# MPLS for Metropolitan Area Networks

# MPLS for Metropolitan Area Networks

Nam-Kee Tan

## AUERBACH PUBLICATIONS

A CRC Press Company

Boca Raton   London   New York   Washington, D.C.

## Library of Congress Cataloging-in-Publication Data

Tan, Nam-Kee.
    MPLS for metropolitan area networks / Nam-Kee Tan.
        p.   cm.
    Includes bibliographical references and index.
    ISBN 0-8493-2212-X
    1. MPLS standard. 2. Metropolitan area networks (Computer networks) I. Title.

TK5105.573.T36 2004
004.6'2—dc22                                                                      2004054874

This book contains information obtained from authentic and highly regarded sources. Reprinted material is quoted with permission, and sources are indicated. A wide variety of references are listed. Reasonable efforts have been made to publish reliable data and information, but the author and the publisher cannot assume responsibility for the validity of all materials or for the consequences of their use.

## Visit the Auerbach Publications Web site at www.auerbach-publications.com

© 2005 by CRC Press
Auerbach is an imprint of CRC Press

No claim to original U.S. Government works
International Standard Book Number 0-8493-2212-X
Library of Congress Card Number 2004054874
Printed in the United States of America  1  2  3  4  5  6  7  8  9  0
Printed on acid-free paper

# Dedication

*To my precious, Chai Tee*

# Preface

**INTRODUCTION**

The demand for bandwidth in metropolitan area networks (MANs) and the increased availability of fiber has created a new breed of service providers—the metro service providers (MSPs). The high-speed networks deployed by the MSPs allow them to distribute new services such as IP virtual private network (VPN), virtual private LAN services (VPLS), Voice-over-IP (VoIP), data warehousing (off-site storage), and Web hosting within metro areas, thus enabling service delivery closer to the customers—the metro subscribers. Because MSPs are offering customers native data services using Ethernet rather than leased lines or time division multiplexing (TDM) circuits, customers are attracted to these services by the flexibility offered, the low cost of bandwidth, and the fact that these services can be quickly provisioned with ease.

Nevertheless, the MSPs still face a number of inevitable challenges. These include integrating new services with existing ones, maintaining the same quality-of-service (QoS) level along with guaranteed bandwidth as in TDM circuits, and providing reliable and responsive networks. The solution to all these stringent requirements is Multi-Protocol Label Switching (MPLS). MPLS is coming into vogue as the technology of choice to converge disparate networks and its services into a unified network core. On top of this MPLS network, value-added capabilities such as traffic engineering, network protection, any-to-any connectivity, user differentiation, service differentiation, and bandwidth/QoS guarantee can be delivered without a glitch. Put simply, MPLS has become the "Swiss Army Knife" of networks, including MANs.

This book addresses the above-mentioned challenges with solutions provided by key MPLS features such as MPLS traffic engineering, MPLS fast reroute, MPLS VPNs, and MPLS QoS. Examples and case studies are used extensively and progressively to illustrate the conceptual aspect of these solutions. This approach facilitates learning by example, which further enhances the ability of the reader to relate and apply what he or she has learned in the examples and case studies to his or her own network environment.

## INTENDED AUDIENCE

This book is most relevant to network engineers, architects, and managers who need to propose, design, implement, operate, administer, or maintain value-added metro network services with MPLS. Besides catering to inter-networking professionals who are interested in knowing more about implementing sophisticated metro network services with MPLS, this book is also appropriate for sales and account managers who are required to justify and propose a foolproof MPLS-based metro solution to their customers.

In order for the audience to gain a better understanding of the various MPLS features discussed in this book, they are expected to have some basic knowledge of MPLS technology, VLAN concepts, end-to-end QoS deployment, and a good grasp of advanced routing protocols such as OSPF, Integrated ISIS, and BGP. These prerequisites are preferred but not mandatory, because the book will attempt to tackle these areas progressively through each chapter.

## ORGANIZATION OF THIS BOOK

### Part 1: Metropolitan Area Networks and MPLS

- Chapter 1 provides a brief overview on the current business prospects and challenges that MSPs are facing. It emphasizes the various service aspects and requirements that need to be incorporated into the present-day MANs so that new revenue opportunities can be realized.
- Chapter 2 gives a concise primer on MPLS and explains how MPLS applications such as traffic engineering, label switched path (LSP) reliability, virtual leased line (VLL) services, virtual private LAN services (VPLS), IP VPNs, and classes of service (CoS) can be deployed in MANs.

### Part 2: Traffic Engineering Aspects of Metropolitan Area Networks

- Chapter 3 focuses first on the fundamental concepts of traffic engineering (TE). After that, it explains the causes and effects of network congestions and the hyperaggregation phenomenon. TE is then introduced as the most effective way to counter network congestion and to gain total network control. This is followed by a comparison between tactical TE and strategic TE. The chapter next discusses the applicability of the IP/ATM overlay model and MPLS in relation to TE. The concluding section describes the three primary tasks that are involved during the information distribution and path selection phases of MPLS-TE.
- Chapter 4 describes the various common trunk attributes that are used to define and modulate the behavioral characteristics of traffic

trunks for MPLS-TE purposes. These attributes include traffic parameters, policing attributes, priority attributes, preemption attributes, resilience attributes, generic path selection attributes, and dynamic path maintenance parameters.

- Chapter 5 gives concise coverage on constraint-based routing (CBR), which is responsible for information distribution and path selection during the MPLS-TE process. It first discusses the limitations of traditional topology-driven IGPs such as OSPF and Integrated ISIS, which lack the awareness of constraints such as resource availability and traffic characteristics during path calculation. The TE extensions to these routing protocols, which include link resource attributes such as maximum bandwidth, maximum reservable bandwidth, unreserved bandwidth, and administrative group (resource class or link color), are brought in to address the limitations and to convert these conventional routing protocols into CBR protocols. The chapter further describes the operational model of CBR and the constrained shortest-path-first (CSPF) algorithm and wraps up with two CSPF path selection examples.

- Chapter 6 presents comprehensive coverage on the Resource Reservation Protocol with TE extension (RSVP-TE). It examines the messages and objects of RSVP-TE and explains how RSVP-TE constructs explicitly routed LSP tunnels, distributes label bindings, supports resource reservations along an LSP tunnel, and tracks the physical route traversed by the LSP tunnel. The chapter further illustrates the path setup operation of RSVP-TE, along with its admission control and preemption capabilities. It then covers the various alternatives to forward traffic across an established LSP tunnel. The chapter later describes how RSVP-TE supports the concept of make-before-break when reoptimizing or rerouting an existing LSP tunnel. The concluding section looks at how RSVP-TE overcomes the scalability, latency, and reliability issues posed by the soft-state model of standard RSVP.

- Chapter 7 incorporates what has been covered in chapters 3 to 6 into a series of case studies so that the readers can relate what they have learned so far to their MAN environments. The four case studies discussed thoroughly in this chapter all focus on the two important aspects of TE—traffic control and resource optimization. Case study 7.1 looks at hop-by-hop routed LSPs, case study 7.2 illustrates the setup of explicitly routed LSPs with loose explicit routes, case study 7.3 demonstrates how CBR can be used to manipulate the desired bandwidth, and case study 7.4 shows how link affinities can be used to construct session attribute filters.

## Part 3: Reliability Aspect of Metropolitan Area Networks

- Chapter 8 discusses how the MPLS fast reroute and path protection features can bring increased reliability to MANs and speed up the restoration time of broken LSPs or TE tunnels. The two case studies in this chapter examine the path protection and the fast reroute schemes, respectively. Case study 8.1 illustrates the deployment of path protection, while case study 8.2 demonstrates fast rerouting using detour LSPs.

## Part 4: Service Aspect of Metropolitan Area Networks

- Chapter 9 provides extensive coverage of layer-3 (L3) and layer-2 (L2) MPLS VPNs. It begins by examining the architectural components of L3 MPLS VPN, followed by a thorough description of the operations of L3 MPLS VPN, such as the propagation of VPN routes along with the distribution of MPLS labels and packet forwarding. The chapter then discusses the implementation of L2 MPLS VPNs with Martini's point-to-point tunneling approach, which involves various constructs such as the VC label and the tunnel label. The next section covers the interaction between OSPF and L3 MPLS VPN, as well as the special handling of OSPF routes across the MPLS VPN backbone (which acts as pseudo-Area 0). The chapter adopts three different case studies to illustrate the actual setup of L3 MPLS VPNs using various PE-CE routing protocols such as static routes, RIPv2, EBGP, and OSPF. Case study 9.1 illustrates the implementation of L3 MPLS VPNs using static routes and OSPF as the PE-CE routing protocols, case study 9.2 demonstrates the setup of L3 MPLS VPNs using RIPv2 and EBGP as the PE-CE routing protocols, and case study 9.3 looks at the deployment of OSPF as the sole PE-CE routing protocol at all the customer sites during the setup of L3 MPLS VPNs.
- Chapter 10 initially cites virtual private LAN services (VPLS) as the approach that allows MSPs with existing MPLS infrastructure to offer geographically dispersed Ethernet multipoint services (EMS). It then provides a brief overview of VPLS as a multipoint L2 VPN service enabler and the service offerings it supports. The functional components of VPLS and its frame forwarding operation are also presented. This is followed by a comparison between VPLS and Martini's point-to-point L2 VPN services. The LDP-based approach to VPLS implementation is covered next. The chapter later discusses hierarchical VPLS and how it can be used to scale LDP-based VPLS. The VPLS discussion rounds off with a comparison between VPLS and L3 MPLS VPN. The two case studies in this chapter examine the setup of a point-to-point L2 VPN service (case study 10.1) and a multipoint L2 VPN service (case study 10.2).

- Chapter 11 first explains the rationale behind the requirement for different service classes. It then gives a brief introduction to the two traditional QoS models: integrated services (IntServ) and differentiated services (DiffServ). This is followed by the definition of the DiffServ field and per-hop behavior. The QoS components of the DiffServ model, such as classification and marking, policing and shaping, queue scheduling, and queue management, are described next. On top of these, a comparison among best-effort services, IntServ and DiffServ, is presented. The chapter then addresses the association between DiffServ and MPLS, particularly focusing on E-LSPs and L-LSPs, and rounds off with a comparison between these two different types of LSPs, followed by a discussion on the various DiffServ tunneling models, such as the uniform, pipe, and short-pipe models. The chapter also discusses DiffServ-aware MPLS-TE, which introduces per-class-type-based TE. Per-VPN QoS service models, which are composed of the point-to-cloud and point-to-point QoS models, are covered as well. The two case studies in this chapter adopt the point-to-cloud model and implement the E-LSP approach. Case study 11.1 uses an L2 MPLS VPN scenario, while case study 11.2 takes on an L3 MPLS VPN environment.

## APPROACH

*What I hear, I forget. What I read, I remember. What I do, I understand.*

**—Confucius**

The entire content of the book adopts a simplify-and-exemplify approach. It packs a whole series of real-life case studies and utilizes representative topologies as a basis for illustrating the concepts discussed in each chapter. This learning-by-example approach will help the readers remember and understand the difficult concepts and technologies much better. The readers can then apply what they have learned from these examples and scenarios to their specific environments.

**Nam-Kee Tan**

# About the Author

**Nam-Kee Tan** ( CCIE #4307) has been in the networking industry for ten years. He is currently the principal consultant of Couver Network Consulting, where he provides consulting and training services to corporate and government clients throughout Asia Pacific. His area of specialization includes advanced IP services, network management, Internet traffic engineering, MPLS technologies, L2/L3 VPN implementations, Internet security solutions, and intrusion detection systems.

Pertaining to MPLS, he has designed and implemented for his clients various mission-critical networks that incorporate MPLS applications such as traffic engineering, fast restoration, L3 VPNs, VPLS, and QoS. Nam-Kee is also a seasoned certified Cisco Systems instructor (CCSI #98976) who can deliver a wide range of Cisco courses. As a veteran in Cisco training, he has taught hundreds of Cisco classes to networking professionals from the Fortune 500 companies.

He writes actively and has three book titles published by McGraw-Hill: *Configuring Cisco Routers for Bridging, DLSw+ and Desktop Protocols*; *Building Scalable Cisco Networks* (coauthor); and *Building VPNs with IPSec and MPLS*. In addition, he holds an M.S. in data communications from the University of Essex, U.K., and an M.B.A. from the University of Adelaide, Australia.

# Acknowledgments

My first acknowledgment goes to my publisher, Rich O'Hanley, who gave me the unique opportunity to write this challenging book. I probably owe the greatest thanks to those who reviewed the manuscript (or a large part of it) looking for obscurities and errors: Ann Chua, Benjamin Chew, Brian Khoo, Charles Yong, Emily Lim, Frank Tan, Jenny Kim, Jeremy Lee, John Kang, Josephine Tang, Tommy Jones, Teresa Lum, and Wesley Khoo.

I also extend my thanks to my elder sister, Sok Ai, and my kid sister, Sok Thian, together with her husband, Yuji Suzuki, and my adorable niece, Emmy Suzuki, for all the support they gave me during my writing stint. In addition, I would like to take this opportunity to pay tribute to my mom for everything she has done for me. She will always be in my heart.

Most of all, I thank my partner, Chai Tee, who helped to finish all the drawings that are associated with this book. Besides putting up, for over a year, with someone who maintains a day job and writes a book in the evenings and weekends, she also had to deal with my making a countless number of changes and amendments to more than a hundred figures and diagrams. I know a few lines of praises and thanks are not enough to make up for the lost time and hard work, but, Chai Tee, I really appreciate your support, patience, and love. Thanks for everything.

Last but not least, I thank my favorite fish, Bi-Bi, and my three pet hamsters, Ham-Ham, Pe-Pe, and Hammida, for keeping my stress level down with their entertaining stunts while I wrote this book.

# Figures

# Tables

# Contents

## PART 3 RELIABILITY ASPECT OF METROPOLITAN AREA NETWORKS

## PART 4  SERVICE ASPECT OF METROPOLITAN AREA NETWORKS

## PART 5    QUALITY-OF-SERVICE ASPECT OF METROPOLITAN AREA NETWORKS

# Part 1
# Metropolitan Area Networks and MPLS

# Chapter 1
# Requirements of Metropolitan Area Network Services

## 1.1 METROPOLITAN AREA NETWORK OVERVIEW

A metropolitan area network (MAN) generally spans within a network radius of 10 to 120 kilometers, and the typical network topology that it adopts can be point to point, ring, or mesh. MANs are used to extend distances outside the range of local area networks (LANs) where there is no need for the long-haul distances characterized by wide area networks (WANs). A MAN is usually segmented into three tiers: core, aggregation, and access, as shown in Figure 1.1. The term *last mile* (or *first mile*), which is considered by many to be the bandwidth bottleneck, refers to the connection between the core and access networks (inclusive of the aggregation).

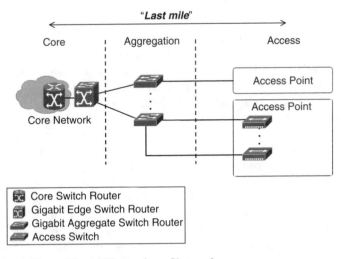

**Figure 1.1   A Three-Tiered Metro Area Network**

## 1.2   THE BANDWIDTH DEMAND

As the bandwidth demand and data processing capability of enterprise networks increasingly surpass the throughput of MANs, relying on legacy access technologies such as time division multiplexing (TDM) circuits (for example, T1/E1 leased lines), providing a faster local access to the customer premise equipment (CPE), has become a top priority for the service provider (SP). SPs are extending their broadband infrastructure via asymmetric digital subscriber lines (ADSLs), cable modems, and 10/100 Ethernet into the customer premises to ease any existing bandwidth bottleneck and to provide bigger bandwidth pipes to the customers.

High-speed access not only encourages the customers to converge their data, voice, and video applications, but also helps the SP to bring in new demands for value-added, high-data-rate (HDR) services such as "constantly on" high-speed Internet access, online gaming, video on-demand (VoD), videoconferencing, e-commerce, virtual private networks (VPNs), Web hosting, and data warehousing (off-site storage). For many SPs, the limelight is on the MAN, which has become the most crucial point of service creation and delivery. Most importantly, MAN spells profitability.

## 1.3   THE METRO SERVICE PROVIDER'S BUSINESS APPROACHES

In the current turbulent economic climate, SPs are striving to reduce costs. Capital investment in new technologies and equipment must be painstakingly considered and well justified to fulfill operational efficiency and realize new revenue opportunities.

In the metro networking arena, the SP can implement two different approaches:

- Provide commoditized transport services such as plain old telephone services (POTS), private leased lines, and Frame Relay. This approach requires the SP to reach out to the mass market. High sales volume is very important here; thus, the pricing of the services has to be very competitive and attractive to the consumers. The main objective is to curtail operational costs significantly and reduce cost per bit. To achieve this goal, the SP can turn to high-speed transport technologies such as next-generation synchronous optical network/synchronous digital hierarchy (SONET/SDH) or Gigabit/10-Gigabit Ethernet as a means of reducing cost per bit and provisioning time, which will help the SP to win the price war. Nevertheless, all price wars will eventually evolve into a free competition market in which profit margin becomes meager. On the other hand, the consumers (end users) will welcome this approach readily because they stand to gain more than the SP.

- Opt for differentiated application-oriented services such as storage area network (SAN), Voice-over-IP (VoIP), VoD, IP VPNs, virtual leased line (VLL), and virtual private LAN services (VPLS). This niche market approach requires the SP to develop a wide range of application-oriented services that can add value, stir up consumer interest, and allow corporate customers to outsource more of their in-house IT applications and telecommuting activities, therefore making way for new sources of revenue. The decisive factor for this approach is to be fully equipped with the capabilities to provide differentiated services. For many SPs, this seems to be the preferred approach. However, SPs have to be cautious when pricing these services to prevent cannibalizing their existing commoditized services.

What are the capabilities SPs need to provide differentiated services to their customers? These service capabilities really depend on the customer's application requirements, which ultimately drive the metro networking solution. Therefore, it is of utmost importance for SPs to understand in the first place what are the emerging metro customer expectations and needs.

## 1.4 THE EMERGING METRO CUSTOMER EXPECTATIONS AND NEEDS

*Flexible* is the keyword that can be used to describe the type of services the current metro subscribers—the customers—want. The service must be flexible in terms of bandwidth provisioning (the required bandwidth can be in any denomination), class of performance (guaranteed versus best effort), and usage (just pay for what you have used). The general expectation is that the new services should maintain the same quality-of-service (QoS) level and guaranteed bandwidth as in TDM circuits.

On top of these, latency-sensitive applications such as videoconferencing and fast service response time, for instance, 200 milliseconds, are two other needs that the SP is required to take into consideration. The wish list also includes achieving transparent LAN interconnections and reducing unit costs through the MAN with Ethernet, because the Ethernet interface is straightforward, cost-effective, and universal. Not only does Ethernet have a low cost of ownership, but it also gives the customers the option of subscribing to the exact amount of bandwidth they need without the hassle of deploying additional equipment or changing the existing network setup. On the other hand, Ethernet helps the SP to reduce costs through bandwidth oversubscription up to a ratio of 1:10 in the metro core, something that is not possible with TDM circuits, which has an oversubscription ratio of only 1:1.

In addition, metro subscribers are an application-inclined lot who are constantly looking for opportunities to use outsourced IT services (for example, storage consolidation, content hosting services, VPNs, and integrated voice/data service delivery) provided by SPs to reduce management costs, which is a big break for SPs who can meet this demand. However, these subscribers are also more concerned about applications performance than anything else. Hence, SPs are anticipated to offer an acceptable service level agreement matching that of TDM circuits to the metro subscribers and build facilities that are reliable, highly scalable, and well engineered to deliver high-speed data, telephony service, and QoS guarantees to multiple subscribers in unison.

Accommodating all these requirements will pose a great challenge to the SP's existing service capability. The greater challenge will be on how to meet the demand of these new-age customers through the SP's present technologies and network facilities.

## 1.5 SOME PREVAILING METRO SERVICE OPPORTUNITIES

Before we examine the wide range of service aspects and requirements, let us look at some of the predominant metro service opportunities. The metro market also renders both appealing and novel service opportunities for data traffic staying exclusively within the metro area. Some of the most prevailing applications include VPLS, IP VPN, VoIP, and storage networking. In this section, we only cover VoIP and storage networking as viable metro services. The detailed inner workings of VoIP and storage networking are beyond the scope of this book. VPLS and IP VPN are discussed in chapter 2.

### 1.5.1 Delivering Premium Voice Service with VoIP

Enterprise customers are migrating voice traffic to IP so that it can be fully incorporated with their prevalent IP traffic to form a single data network, which can then carry both voice and data traffic for interoffice communications, thus avoiding long-distance tolls. From the SP perspective, transporting voice traffic across a data- or packet-switched infrastructure is cheaper than using a fixed circuit-switched infrastructure, as it will be more cost-effective for the SP to unite these infrastructures together to provide integrated voice and data services. In short, the adoption of VoIP is driven by cost savings. Offering VoIP over a MAN will promise good returns for the SP if the quality and reliability of VoIP can be assured to the subscribers.

End-to-end delay is the major factor that decides the quality of VoIP. Most users will not notice the delay if it is under 150 milliseconds. Slight pauses will result in timings between 150 and 400 milliseconds. Beyond 400 milliseconds, the delay will become very obvious. Refer to Table 1.1 for the maximum delay recommendations from G.114. Delay will always be present due to the physical constraints of coder/decoder (codec) processing time, packetization time, and propagation time, which includes interface queuing, MAN ingress serialization, MAN cloud transmission and buffering, MAN egress serialization, and playback (de-jitter) buffer. Even

**Table 1.1   One-Way Delay Considerations from International Telecommunication Union-Telecommunication Standardization Sector (ITU-T) Recommendation G.114**

| One-Way Delays | Acceptable Conditions |
| --- | --- |
| 0–150 milliseconds | Acceptable for most user applications. Recommended delay for intracountry calls. |
| 150–400 milliseconds | Acceptable, provided the impact to applications is assessed. Recommended delay for international calls. |
| >400 milliseconds | Unacceptable for general planning; may be required in specific cases. Recommended delay for calls with a satellite hop. |

though these timings cannot be made smaller, they can definitely be limited to an unvarying high watermark.

Other substantial delay elements include network congestion, de-jitter buffer, packet loss, and network link/node failure. Voice traffic is time sensitive. It has a consistent delay and thus requires a constant available bandwidth. Network congestion is the result of the depletion of network resources such as bandwidth at a specific network path and constitutes a considerable delay factor for voice traffic. To achieve the QoS in premium-class voice, it is mandatory to ensure that there is enough bandwidth throughout the network path that is traversed by the voice traffic. We can use the following equation to calculate amount of bandwidth per call:

$$\text{Bandwidth} = (\text{Codec bandwidth}) \times [(\text{Payload length} + \text{Header length})/(\text{Payload length})]$$

The various codec bandwidth values are shown in Table 1.2 and the header lengths for VoIP implementation over metro Ethernet are listed in Table 1.3.

**Table 1.2   Bandwidth Requirement per Call for Various Codecs**

| Codec | Bandwidth (kilobits per second) |
| --- | --- |
| G.711 | 64 |
| G.723.1 | 6.3/5.3 |
| G.726 | 16/24/32/40 |
| G.728 | 16 |
| G.729 | 8 |
| G.729A | 8 |

**Table 1.3    Header Lengths (in Bytes) for VoIP over Metro Ethernet**

| Ethernet Frame Assortments | Link Layer | IP/UDP/ RTP | CRTP (with UDP Checksums) | CRTP (without UDP Checksums) |
|---|---|---|---|---|
| No 802.1q trunking | 18 | 40 | 4 | 2 |
| 802.1q trunking | 22 | 40 | 4 | 2 |
| Layer-3 MPLS VPN pass-through with two 4-byte labels | 26 | 40 | 4 | 2 |
| 802.1q trunking + layer-3 MPLS VPN pass-through with two 4-byte labels | 30 | 40 | 4 | 2 |
| 802.1q trunking + layer-2 MPLS VPN pass-through with two 4-byte labels | 44 | 40 | 4 | 2 |

*Note:* UDP = User Datagram Protocol; RTP = Real-Time Transport Protocol; CRTP = Compressed RTP.

To avoid speech distortion, voice packets must be buffered at the receiving end to offset the variation in packet interarrival times, which is also commonly known as jitter or delay variation. This de-jitter buffer assures a smooth playback of voice at the receiving end, but also constitutes another delay factor. Using QoS mechanisms that prioritize voice traffic over other traffic in the network can help reduce jitter and shorten buffering times. However, a balance needs to be struck so that data traffic performance will not be adversely affected by large amounts of priority voice traffic.

Packet loss is associated with the reliability aspect of VoIP, and it can be due to network link congestion in which router buffers are exhausted by many traffic streams and start to drop packets. Packet loss also occurs when a network link/node fails. For non-real-time applications such as file transfers and e-mail, packet loss is nonvital because these data applications use Transmission Control Protocol (TCP) to provide retransmission to recover dropped packets. However, in the case of real-time voice applications, voice packets have to reach the receiver end within a rather small time window to remain valid in the reconstruction of the original voice signal. Retransmissions in the voice situation would add a sizeable delay to the reconstruction and would result in clipped or unintelligible speech. To prevent packet loss for VoIP from happening, QoS prioritization schemes and flow control mechanisms must be deployed accordingly along the network path that is traversed by the voice traffic. In addition, the routing environment must be able to provide fast restoration to any link/node failures and topology changes to bolster network reliability, which in turn prevents unnecessary packet loss. Table 1.4 gives the reader a rough guide of the QoS requirements for some multi-service applications in a MAN setting.

**Table 1.4    QoS Requirements for Some Multi-Service Applications over MAN**

| Application Type | Packet Loss Limit | One-Way Latency Limit | Jitter Limit |
|---|---|---|---|
| Data | Variable | Variable | Variable |
| Videostreaming | 2% | 5 seconds | Not applicable |
| Videoconferencing | 1% | 200 milliseconds | 30 milliseconds |
| VoIP | 1% | 200 milliseconds | 30 milliseconds |

To sum up, VoIP can provide good-quality voice if the SP network infrastructure delivers the required performance. This is possible using a VPN with QoS guarantees, but dubious in the best-effort public Internet. Resolving the QoS issues mentioned earlier will give the SP the much needed assurance to offer premium-grade voice to the metro customers using VoIP.

### 1.5.2    Storage Networking

The need for distributed network storage has been driven by data-intensive applications, content-based applications, and the huge growth of Internet traffic. There are two existing network storage models: storage area network (SAN) and network-attached storage (NAS).

A SAN is a network dedicated to providing and managing storage as well as administering backup for enterprise businesses or data centers. It enables the extension of the Small Computer Systems Interface (SCSI) Protocol over longer distances. SANs are typically based on fiber channel switches configured to occupy the back-end portion of a network behind the data center or server farm (see Figure 1.2). Fiber channel's support of block-level data enables very efficient transfer of large quantities of data over longer distances with minimal server intervention.

NAS devices allow users to attach scalable storage directly to existing LAN infrastructure instead of a separate SAN, which simplifies installation and maintenance (see Figure 1.3).

The main difference between these two approaches is that SAN is data centric while NAS is network centric. However, the emerging Internet SCSI (ISCSI) combines the best of both worlds. ISCSI (see [DRAFT-ISCSI]) incorporates Ethernet and IP-based NAS file-level access with the performance of SAN-based block-level access.

ISCSI is a networking protocol for SCSI-3 traffic over TCP/IP Ethernet. This protocol allows block-level storage data to be transported over common IP networks, enabling end users to access the storage network anywhere in the enterprise. ISCSI extends over unlimited distances and can be implemented as a converged storage network in the metro environment

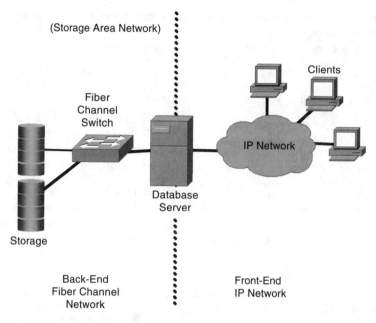

Figure 1.2   SAN Storage Approach

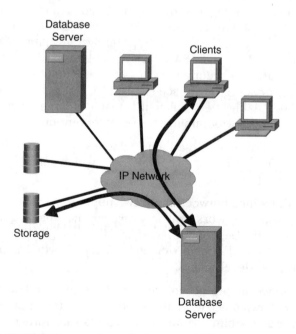

Figure 1.3   NAS/ISCSI Storage Approach

through the use of VPNs to support multiple-user data applications within a single network infrastructure.

There is the tendency to associate storage with pure data applications, but in reality, most data applications have diverse needs as well as specific performance goals. Hence, the metro subscribers might want to have different types of storage services that are catered to their specific application needs. To satisfy this demand, the SP would have to come out with a network storage service specification focusing on two aspects: traffic profile and performance. The traffic profile defines the amount of storage capacity to be reserved, the period of reservation, and the allocation of data accesses. Parameters such as data access latency, data access jitter, and acceptable loss rate can be used to specify the performance requirements of a particular data application.

With the service specification in place, the SP can then classify applications into three main service categories: deterministic guarantee, statistical guarantee, and best effort. Deterministic guarantee service is well suited for nonbursty applications that require constant available bandwidth and fixed-delay bound. Applications that exhibit stochastic behaviors or burstiness will match the service profile of statistical guarantee, whereas applications that do not require any guarantees can use best-effort service at a more affordable price than the other services.

In order for the metro SP to tap into the lucrative network storage market, the SP distributed network storage infrastructure must have the capability to incorporate new service classes and performance attributes when the market demands them.

## 1.6 SERVICE ASPECTS AND REQUIREMENTS

What are the various types of service features or capabilities required to satisfy the needs of the metro subscribers? This is a tough question, but responding to this issue with the right solution will certainly propel the SP ahead of its competitors. The SP has to convert the overall needs of the metro subscribers into specific functional attributes and embed these features into its existing network infrastructure.

First and foremost, the SP would need plenty of bandwidth and speed on its transport media. Today, with Ethernet speeds going progressively from 1 gigabit per second (IEEE 802.3z) to 10 gigabits per second (IEEE 802.3ae), the SP should not have any problem in achieving a high-speed and high-bandwidth core transport layer. Due to its simplicity, Ethernet helps to speed up the service provisioning process, and the cost of bandwidth offered by Ethernet is also cheaper.

To meet the performance requirements of new application-oriented services and to manage the large streams of traffic generated by these services as a result, the SP would need to ensure that the utilization of its network infrastructure can be optimized and enhanced. This means the network path designated for a specific application can not only be optimized in terms of the cost of the path but also take into consideration available path resources such as bandwidth during path determination.

Put another way, network paths can be selected explicitly—you choose which path you want—and path determination is no longer based on just path cost metric. Instead, path resource information has taken over as the main prerequisite when selecting the best path. Hence, packet delivery between any two nodes on the SP network will become more deterministic and predictable than before, which is crucial for real-time applications. Moreover, service revenue is directly related to uptime. So the responsiveness and reliability of the SP network has to be improved and reinforced to prevent or preempt any considerable downtime during link or node failures.

The SP should also bear in mind the scalability issue. One factor to consider is the any-to-any connectivity between different access networks such as private leased lines, Frame Relay, Asynchronous Transfer Mode (ATM), digital subscriber line (DSL), Ethernet, and IP. The other is the convergence of multi-service (audio, video, and integrated data) applications running on top of these access networks with the SP network. More concisely, the SP should have a common carrier infrastructure for multiple disparate networks and applications. There should also be a feasible option for the migration of existing TDM services to new-age services.

As the metro subscribers are only interested in the performance of their applications running on an end-to-end basis, the performance of the SP network will need to qualify in stipulations that are useful to the subscribers. To cut costs, the SP might want to oversubscribe (see section 1.4) the bandwidth in the metro core while at the same time selling a premium service to subscribers, guaranteeing their traffic will always get priority when forwarding through the SP network. This is what the MAN is all about—providing differentiated and guaranteed services.

With differentiated services, the SP can offer a wide range of options on attributes such as latency, jitter, committed bit rate, and peak burst rate to meet the requirements of a specific application. This is the competitive edge and service differentiation that the SP has always wanted. As such, the SP can now achieve better margins along with higher customer satisfaction and retention.

Succinctly, meeting the diverse service requirements imposed by the new-age metro subscribers would require the SP network to incorporate service aspects such as network path control and optimization, a responsive and reliable carrier network, a scalable technology to converge multiple networks and user applications within a single network infrastructure, and the capability to implement differentiated and guaranteed services.

The solution to all these stringent requirements is Multi-Protocol Label Switching (MPLS). MPLS can be regarded as a multipurpose internetworking technology that consolidates disparate networks (such as ATM, Frame Relay, Ethernet, and IP networks), delivers VPN services, guarantees QoS, optimizes network routes (with path resource information), and provides fast restoration during path/link/node failures.

Ethernet is a very efficient aggregator of traffic in native format, but when it comes to the consolidation of dissimilar networks and providing transparent (or virtual) connectivity across the metro network, MPLS is the way out.

Adopting MPLS in the SP networks not only makes differentiated services a breeze, but also supplements the SP networks in the many aspects of performance, reliability, and scalability. MPLS also helps SPs to leverage on their existing networks to provide high-grade multi-service applications and accommodate any-to-any connectivity.

In other words, MPLS has become the key service enabler for MANs. For many SPs, providing first-rate value-added services is the primary differentiator over other competitors in the metro market segment, and deploying MPLS in the SP MANs makes this a reality.

## 1.7  SUMMARY

To be at the top of the metro service value chain, SPs must offer value-added applications-oriented services to customers and help them to outsource their IT applications as they expand their LANs into the MANs. Even though application-oriented services are more sophisticated in terms of provisioning, they create new sources of revenue in the emergent MAN environments.

Metro customers will become more dependent on the SPs to support mission-critical applications such as VoIP, videoconferencing, videostreaming, and storage networking. The performance of these applications will be judged in terms of throughput, latency, jitter, and errors. To deliver premium applications, SPs will need to enhance the performance, reliability, and scalability aspects of their networks. The good news is that there is one versatile internetworking technology that can be used to resolve all these demanding requirements, and that intriguing technology is MPLS. MPLS, together with IP, provides a unified packet-based infrastructure where disparate networks and distinct services can all converge.

Metro SPs need to take on a service-oriented positioning in the new Internet era, and this book basically equips readers with the key technologies and applications to achieve this. The subsequent chapters of this book address the above-mentioned challenges with solutions provided by key MPLS features such as MPLS traffic engineering (TE), MPLS fast reroute, MPLS VPNs, and MPLS QoS.

# Chapter 2
# Roles of MPLS in Metropolitan Area Networks

## 2.1 INTRODUCTION

One of the major benefits of Multi-Protocol Label Switching (MPLS) is that it allows metro service providers (SPs) to deliver new services that cannot be readily supported by conventional IP routing techniques. MPLS augments traditional IP routing by supporting more than just destination-based forwarding. Through separating the control component [REF01] from the forwarding component [REF01] in legacy routing, MPLS is designed with the ability to evolve control functionality without changing the forwarding mechanism. With this newly acquired flexibility, MPLS can be distinctively positioned to support the deployment of enhanced control and forwarding capabilities that are required in new value-added and highly profitable metro services such as traffic engineering (TE), classes of service (CoS), virtual private networks (VPNs), virtual leased lines (VLLs), and virtual private LAN services (VPLS).

MPLS provides a set of powerful constructs that allow metro SPs to conquer all the challenges set forth in chapter 1. So it is important to understand the fundamentals of MPLS up front before covering its metro applications in detail. This chapter focuses on the relevant MPLS foundation that the reader can build on when the actual MPLS metro service implementations are discussed in the later chapters.

## 2.2 MPLS PRIMER

MPLS primarily performs flow aggregation that incorporates a label-swapping framework with layer-3 (L3) routing. This relatively simple paradigm involves assigning short fixed-length labels to packets at the ingress of an MPLS network based on the concept of forwarding equivalence classes (FECs). Within the MPLS network, instead of using address information found in the original L3 packet headers, labels attached to the packets are used to make the forwarding decisions. The basics of MPLS are discussed in the subsequent sections.

### 2.2.1 Forwarding Equivalence Classes

As defined in [RFC3031], an FEC is a group of IP packets that is forwarded over the same path with the same forwarding treatment. Specifically, an FEC can be regarded as a traffic policy that examines and classifies traffic flow according to a set of conditions or attributes. In a non-MPLS network, the assignment (or binding) of a packet to an FEC is based solely on destination subnet derived from the destination IP address in the packet header. In the MPLS case, besides the information found in the packet header, the binding of a packet to an FEC could be influenced by other packet classification criteria such as:

- Combination of source and destination subnet
- Combination of destination subnet and application type
- IP multicast group
- TE tunnel
- Virtual LAN (VLAN) identifier
- VPN identifier
- Quality-of-service (QoS) requirement and class of service (CoS)

### 2.2.2 Architecture Overview

Unlike traditional hop-by-hop routing, a path through the MPLS network, known as a label switched path (LSP), must be established prior to the forwarding of packets in a given FEC. An LSP is functionally analogous to a virtual circuit (VC) because it defines an ingress-to-egress path through a network to be followed by all packets assigned to a particular FEC. Put another way, an LSP is really a series of label switch routers (LSRs) that forward packets for a particular FEC. More than one FEC can be mapped to a single LSP. LSPs are unidirectional (or simplex), which implies that the return traffic will have to use another LSP set up in the reverse direction. Through the use of LSPs, MPLS converts connectionless networks such as IP to connection-oriented ones. Note that the terms *LSP* and *LSP tunnel* are interchangeable throughout the book.

There are two types of LSRs: core and edge. Core LSRs (or transit LSRs) reside within the MPLS network and primarily forward packets based on labels (see section 2.2.3). They do not examine the packet's header. Labels play a very important role in MPLS because they are used to map packets classified under a particular FEC to its corresponding LSP. The LSP can be provisioned to satisfy specific FEC requirements such as minimizing the number of hops, conforming to specific bandwidth requirements, and diverting traffic across certain links in the network. Edge LSRs, also known as label edge routers (LERs), reside at either the ingress or the egress of

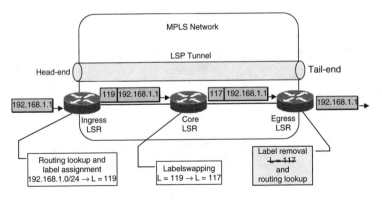

**Figure 2.1   MPLS Forwarding Operation**

the MPLS network. The ingress LSR (or head-end LSR) receives IP packets, does an L3 table lookup, performs packet classifications by grouping packets into FECs, assigns labels, and forwards the labeled IP packets into the head-end of the LSP. The egress LSR (or tail-end LSR) removes the labels from the IP packets before forwarding them out from the tail end of the LSP based on the destination IP addresses contained in their packet headers.

Put another way, only the ingress or egress LSR performs a routing lookup and label assignment/removal, while the core routers swap labels (label swapping) and forward packets based on simple label lookups. As such, MPLS not only simplifies the forwarding process and reduces the forwarding overhead on the core LSRs, but also speeds up the data transfer time. Figure 2.1 shows an example of the MPLS forwarding operation. In this example, the ingress LSR receives an unlabeled IP packet with a destination address of 192.168.1.1, does a routing lookup (in this example the derived FEC is destination network 192.168.1.0/24), assigns a label value of 119 to the packet, and forwards it to the core LSR. When the core LSR receives the labeled packet, it simply does a label lookup, swaps label value 119 with 117, and forwards the relabeled packet to the egress LSR. The egress LSR removes the label value 117 upon receiving the packet, does a routing lookup, and forwards the unlabeled packet using the longest-match route found in the routing table.

## 2.2.3   Labels

The MPLS label, a field found in the MPLS shim header (see Figure 2.2), is a locally significant 20-bit integer (between 0 and 1,048,575) that is used to identify a particular FEC. The label that is imposed on a particular packet indicates the FEC to which that packet is assigned.

The MPLS label is analogous to the data-link connection identifier (DLCI) of Frame Relay or the virtual path identifier/virtual channel identifier

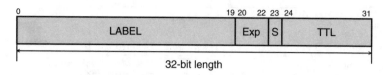

**Figure 2.2   MPLS Shim Header**

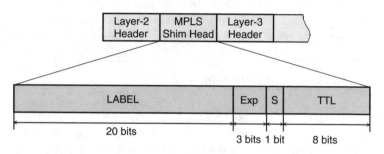

**Figure 2.3   Location of the MPLS Shim Header in a Packet**

(VPI/VCI) of Asynchronous Transfer Mode (ATM). In fact, labels are derived from VPI/VCI for MPLS over ATM (cell-mode MPLS). This book focuses only on frame-mode MPLS (for example, MPLS over Ethernet). Cell-mode MPLS or label-controlled (LC) ATM is beyond the scope of this book. Both DLCI and VPI/VCI are used to identify a particular virtual circuit (VC), whereas an MPLS label is used to identify a specific LSP. The MPLS shim header (or MPLS label header) is inserted (or encapsulated) between the layer-2 (L2) header and the layer-3 (L3) header (see Figure 2.3).

Table 2.1 describes all the fields in the MPLS shim header, and Table 2.2 illustrates the reserved label values 0 to 15 and their respective meanings.

Typically, only one label entry is assigned per packet. However, additional label entries may be added to packets, organized as a last-in, first-out buffer, by MPLS applications such as TE, VLL, VPLS, and IP VPNs (see section 2.3). This last-in, first-out buffer is referred to as an MPLS label stack [RFC3032].

In other words, the top label entry of the label stack appears first in the packet, and the bottom label entry appears last. In Figure 2.4, the packet's label stack has a depth of 3. In this case, the label at the bottom of the stack is referred to as the level-1 label (label 1), the label above it is referred to as the level-2 label (label 2), and the label at the top of the stack is referred to as the level-3 label (label 3). The L3 header immediately follows the last label entry (that is, label 1 with bottom-of-stack indicator set, S = 1) in the

**Table 2.1  MPLS Shim Header Fields**

| MPLS Shim Header Fields | Length | Description |
|---|---|---|
| Bottom of stack (S) | 1 bit | This bit supports a hierarchical label stack (typically used for nested LSPs) and denotes whether this is the last label in the label stack before the L3 header. It is set to 1 for the last entry in the label stack (bottom of the stack) and 0 for all other label stack entries. |
| Time to live (TTL) | 8 bits | This field provides traditional IP TTL functionality within the MPLS network. The TTL field is used to prevent forwarding loops and is decremented by a value of 1 on every hop. For more information on MPLS TTL processing, see [RFC3031]. |
| Experimental use (Exp) | 3 bits | This field is used to define different classes of service (CoS) that will in turn influence the queuing and discard algorithms applied to the packet as it traverses the MPLS network. |
| Label value | 20 bits | This field defines the actual value of the label used, which can range from 0 to 1,048,575 ($2^{20}-1$). Just like Frame Relay's DLCI or ATM's VPI/VCI, a label typically has only local significance and changes on every hop. Not only do globally unique labels limit the number of usable labels, but they are also difficult to manage. |

**Table 2.2  Reserved Label Values and Their Meanings**

| Label Value | Description |
|---|---|
| 0 | IPv4 explicit null label. This label value is only valid at the bottom of the label stack. It indicates that the label stack must be popped, and the forwarding decision of the packet must then be based on the IPv4 header. |
| 1 | Router alert label. This label value is valid anywhere in the label stack except at the bottom. When a received packet contains this label value at the top of the label stack, it is delivered to the local operating system for processing. |
| 2 | IPv6 explicit null label. This label value is only valid at the bottom of the label stack. It indicates that the label stack must be popped, and the forwarding decision of the packet must then be based on the IPv6 header. |
| 3 | Implicit null label (see penultimate popping in section 2.2.6). When an LSR swaps the label at the top of the stack with a new label, and the new label is implicit null, the LSR will pop the stack instead of performing the label swap. Even though this label can be assigned and distributed by an LSR, it will never appear in the encapsulation. |
| 4–15 | Reserved for future use. |

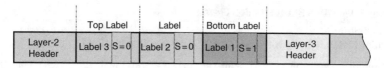

**Figure 2.4   MPLS Label Stack**

label stack. The label stack mechanism enables MPLS to support a hierarchy of LSPs by allowing the nesting of LSPs (or LSPs within LSPs) to any depth. The label stack is by far the most powerful construct in MPLS.

LSRs forward a packet based on the top label entry from the packet's label stack. By examining the top label entry of an incoming packet, each LSR along the LSP can determine:

- The next-hop to which the packet is to be forwarded.
- The operation to be performed on the packet's label stack before forwarding. It can be any one of the following tasks:
  — Swap (replace) the label entry at the top of the label stack with a new label.
  — Pop (remove) the label entry at the top of the label stack.
  — Swap the label entry at the top of the label stack, and then push (add) one or more new label entries onto the label stack.

The packet's label stack is considered empty when the last (or bottommost) label entry is popped. Further processing of the packet is then based on the packet's L3 header. Therefore, the LSR that pops the last label off the label stack must be able to identify the packet's L3 address and perform a routing lookup.

Besides learning the next-hop and performing the various label stack operations such as push, swap, or pop, the LSR may also need to know the type of outgoing data-link encapsulation (for instance, an Ethernet frame). The MPLS label does not contain any information about the L3 protocol being carried in a packet. Therefore, a protocol identifier (PID) for the MPLS-capable L3 protocol needs to be specified in the L2 header to signify that the payload is preceded with a label or labels, and is followed by the L3 protocol header. In the case of IP over Ethernet, the PID is the EtherType value. The EtherType value is $0 \times 8847$ for an Ethernet frame carrying an MPLS unicast IP packet and $0 \times 8848$ for an Ethernet frame carrying an MPLS multicast IP packet.

### 2.2.4   Label Bindings

Label bindings can be associated either with an interface (per-interface basis) or with the router as a whole (per-platform basis). A per-interface label binding uses a per-interface label space and a per-router binding uses a per-platform label space. A per-interface label space is a separate pool of label values defined for each interface on which MPLS is enabled, whereas a per-platform label space is a single global pool of label values defined for the entire router.

With per-interface label space, interface-unique labels are assigned to an FEC on a per-interface basis. In this case, the label assigned to an interface can be reused on another interface with a different FEC. Nevertheless, these labels must still be unique for a specific interface. With per-platform label space, a platform-unique label is assigned to any particular FEC and announced to all neighbors. This label can be used on any interface, but labels that are assigned to different FECs cannot have the same value.

Whether label bindings are assigned on a per-interface or per-router basis really depends on the label distribution protocol that is being deployed (see section 2.2.5.1).

### 2.2.5 Label Distribution and Management

There must be an LSP to a specific FEC before the ingress LSR can label and forward incoming packets to that FEC. To set up the LSP, the binding of the label to the FEC is advertised to adjacent LSRs. As mentioned in [RFC3031], the decision to bind a specific label to a particular FEC is made by the LSR that is downstream with respect to that binding. The downstream LSR then informs the upstream LSR of the binding. So labels are downstream-assigned, and label bindings are distributed in the downstream-to-upstream direction.

The terms *upstream* and *downstream* are relative to a particular FEC. Packets that are bound for a particular FEC travel from the upstream LSR to the downstream LSR. In Figure 2.5, for FEC-1, Kastor-R13 is the downstream neighbor of Iris-R12, and Iris-R12 is the downstream neighbor of Aeolus-R11. An LSR learns about its downstream neighbor through the IP routing protocol, and the next-hop for a particular FEC is the downstream neighbor. From the upstream perspective for FEC-1, Iris-R12 is the upstream neighbor of Kastor-R13, and Aeolus-R11 is the upstream neighbor of Iris-R12.

Likewise for FEC-2, Aeolus-R11 is the downstream neighbor of Iris-R12, and Iris-R12 is the downstream neighbor of Kastor-R13. From the upstream perspective for FEC-2, Iris-R12 is the upstream neighbor of Aeolus-R11, and Kastor-R13 is the upstream neighbor of Iris-R12.

The downstream LSR determines the binding of a label to an FEC and notifies the upstream LSR of the binding through a label distribution protocol (see section 2.2.5.1). There are two techniques that an LSR can use to distribute label bindings: downstream on-demand and downstream unsolicited. In downstream on-demand label distribution, an LSR explicitly

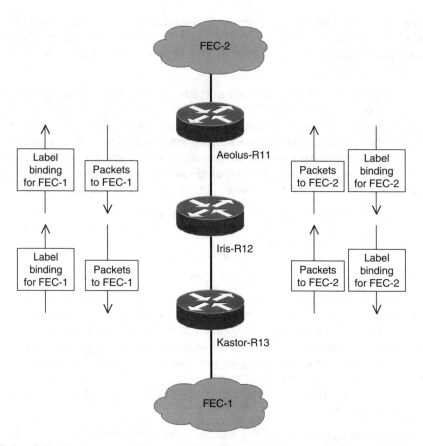

**Figure 2.5   Upstream and Downstream LSRs with Respect to Different FECs**

requests a label binding for a particular FEC from its downstream neighbor (or next-hop), and the downstream neighbor distributes the label upon request. In downstream-unsolicited label distribution, an LSR distributes bindings to other LSRs that have not explicitly requested them. In this case, the label for an FEC is asynchronously allocated and advertised to all neighbors, regardless of whether the neighbors are upstream or downstream LSRs for a particular FEC. Whether downstream on-demand or downstream unsolicited, the upstream LSR and the downstream LSR must agree on which one to use.

**2.2.5.1  Label Distribution Protocols.** An LSP is defined by a set of labels from the ingress LSR to the egress LSR (inclusive of the transit LSRs). When an LSR allocates a label to an FEC, it must propagate this label information to other LSRs in the path so that these LSRs are aware of the label and its meaning. A label distribution protocol helps to establish the LSP by providing a set of procedures by which one LSR informs another of the label-to-FEC bindings (next-hop labels) it has made. The MPLS architecture allows the use of multiple label distribution protocols, such as the following:

- Resource Reservation Protocol with TE extensions (RSVP-TE [RFC3209]) allows channels or paths to be reserved for high-band-width transmissions. The main application of RSVP-TE is traffic engineering the backbone or core networks where resources might not be available at all times. RSVP-TE assigns labels on a per-interface basis and it supports downstream on-demand label distribution. (More information on RSVP-TE can be found in the subsequent chapters.)
- Label Distribution Protocol (LDP [RFC3036]) allows LSRs to distribute labels and their meaning to LDP peers. LDP assigns labels on a per-platform basis and it supports downstream-unsolicited label distribution. Because LDP does not support traffic engineering, it can only set up best-effort LSPs. To establish an LSP, LDP does not need to rely on routing protocols at every hop along the path. LDP is required for the tunneling of L2 frames across MPLS networks. Note that throughout this book the term *LDP* refers specifically to the protocol defined in [RFC3036]; this term is avoided when discussing label distribution protocols in general. (More information on LDP can be found in the subsequent chapters.)

Two LSRs become label distribution peers when they exchange label-to-FEC binding information via a label distribution protocol, and a label distribution adjacency is established between them. The concept of label distribution peers is subjective to the set of label-to-FEC bindings. For example, the two LSRs might be LDP peers but not RSVP peers. The label distribution protocol also includes any negotiations in which two label distribution peers need to engage to be aware of each other's MPLS capabilities.

**2.2.5.2 Label Distribution Control.** An LSR can distribute label bindings to its label distribution peers in two ways:

- Independent control mode: In this control mode, each LSR, upon recognizing a particular FEC, makes a decision to bind a label to that FEC and distribute that binding to its label distribution peers independent of any other LSR. With the independent control mode, an LSR might encounter an incoming labeled packet where there is no corresponding outgoing label in the next-hop label forwarding entry (NHLFE) table. Thus, an LSR using independent control mode must be able to perform full L3 (routing) lookups.
- Ordered control mode: In this control mode, an LSR only binds a label to a particular FEC if it is the egress LSR for that FEC, or if it has already received a label binding (or next-hop label) for that FEC

from its downstream neighbor (or next-hop); otherwise, it must request a label from its next-hop. This results in an ordered sequence of downstream requests until an LSR is found to have a next-hop label or an LSR is reached that uses the independent control mode (for instance, an egress LSR). The ordered control mode can ensure that traffic in a particular FEC follows a specific path with a defined set of attributes, something that cannot be achieved through the independent control mode.

**2.2.5.3 Label Retention Modes.** There are two available label retention methods an LSR can use to store label bindings:

- Conservative retention mode: In this retention mode, an LSR retains labels only from next-hops and discards all labels received from neighbors that are not the downstream peers of any particular FEC. If any of these neighbors later becomes the next-hop for a particular FEC, the label binding will have to be reacquired from that particular next-hop neighbor.
- Liberal retention mode: In this retention mode, an LSR retains labels from all neighbors even if these neighbors are not the downstream peers of any particular FEC. In this case, the LSR may straightaway use the stored labels if one of the neighbors eventually becomes the next-hop for a particular FEC.

While an LSR using the conservative retention mode has a longer convergence time, it uses less memory because it maintains fewer labels. Even though the liberal label retention mode allows for a shorter convergence time, it also requires more memory.

## 2.2.6 Penultimate Hop Popping

The word *penultimate* means second to last, and penultimate hop refers to the second-to-last LSR in an LSP, or simply the LSR just before the egress LSR. Penultimate hop popping (PHP) allows the label stack to be popped at the penultimate LSR of the LSP rather than at the egress LSR.

In a typical single-level label stack scenario, the egress LSR is required to do two lookups, a label lookup followed by a routing lookup, if PHP is not enabled. This dual lookup is not the most optimal way of forwarding labeled packets. Penultimate hop popping eases the requirement for a dual lookup on the egress LSR. With PHP enabled, only one lookup is performed at the penultimate LSR and the egress LSR, respectively.

As illustrated in Figure 2.6, the penultimate node (Core-2) pops the stack and forwards the packet to the egress LSR based on the information provided by the label that was previously at the top of the stack. In the case of nested LSPs (where the packet is carrying more than one label), when the egress LSR receives the packet, the label that is now at the top of the stack

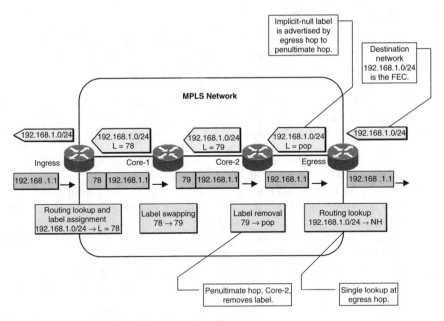

**Figure 2.6    Penultimate Hop Popping Operation**

will be the label the egress LSR needs to look up to make its own forwarding decision. If the packet is only carrying a single label (see Figure 2.6), the egress LSR will simply do a routing lookup and make its forwarding decision based on the longest-match route found in the routing table. As PHP might not be desirable in all situations, the penultimate LSR will only pop the label stack if this is specifically requested by the egress LSR through an implicit null label (see Figure 2.6 and Table 2.2).

## 2.3  MPLS APPLICATIONS

Besides reducing the amount of per-packet processing time required at each router in an IP network, MPLS also offers new service capabilities in applications, such as:

- Traffic engineering
- LSP reliability
- Virtual leased line and virtual private LAN services
- IP VPNs
- Classes of service

These MPLS applications are briefly discussed in the following sections. In-depth discussions of these intriguing topics are presented in the subsequent chapters.

### 2.3.1 Traffic Engineering

From the metro SP perspective, the outcome of traffic engineering (TE) is to have a responsive and efficiently operated network that can provide more predictable services and meet service-level agreements. An effective TE solution can balance a network's aggregate traffic load on the many links and nodes in the network so that none of these network components is over- or underutilized. This is where MPLS-TE comes into the picture because it reinforces the TE capabilities of IP networks by providing:

- Support for explicit paths that can be used to define the exact physical path taken by an LSP through the MAN.
- Constraint-based routing capabilities in link-state routing protocols, such as open shortest path first with TE extensions (OSPF-TE) and integrated intermediate system to intermediate system with TE extensions (ISIS-TE), which provides additional link-state information such as reserved bandwidth, available bandwidth, and link affinity, along with routing updates. This information is used to ensure that an LSP can meet specific performance (or QoS) requirements before it is set up.

### 2.3.2 LSP Reliability

The successful delivery of metro services depends very much on whether protections are in place against network link or node failures. The last thing metro SPs want is the effect of disruptions caused by network outages, which can easily lead to costly rectification and customer dissatisfaction. This becomes even more drastic as the amount of mission-critical data involved in new-age applications increases.

In an MPLS environment, metro SPs need to have a protection scheme in place against LSP outages. Therefore, to improve overall service reliability for the metro subscriber, protection paths or backup LSPs can be configured to speed up the restoration time of an LSP failover during an outage.

One way to achieve this is to dynamically set up the backup LSP upon the failure of the primary LSP. The other is to have a preestablished (or hot standby) backup LSP. The third alternative is to enable the MPLS fast reroute option during the setup of an LSP. This option creates detour LSPs around each point of failure in the path. During a link or node failure, the ingress LSR will be notified and the corresponding detour LSP or bypass tunnel will be used.

### 2.3.3 Virtual Leased Line and Virtual Private LAN Services

A virtual leased line (VLL) is formed by two MPLS LSPs set up in reverse directions. These LSPs can be statically preconfigured or dynamically created using MPLS signaling protocols. The particular route taken by an LSP can be defined explicitly to meet specific QoS requirements and traffic properties. In this case, the metro subscriber's network information, such as media access control (MAC) addresses and VLAN identifiers, is transparent to the SP core network because it is all mapped to MPLS labels at the network ingress. In addition, LSPs created at the edge can be nested hierarchically within LSP tunnels set up in the core network based on the MPLS label stack.

Metro SPs can extend the VLL model to offer virtual private LAN services (VPLS) when more than two sites must be interconnected. VPLS (formerly known as transparent LAN services) delivers a multipoint-to-multipoint LAN service that provides connectivity between multiple sites as if these sites were attached to the same LAN. MPLS supports the broadcasting and address learning capabilities required for emulating such a LAN.

In VLL/VPLS implementations, metro subscribers' VLANs are mapped to specific virtual channel LSPs signaled via LDP, which in turn are bundled in the SP core network within tunnel LSPs that carry traffic between multiple points of presence (POPs). These tunnel LSPs are usually signaled via RSVP-TE because traffic engineering is frequently required in the core network. When a virtual channel LSP is assigned to a subscriber, it can carry all the traffic from the subscriber despite the VLAN topology configured at the subscriber's site, and no further provisioning is required.

### 2.3.4  IP VPNs

Metro SPs can use MPLS to deliver state-of-the-art IP VPN services based on the latest draft of [RFC2547bis] with the help of Border Gateway Protocol (BGP) multi-protocol extensions [RFC2858]. The MPLS VPN implementation requires metro SPs to assign a VPN identifier for each subscriber. To ensure that the IP addresses of one metro subscriber do not overlap with the others, a combination of VPN identifiers and IP addresses is used in the forwarding tables. Thus, MPLS VPN supports the use of nonunique IP addresses within a VPN. Because the VPN information is distributed by multi-protocol BGP (MP-BGP) to members of the same VPN only, traffic isolation is also achieved.

With MPLS VPN, metro SPs can offer tiered IP VPN services coupled with different QoS and reliability guarantees for each individual metro subscriber. These simple and yet robust VPN services not only fulfill the metro subscribers' common VPN needs, but also become powerful service-delivery constructs for the metro SPs.

### 2.3.5  Class of Service

Bandwidth has become such a commodity in the metro space that it is no longer enough for metro SPs to offer just sheer bandwidth. For the customer to favor their higher-grade services, metro SPs have to differentiate their offerings with value-added features, and MPLS provides the means to deploy such offerings through the use of service classes.

Service classification is achieved through the various priority values indicated by CoS field types such as 802.1P priority bits, IP type-of-service (ToS) precedence bits, and differentiated services code point (DSCP) bits, which are in turn mapped to the Exp bits in the MPLS shim header. QoS and reliability guarantees can then be implemented based on these priority markings and classifications. There are two approaches to MPLS-based class-of-service (CoS) forwarding:

- With a single LSP, each LSR along the path will conform to a packet's priority by placing it into the proper outbound interface queue for transmission based on the priority values set on the Exp bits' field of the MPLS shim header. The packet is then serviced accordingly.
- With multiple LSPs, the LSPs are traffic engineered to meet the various performance (or QoS) as well as bandwidth guarantee requirements and are provisioned between each pair of edge LSRs (ingress and egress). The CoS-based forwarding decision is made at the ingress point of the path, and a packet will traverse the LSP that has been set up for its corresponding service class. For instance, the ingress LSR can put packets marked as high priority in one LSP, packets marked as medium priority in another, packets marked as best effort in a third LSP, and so forth.

## 2.4 SUMMARY

The first part of this chapter gives the reader a concise overview of MPLS by going through the MPLS architecture, the concept of MPLS label and label stacking, the distribution and management of MPLS labels, and the penultimate hop popping operation.

The second part of the chapter briefly describes how MPLS applications such as traffic engineering, LSP reliability, VLL, VPLS, IP VPNs, and CoS-based forwarding can present a whole series of service enablers that not only bolster the metro SP's ability to offer more value-added services, but also give the metro SP better control over its metro network. We shall take a closer look at these MPLS applications in the subsequent chapters.

# Part 2
# Traffic Engineering Aspects of Metropolitan Area Networks

# Chapter 3
# Traffic Engineering Concepts

## 3.1 INTRODUCTION

To get ready for the challenges posed by new-age metro applications (see chapter 1), metro service providers (SPs) have to ensure that their network operations are both efficient and reliable while utilizing available network resources and optimizing traffic performance. Traffic engineering (TE) helps metro SPs to optimize the performance of their operational networks at both the traffic and resource levels.

The traffic optimization aspect of TE can be achieved through capacity management [RFC3272], which consists of capacity planning, routing control, and resource management. The main objective of traffic-oriented performance is to enhance the quality of service (QoS) of traffic streams. For the best-effort network service model, the performance parameters include minimization of packet loss, minimization of delay, maximization of throughput, and conformance with service-level agreements. In the case of the differentiated network services concerning statistically bounded traffic, the performance attributes include peak-to-peak packet delay variation, packet loss ratio, and maximum packet transfer delay.

The resource optimization aspect of TE can be achieved through traffic management [RFC3272], which includes nodal traffic control functions such as traffic conditioning, queue management, scheduling, and other functions that regulate traffic flow through the network or control access to network resources among various packets or among various traffic streams. The main objective of resource-oriented performance is the efficient management of network resources, ensuring that subsets of network resources do not become overutilized and congested while other subsets along alternate feasible paths remain underutilized. Bandwidth is an essential resource in metropolitan area networks (MANs). Therefore, one of the main functions of TE is to efficiently manage bandwidth resources.

Another important goal of TE is to facilitate reliable network operations [RFC2702]. Reliable network operations can be facilitated by providing mechanisms that improve network integrity and by implementing policies

31

that reinforce network survivability [RFC3272]. This minimizes the susceptibility of the metro network to service outages due to errors, faults, and failures occurring within the network infrastructure.

## 3.2  NETWORK CONGESTION

Multiple traffic streams vie for the use of network resources such as bandwidth when they are transported through a network. Network congestion occurs when the arrival rate of data packets exceeds the output capacity of the network resources over a period of time. Congestion is highly undesirable because it increases transit delay, delay variation, and packet loss and lowers the predictability of network services. The fundamental challenge in network operation is to increase the efficiency of resource utilization while minimizing the possibility of congestion.

Reducing congestion is the main objective of TE, and it has also become a primary traffic and resource-oriented performance goal. The key focus is on prolonged congestion problems rather than on ephemeral congestion caused by instantaneous bursts. Congestion normally transpires under two circumstances:

- When network resources do not have enough capacity to contain the offered load
- When traffic streams are inefficiently allocated to available resources, resulting in subsets of network resources becoming over-utilized while other subsets remain underutilized

The first congestion scenario can be addressed by:

- Increasing the capacity of the network resources
- Applying conventional congestion control techniques such as rate limiting, window flow control, router queue management, and schedule-based control to regulate the resource demand so that the traffic is allocated to resources with the right capacity

The second congestion scenario caused by inefficient resource allocation can be resolved through TE.

## 3.3  HYPERAGGREGATION PROBLEM

What causes the inefficient allocation of network resources? The problem resides in the contemporary destination-based interior gateway protocols (IGPs) that we use for routing. IGPs such as open shortest path first (OSPF) and integrated intermediate system to intermediate system (ISIS) refer to a link-state database (LSDB) to realize the entire network topology and select the best next-hop based on the lowest cost. Consequently, all lowest equal-cost paths are selected and used to forward data. Because the lowest

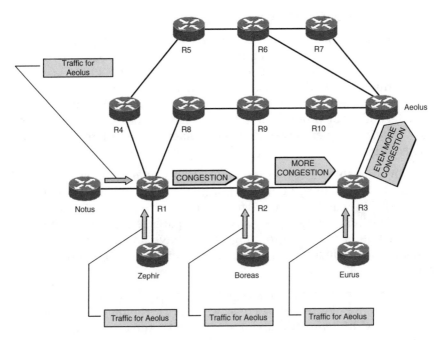

**Figure 3.1  Hyperaggregation Scenario**

equal-cost paths may not be the only paths around, alternate paths may become underutilized while the lowest equal-cost paths become overutilized. The full utilization of single or multiple equal-cost routes when other routes are underutilized, or not used in any way, is known as hyperaggregation. Figure 3.1 illustrates a hyperaggregation scenario. The network link R1–R2–R3-Aeolus happens to be the shortest path or part of the shortest path to Aeolus from Notus, Zephir, Boreas, and Eurus. As illustrated in this example, traffic streams from Notus, Zephir, Boreas, and Eurus are traversing and congesting link R1–R2–R3-Aeolus to reach Aeolus, while links R1–R4–R5–R6–R7-Aeolus, R1–R8–R9–R10-Aeolus, and R2–R9–R6-Aeolus remain underutilized.

Hyperaggregation is an unfavorable phenomenon that usually leads to chronic local congestion. Through the adoption of load-balancing policies, such congestion situations can be alleviated. Adopting load-balancing policies can reduce congestion resulting from inefficient resource allocation. The objective of such strategies is to minimize congestion or to avoid maximum resource utilization, through more efficient resource allocation, which in turn reduces packet loss, decreases transit delay, and increases aggregate throughput. In doing so, the network service quality experienced by metro subscribers will be notably improved.

## 3.4 EASING CONGESTION

As mentioned in the previous section, we can avoid a congestion situation through more efficient resource allocation or load balancing. The rudimentary way to achieve load balancing is to manipulate the IGP metrics hop by hop so that all routes to the same destination will have the same aggregate path cost. Nevertheless, adjusting IGP metrics hop by hop can be a daunting trial-and-error process, and its outcome can become unpredictable, especially in complex networks with long haul links of different speeds and bandwidths.

In an effort to utilize lesser-preferred routes, some vendor-specific IGPs (such as Cisco's Interior Gateway Routing Protocol (IGRP) and Enhanced IGRP (EIGRP)) employ a user-defined tolerance or variance to implement load balancing with multiple unequal-cost paths. Nevertheless, networks with partial- or full-mesh topologies may have unequal-cost paths that remain underutilized because they are outside the defined tolerance ranges.

As it can be an uphill task to get conventional routing protocols to make good use of unequal size links without overloading the lower speed link, bandwidth overprovisioning seems to be the other way out. Bandwidth overprovisioning is really network engineering in which bandwidth is deployed where the traffic is. For instance, in Figure 3.1, we can overprovision the network link R1–R2–R3-Aeolus so that it can comfortably accommodate all the traffic streams going to Aeolus from Notus, Zephir, Boreas, and Eurus. Not only is bandwidth overprovisioning costly, but certain links will still remain under- or overutilized. In other words, increasing bandwidth may ease congestion temporarily, but it does not guarantee an efficient resource allocation, which is the best way to purge chronic local congestion.

The first step to efficient resource allocation is to gain control of the network and direct customer traffic streams to the relevant (but not necessarily the shortest) network path or paths that meet that specific customer's requirements. This is where TE comes into the picture. Contrary to network engineering, in TE we direct traffic streams to where the bandwidth is. Put another way, traffic streams are diverged from resource-poor network segments to resource-rich network segments in TE.

As depicted in Figure 3.2 (continuation of Figure 3.1), TE attempts to distribute traffic over the currently available but underutilized network resources. To reach Aeolus, traffic streams from Notus are rerouted to link R1–R4–R5–R6–R7-Aeolus, traffic streams from Zephir are rerouted to link R1–R8–R9–R10-Aeolus, traffic streams from Boreas are rerouted to link R2–R9–R6-Aeolus, and traffic streams from Eurus are rerouted to link R3-Aeolus.

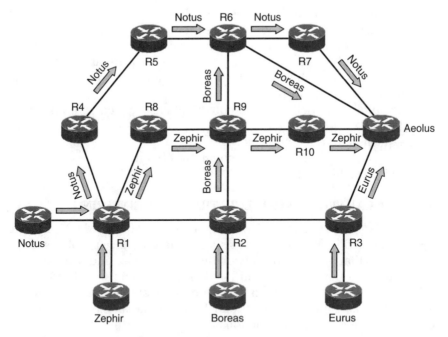

**Figure 3.2   Traffic Engineering Scenario**

## 3.5   NETWORK CONTROL

Due to changing business needs, capabilities, and operating limitations, different networks will have different optimization requirements. No matter what the optimization goals (such as proper resource allocation) are, the optimization aspects of TE can be either proactive or reactive network control. In the proactive case, the TE control system takes preemptive measures to prevent any unfavorable network events from happening, whereas in the reactive case, the TE control system responds correctively and adaptively to events that have already occurred in the network.

Some of the network control actions include:

- Adjustment of traffic management parameters. Traffic management provides TE at a node-level basis, and the tweaking of traffic management parameters provides:
  — Classification and marking
  — Policing, metering, and shaping
  — Buffering and discarding
  — Queuing and scheduling
- Manipulation of parameters associated with routing. The routing of traffic from ingress nodes to egress nodes is one of the most essential

functions performed in any network. In this aspect, TE controls and optimizes the routing function (either online or offline [RFC3272] optimization of routes) by steering traffic through the network in the most effective way using parameters that are associated with bandwidth availability and traffic characteristics to determine the most suitable network path.

- Fine-tuning of attributes and constraints associated with resources [RFC3212] such as:
  — Peak data rate and peak burst size
  — Committed data rate and committed burst size
  — Excess burst size

### 3.6   TACTICAL VERSUS STRATEGIC TRAFFIC ENGINEERING

There are two approaches to TE: tactical and strategic [RFC3272]. The objective of tactical TE is to address specific performance problems (such as hot spots) that occur in the network in an improvised and reactive manner, without proper planning and long-term consideration. Strategic TE tackles the congestion problem from a more systematic and proactive standpoint, taking into consideration the immediate and longer-term outcomes of specific policies and actions.

Even though both techniques can coexist together, the latter is still recommended in the long run because we would rather prevent congestion from evolving than permit it to transpire and then react to it accordingly, which may eventually become too late for any corrective actions.

### 3.7   IP/ATM OVERLAY MODEL

One way to get around the shortcomings of traditional IGPs (see sections 3.3 and 3.4) is through the use of an overlay model such as IP over Asynchronous Transfer Mode (ATM) (or IP over Frame Relay). This L3-overlaying-L2 approach allows virtual topologies to be provisioned on top of the network's physical topology. The virtual topology is constructed from virtual circuits (VCs), which appear to the IGPs as physical links. A full mesh of VCs is provisioned between all L3 nodes (or routers)—each L3 node (or router) has a direct VC to every other L3 node in the mesh.

Figure 3.3 illustrates both the physical and logical topologies of an overlay model. From the physical perspective, L3 nodes Boreas, Eurus, Zephir, and Notus are all interconnected via four L2 ATM switches. These ATM switches are transparent to the four L3 nodes from the logical perspective, whereby each L3 node establishes a logical any-to-any connectivity to every other L3 node with direct VCs.

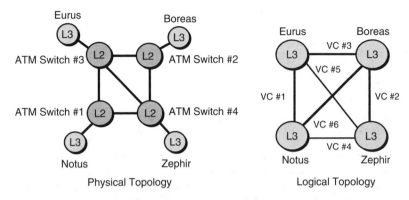

**Figure 3.3  Physical and Logical Topologies of an Overlay Model**

Put another way, the overlay model uses a separate transit L2 layer, for example, ATM (or Frame Relay), to provide TE features (such as the precise control of how traffic uses the available bandwidth) that are inadequately supported by adjusting L3 metrics alone. TE is performed at L2 as shown in Figure 3.4 (continuation of Figure 3.3), transparent to the L3 nodes. In this case, the path SW3–SW2 is utilized for traffic from Eurus to Boreas, the path SW3–SW2–SW4 is utilized for traffic from Eurus to Zephir, and the path SW3–SW1 is utilized for traffic from Eurus to Notus. From the perspective of Eurus, the L2 ATM network appears as point-to-point links between Eurus–Boreas, Eurus–Zephir, and Eurus–Notus.

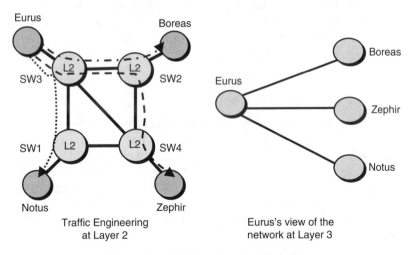

**Figure 3.4  Traffic Engineering with the Overlay Model**

The overlay model supports traffic and resource control by providing essential capabilities such as:

- Configurable explicit VC paths
- Constraint-based routing at the VC level
- Path compression
- Call admission control functions
- Traffic shaping and traffic policing functions
- Survivability of VCs

These capabilities enable the fulfillment of a range of TE policies. For instance, VCs can easily be rerouted to divert traffic from overutilized resources onto relatively underutilized ones. Even though the IP/ATM overlay model provides full traffic control and balanced flow of traffic across network links, it is not without any limitations. Some of the constraints [RFC2702] include:

- Combination of L2 (ATM) and L3 (IP) network devices, which incurs additional operational costs for both control planes
- Two different networks to administer, resulting in more complex network management
- N-squared VCs (based on full mesh formula $n*\{n-1\}/2$, where n is the number of L3 nodes) to be provisioned
- IGP scalability issue for meshes (again requiring N-squared routing adjacencies)
- Additional bandwidth overhead due to cell tax (ATM encapsulation overhead), especially for short packets

From the above-mentioned list, the limitations of the IP/ATM overlay model clearly outweigh its benefits. The overlay model is a good approach, but the IP and ATM combination needs more enhancements. The challenge is how to merge L2 and L3 together as a single network supported by a single entity or device.

### 3.8 MPLS AND TRAFFIC ENGINEERING

The many restrictions faced by the IP/ATM overlay model can be resolved by Multi-Protocol Label Switching with TE extensions (MPLS-TE), which to a certain extent has a similar level of functionality. With MPLS-TE, most of the capabilities available from the IP/ATM overlay model are provided in an integrated manner and at a lower cost. MPLS-TE can also automate the various aspects of the TE function, thus minimizing any manual intervention in the TE process. The value of MPLS-TE can be attributed to the following factors:

- LSPs can be efficiently maintained.
- Traffic trunks can be instantiated and mapped onto LSPs. A traffic trunk [RFC3272] is a unidirectional aggregation of traffic flows

belonging to the same FEC, which are forwarded through a common LSP from an ingress LSR to an egress LSR. It is also associated with a set of attributes that determine its behavioral characteristics and requirements from the network. Traffic trunks are routable entities, and the LSP through which a traffic trunk traverses can be changed. In this aspect, traffic trunks are analogous to VCs in ATM and Frame Relay networks.

- Both traffic aggregation and disaggregation (or segregation) are permissible, whereas classical destination-based IP forwarding allows only aggregation. Destination-based routing aggregates traffic, which can lead to congestion. Segregating traffic instead of aggregating it helps to redirect (or reroute) flows from overutilized segments to underutilized segments.
- Explicit LSPs that are not restricted by the destination-based forwarding paradigm can be created manually through administrative action or automatically through underlying protocols. Segregating traffic requires explicit routing (or non-destination-based routing), which can be implemented with explicit LSPs. One of the most significant functions performed by TE is the control and optimization of the routing function to maneuver traffic through the network in the most effective way. Through explicit LSPs, MPLS allows a quasi-circuit-switching capability to be superimposed on the current IP forwarding paradigm.
- Constraint-based routing and trunk protection are supported.
- A set of attributes can be associated with:
  — Traffic trunks to modulate their behavioral characteristics
  — Resources to constrain the placement of LSPs and traffic trunks across them
- It offers significantly lower overhead than competing TE options.

In the end, what truly matters is the performance of the network services as experienced by metro subscribers. As such, the common goal of metro SPs is to enhance the QoS delivered to end users of network services with MPLS-TE while at the same time taking into consideration the economic factors involved. The next section provides an overview of the MPLS-TE tasks.

### 3.9 MPLS TRAFFIC ENGINEERING TASKS

There are primarily three tasks [RFC2702] to be executed sequentially throughout the MPLS-TE process:

- *Task 1—Mapping packets onto FECs (defining flows)*: The first task of MPLS-TE is the classification of packets into separate flows or FECs. (See section 2.2.1 for a list of packet classification criteria.)

- *Task 2—Mapping FECs to traffic trunks (defining behavior)*: Once packets belonging to a particular FEC have been identified, they need to be mapped to traffic trunks and associated with resource requirements. This is normally a one-to-one mapping process in which one FEC is mapped to one trunk, although in some instances multiple FECs could be mapped to a single trunk (for example, the forwarding of bundled traffic through a metro SP backbone). At this juncture, the behavior of the traffic trunk, such as priority, bandwidth, and protection, must be administratively defined for the third task through the implementation of MPLS-TE policies.
- *Task 3—Mapping trunks to physical network topology (implementing constraint-based routing)*: This is the most important task of MPLS-TE. The traffic trunks defined in the previous tasks need to have appropriate paths through the physical network. One way to achieve this is through constraint-based routing protocols, which distribute the sets of attributes associated with the traffic trunks and the resources. Based on the exchanged information, the constraint-based routing protocols can then dynamically compute and determine the most suitable routes through the network for the traffic trunks according to their QoS needs. These constrained routes can also be determined administratively with manual configurations. Constraint-based routing is covered in more detail in the later chapters.

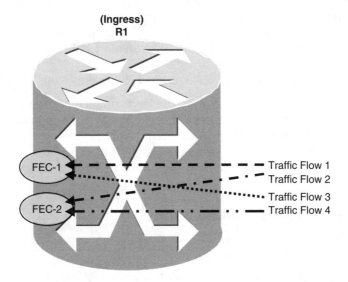

**Figure 3.5    Defining Traffic Flows**

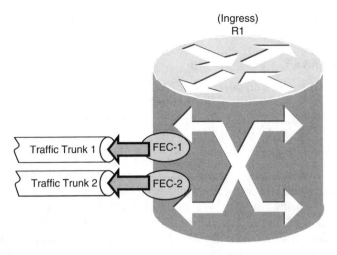

**Figure 3.6  Mapping FECs to Traffic Trunks**

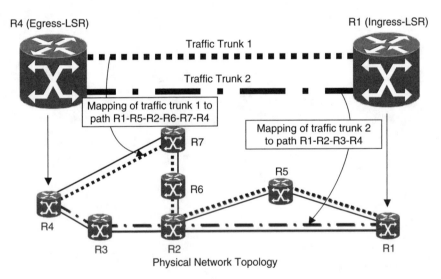

**Figure 3.7  Mapping Traffic Trunks to Physical Network Topology**

Figures 3.5 to 3.7 illustrate the three MPLS-TE tasks in order. In Figure 3.5, traffic flows 1 and 3 are assigned to FEC-1, whereas traffic flows 2 and 4 are assigned to FEC-2. Figure 3.6 is a continuation of Figure 3.5 in which FEC-1 is mapped to traffic trunk 1 and FEC-2 is mapped to traffic trunk 2. Finally, Figure 3.7 shows how traffic trunks 1 and 2 from Figure 3.6 are mapped to the actual physical network topology. In the example, R1 is the ingress LSR and R4 is the egress LSR. Traffic trunk 1 traverses path R1–R5–R2–R6–R7–R4, while traffic trunk 2 traverses path R1–R2–R3–R4.

## 3.10 SUMMARY

This chapter provides an overview of TE concepts. We first look at the causes and effects of network congestions and the hyperaggregation phenomenon. TE is then introduced as the most effective way to counter network congestion and gain network control. This is followed by comparing and contrasting tactical TE and strategic TE, stating which is the more preferred approach.

After that we discuss the applicability of the IP/ATM overlay model and MPLS to TE. The concluding section describes the three different tasks that are involved during the MPLS-TE process. Implementing a good TE policy requires a better understanding of the basic TE attributes associated with traffic trunks. Traffic trunk attributes and its characteristics will be covered in the subsequent chapter.

# Chapter 4
# Functions of Trunk Attributes

## 4.1 INTRODUCTION

As mentioned earlier in chapter 3, a trunk attribute is an administratively adjustable parameter that can be used to influence the behavioral characteristics of a traffic trunk. Trunk attributes can be either manually assigned to traffic trunks through administrative action or automatically assigned by the underlying protocols when packets are classified and mapped into forwarding equivalence classes (FECs) at the ingress to a Multi-Protocol Label Switching (MPLS) network.

The rest of this chapter describes the following basic trunk attributes (see Figure 4.1), which are used to define and modulate the behavioral characteristics of traffic trunks for traffic engineering (TE) purposes:

- Traffic parameters
- Policing attributes
- Priority attributes
- Preemption attributes

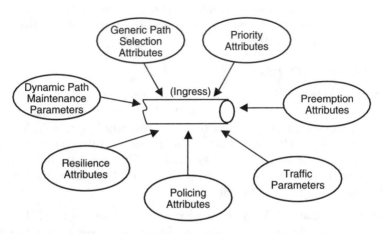

**Figure 4.1 The Taxonomy of Trunk Attributes**

- Resilience attributes
- Generic path selection attributes
- Dynamic path maintenance parameters

Because most of the traffic trunk attributes listed above have analogies in well-established technologies, they can be mapped onto the different types of existing switching and routing architectures rather easily. For instance, the traffic parameter attribute and the policing attribute together provide usage parameter control (UPC) similar to those in Asynchronous Transfer Mode (ATM) networks. UPC is a set of actions taken by the ATM network to monitor and control traffic offered by the end user. It detects violations of negotiated traffic contract parameters, and nonconforming cells may be either dropped immediately or tagged for disposal later. The priority attribute and the preemption attribute also define the way in which traffic trunks interact with each other as they contend for network resources during path establishment and path maintenance. We shall cover functions of trunk attributes in more detail in the subsequent sections.

## 4.2   TRAFFIC PARAMETERS

Traffic parameters are used to define the characteristics of the traffic streams or the FEC traversing the traffic trunk. Such characteristics may include peak rates, average (committed) rates, permissible burst size, and excess burst size. Because traffic parameters indicate the resource requirements of the traffic trunk, they are useful for resource allocation and congestion avoidance through the use of anticipatory TE policies. Traffic parameters are typically exchanged by a signaling protocol (such as Resource Reservation Protocol with TE extensions (RSVP-TE)) and are used by the ingress label switch router (LSR) to inform the network about quality-of-service (QoS) requirements or service-level agreements (SLAs). For instance, the bandwidth requirements of a traffic trunk can be derived from its associated traffic parameters, and the required bandwidth can then be reserved accordingly.

## 4.3   POLICING ATTRIBUTES

Before going into policing attributes, it is worth mentioning the generic cell rate algorithm (GCRA) as proposed by the ATM forum because adaptations of the GCRA can be used to perform the policing function in MPLS-TE. GCRA defines cell rate conformance using a set of traffic parameters. It is also known as the leaky bucket algorithm. The leak rate of the bucket defines a particular rate, while the bucket depth determines the tolerance level for accepting bursts of cells.

The functionality of the policing attribute is analogous to ATM GCRA. It determines the actions that should be taken by underlying protocols when a traffic trunk goes beyond its committed/peak rate as specified by the traffic parameters (see section 4.2). Besides performing similar GCRA functions, other possible actions for a noncompliant traffic trunk include:

- Rate limiting
- Tag (for disposal if a congestion arises later) and forward
- Forward without any policing action

## 4.4 PRIORITY ATTRIBUTES

The priority attribute defines the relative weight of traffic trunks and is used to determine the order in which path selection is done for traffic trunks during label switched path (LSP) establishment and under LSP fault scenarios. Priority values range from 0 (highest) to 7 (lowest). There are two types of priority: the setup priority, which is valid at the time the LSP is set up, and the holding priority, which is valid throughout the lifetime of the LSP. Priority attributes can also be used together with preemption attributes to enforce a preemptive policy on a set of traffic trunks when there are no available resources to establish a new LSP. Holding priorities are used to rank existing LSPs, and setup priorities are used to determine if the new LSP can preempt an existing LSP. A higher setup priority will mean that an LSP can preempt other LSPs if resources are unavailable, while a higher holding priority will imply a smaller possibility of being preempted.

Take a tiered service model for example. In this instance, LSPs can be assigned different priorities that correspond to the SLAs of different customers. If a higher-priority LSP fails and no resources are available to establish a new LSP, lower-priority LSPs' resources can be preempted. This ensures that customers with higher-priority SLAs will continue to receive guaranteed service at the expense of customers with best-effort services that have no guarantee. More details of preemption attributes are covered in the next section.

## 4.5 PREEMPTION ATTRIBUTES

Preemption plays a crucial role in achieving both traffic-oriented and resource-oriented performance objectives. In the best-effort service model, preemption is not a must. Conversely, in the differentiated services environment, the need for preemption becomes more crucial because it ensures that high-priority traffic trunks can always be routed through relatively favorable paths. Preemption can also be used to implement a range of prioritized restoration policies and reduce the time for the restoration of high-priority traffic trunks following fault events.

A preemption attribute determines whether a trunk may preempt another trunk, and it must interact closely with the priority attributes to be effective. There are four preempt modes for a traffic trunk:

- *Preemptable*: The trunk may be released if higher-priority trunks request new resources.
- *Nonpreemptable*: The trunk cannot be preempted by any other trunks, regardless of relative priorities.
- *Allowed to preempt*: The trunk may disconnect lower-priority trunks designated as preemptable to facilitate its own new resource request.
- *Not allowed to preempt*: The trunk cannot perform preemption.

In addition, for a traffic trunk (hereby referred to as trunk 1) to preempt another traffic trunk (hereby referred to as trunk 2), all of the following five conditions must be met:

- Trunk 1 has a relatively higher priority than trunk 2.
- Trunk 1 competes for a resource that is currently utilized by trunk 2.
- The resource does not have the capacity for both trunk 1 and trunk 2 at the same time.
- Trunk 1 is allowed to preempt.
- Trunk 2 is preemptable.

## 4.6 RESILIENCE ATTRIBUTES

The resilience attribute determines the behavior of a trunk under fault conditions and requires close interaction between MPLS and the routing process. This attribute facilitates fault detection, failure notification, and service restoration in the MPLS network with the following possible scenarios:

- Do not reroute the traffic trunk when there is a survivability scheme already in place that ensures service continuity under failure situations. The survivability scheme includes the provisioning of multiple parallel LSPs such that the failure of one LSP is transparently offset by the remaining LSPs.
- Only reroute through a feasible path with enough resources, but do not reroute if there are no available routes with the required resources.
- Reroute through any suitable path regardless of resource constraints.

The resilience attribute can be further classified into two types: standard and extended. A standard resilience attribute is a binary variable that determines whether the affected traffic trunk is to be rerouted when portions of its LSP fail. The extended resilience attributes are used to define the actions to be taken under fault scenarios. For instance, under fault conditions, an extended resilience attribute can be used to specify which set of alternate paths to use, and it also defines the relative preference of each specified path.

## 4.7  GENERIC PATH SELECTION ATTRIBUTES

Generic path selection attributes are used to define the policy for selecting the route taken by a traffic trunk. Paths can be either selected automatically by the underlying routing protocols or defined administratively by a network operator. A topology-driven (link-state) routing protocol can be used to select the paths for a traffic trunk if there are no resource restrictions associated with the trunk. If resource requirements do exist, a constraint-based routing protocol should be used for path selection instead. Constraint-based routing and its related attributes are discussed in the next chapter. Issues regarding explicit paths instantiated through administrative action are discussed in the following subsections.

### 4.7.1  Administratively Specified Explicit Paths

An administratively specified explicit path (or LSP) for a traffic trunk is one that is configured manually and can be fully specified (strict route) or partially specified (loose route). A path is considered fully specified if all of the transit hops in between the path are defined. In other words, the path must go through the specified routers and must not include other unspecified routers. A path is considered partially specified if only a subset of transit hops are defined. In this instance, the underlying protocols are required to determine the rest of the incomplete path through unspecified routers.

### 4.7.2  Preference Value for Parallel Paths

A list of parallel explicit paths can be administratively specified for a given traffic trunk and a preference value can be defined for each path. During path setup, the preference value is used to identify a desired path from the list of paths, whereas under failure circumstances, the preference value is used to determine a preferred alternate path from the list.

## 4.8  DYNAMIC PATH MANAGEMENT PARAMETERS

Dynamic path management parameters are used to define the policy for the maintenance of established paths. In other words, they take care of the maintenance of paths traversed by traffic trunks. From the operation perspective, it is important that an MPLS implementation can dynamically reconfigure itself according to different network state changes. The adaptability and resilience (see section 4.6) attributes are used for dynamic path management. The adaptability attribute is described in the following subsection.

### 4.8.1  Adaptability Attributes

Network characteristics and network state change constantly. Put another way, new resources are made available over time, failed resources are reactivated again after a while, allocated resources are unallocated in due course, and more efficient paths become available after some time. From the TE perspective, it is essential to have administrative control parameters in place so that we can define how traffic trunks react to these changes. On some occasions, it might be useful to dynamically change the paths of certain traffic trunks in response to changes in network state. This process is known as reoptimization.

When a path is less than optimal, it becomes important to reoptimize it. For instance, if a high-priority LSP preempts a medium-priority LSP, a suboptimal medium-priority LSP might be established as a result. Reoptimization ensures that the medium-priority LSP will be reinstated with its initial characteristics once resources become available again.

An adaptability attribute can be used to enable or disable reoptimization for traffic trunks and is part of the path maintenance parameters associated with the trunks. If reoptimization is enabled, then a traffic trunk can be rerouted through different paths by the underlying protocols in response to changes in network state (mostly due to changes in resource availability). On the other hand, if reoptimization is disabled, then the traffic trunk is permanently fixed to its established path and cannot be rerouted in the event when the network state changes. Therefore, reoptimization should be enabled whenever possible.

## 4.9  BASIC OPERATIONS OF TRAFFIC TRUNKS

The operations of traffic trunks function cohesively with the traffic trunk attributes. It is therefore worthwhile to briefly describe the basic operations of traffic trunks as follows:

- Establish operation: This operation creates an instance of a traffic trunk.
- Activate operation: This operation initiates a traffic trunk to begin traffic delivery.
- Deactivate operation: This operation causes a traffic trunk to cease traffic delivery.
- Modify attributes operation: This operation modulates the attributes of a traffic trunk.
- Reroute operation: This operation allows a traffic trunk to change its route either manually through administrative action or automatically by the underlying protocols.
- Teardown operation: This operation removes an instance of a traffic trunk from the network and releases all resources such as label space and available bandwidth allocated to the trunk.

Additional operations, such as policing and traffic shaping, can also be included to enhance the overall efficiency of the traffic trunks.

## 4.10  SUMMARY

The traffic trunk attributes are only covered generically in this chapter. Table 4.1 summarizes the various traffic trunk attributes described in this chapter together with their respective functionalities. Each of these attributes will correspond to a specific underlying protocol attribute, to be discussed in the later chapters, together with their applications.

**Table 4.1  Summary of the Traffic Trunk Attributes and Their Corresponding Functions**

| Description | Functions |
|---|---|
| Traffic parameters | Define the resource requirements of the traffic streams (or the FEC) traversing a traffic trunk. |
| Policing attribute | Determines the actions that should be taken by underlying protocols when a traffic trunk goes beyond its committed/peak rate as specified by the traffic parameters. |
| Priority attribute | Defines the relative importance of traffic trunks and is used to determine the order in which path selection is done for traffic trunks during LSP establishment and under LSP fault conditions. |
| Preemption attribute | Determines whether a trunk may preempt another trunk and it must interact closely with the priority attributes to be effective. |
| Resilience attribute | Determines the behavior of a trunk under fault conditions and requires close interaction between MPLS and the routing process. It facilitates fault detection, failure notification, and service restoration in the MPLS network. |
| Generic path selection attributes | Control how the path for a traffic trunk is selected. Paths can be either selected automatically by the underlying routing protocols or defined administratively by a network operator. |
| Dynamic path maintenance parameters | Define the policy for the maintenance of established paths. These attributes take care of the maintenance of paths traversed by traffic trunks. |
| Adaptability attribute | Enables or disables reoptimization for traffic trunks. It is part of the path maintenance parameters associated with the traffic trunks. |

# Chapter 5
# Constraint-Based Routing

## 5.1 INTRODUCTION

In Multi-Protocol Label Switching with traffic engineering extensions (MPLS-TE), paths need to be determined for traffic trunks and the path selection can be done either administratively by an operator or automatically by underlying routing protocols. For scalability reasons, the automation of the path selection process is preferred. However, most classic interior gateway protocols (IGPs) do not optimize network resources because they lack the capability to assess resource availability. Put another way, the routing criteria of these IGPs are based on shortest paths that are calculated using a simple additive metric, whereas factors such as resource availability and traffic characteristics are not considered.

One way to alleviate this limitation is to deploy quality-of-service (QoS) routing [RFC2386] instead. QoS routing itself contributes another limitation whereby it does path selection for an individual traffic flow only. The need to generalize QoS routing arises because traffic trunks in TE are basically aggregation of traffic flows. Constraint-based routing (CBR) is a generalization of QoS routing. It is applicable to traffic aggregates as well as individual traffic flows and takes a specified set of constraints such as bandwidth, hop count, delay, and policy mechanisms (for example, the resource class attributes) into consideration when making routing decisions. These constraints can be imposed either by the network itself or by administrative policies.

## 5.2 MPLS-TE OPERATION

The entire MPLS-TE operation involves the following processes:

- Information distribution
- Path selection
- Path setup
- Forwarding traffic across the path

The above processes are handled by two separate but interrelated protocols working closely together to achieve the network control (see section 3.5) that is required in MPLS-TE. A CBR protocol is used for dynamic information distribution and path calculation. Information distribution and path calculation are explained further in later sections. Once the path is calculated, it is handed over to a signaling protocol (for example, Resource Reservation Protocol with TE extensions (RSVP-TE)), which then takes care of the path setup. Path setup and traffic forwarding are discussed in the next chapter.

## 5.3  LIMITATIONS OF TRADITIONAL IGPs

Routers running on distance vector IGPs will be not know the full topology of the network domain they are in. By design, this is how a distance vector IGP works. For TE to function properly, these routers must have full network visibility. Only link-state or topology-driven IGPs can satisfy this requirement.

As such, traditional topology-driven IGPs such as open shortest path first (OSPF) and integrated intermediate system to intermediate system (ISIS) are used to disseminate network information among label switch routers (LSRs) during MPLS-TE. However, the routing decisions of these IGPs are based on shortest paths computed using an additive cost metric. Constraints such as resource availability and traffic characteristics are not taken into account. A control mechanism based solely on a single additive link metric is insufficient for MPLS-TE to support a connection-oriented service with guaranteed bandwidth and QoS. Hence, there is a need to enhance these contemporary IGPs with TE extensions so that information related to resource availability and traffic characteristics, which are deemed necessary in MPLS-TE, can be distributed accordingly between the LSRs. These standard link-state IGPs are transformed into constraint-based IGPs with TE extensions that incorporate resource attributes.

## 5.4  RESOURCE ATTRIBUTES

To take resource availability and traffic characteristics into consideration during path determination, topology-driven IGPs need to have link attributes that are associated with network resources. While trunk attributes (see Chapter 4) are configured at the ingress of a traffic trunk, resource attributes are part of the topology state parameters that are used to constrain the routing of traffic trunks through specific resources. Resource attributes are configured on every link in a network and are flooded throughout the network.

### 5.4.1  Traffic Engineering-Specific Metric

The TE-specific metric is used to specify the link metric for TE purposes. Unless explicitly defined, the TE link metric is the standard link metric (for example, the cost metric) used by legacy topology-driven or link-state IGPs. This attribute is advertised throughout the network.

### 5.4.2 Maximum Link Bandwidth

By default, the maximum link bandwidth defines the true link capacity (or physical link speed). This attribute can be used to set the maximum unidirectional (going outward from local LSR) bandwidth (apart from the actual physical capacity) that can be used on a particular link. This attribute is advertised throughout the network.

### 5.4.3 Maximum Allocation Multiplier

As defined in [RFC2702], the maximum allocation multiplier (MAM) of a resource is an administratively configurable attribute that determines the proportion of the resource that is available for allocation to traffic trunks. This attribute modulates the link bandwidth. The concept of MAM is similar to that of subscription ratios in which the values of the MAM can be adjusted so that a resource can be underallocated (undersubscribed) or overallocated (oversubscribed).

A resource is underallocated if the aggregate demands of all traffic trunks (as specified in the trunk traffic parameters) that can be allocated to it are always less than the maximum capacity of the resource. A resource is overallocated if the aggregate demands of all traffic trunks allocated to it exceed the maximum capacity of the resource. Underallocation can be used to curtail the utilization of resources, while overallocation can be used to utilize the statistical behavior of traffic for the implementation of more efficient resource allocation policies. MAM is a nonadvertising attribute.

### 5.4.4 Maximum Reservable Link Bandwidth

The maximum reservable link bandwidth defines the maximum unidirectional bandwidth that can be reserved on a particular link. When a link is oversubscribed, this value will be greater than the maximum link bandwidth. This attribute is advertised throughout the network and can be modulated by the maximum allocation multiplier. Its default value is the maximum link bandwidth.

### 5.4.5 Unreserved Bandwidth

The unreserved bandwidth corresponds to the bandwidth that can be reserved with a setup priority of 0 through 7 (see section 4.4). The initial values (before any bandwidth is reserved) are all set to the maximum reservable link bandwidth. Each value will be less than or equal to the maximum reservable link bandwidth.

### 5.4.6   Resource Class Attributes

A resource class attribute can be associated with both a network link and a traffic trunk. These two different aspects of the resource class attribute are discussed in the following subsections.

**5.4.6.1   Resource Class Attributes and Network Links.** Resource class attributes can be regarded as "colors" that are administratively assigned to resources such that the set of resources with the same color conceptually belong to the same resource class or administrative group. The resources in this aspect are network links. When applied to network links, the resource class attribute is in effect a link-state parameter.

The resource class attributes are used to implement a range of policies with respect to both traffic- and resource-oriented performance optimization. Specifically, resource class attributes can be used to:

- Apply the same policies to a set of resources that are in different topological regions
- State the preferred sets of resources during path placement of traffic trunks
- Explicitly confine the placement of traffic trunks to certain subsets of resources
- Implement universal inclusion or exclusion policies
- Perform traffic control policies that contain local traffic within specific topological regions of the network
- Identify specific network links

Typically, a network link can be assigned more than one resource class attribute. To implement more specific traffic control policies or to manipulate network traffic in a particular way, extra resource class attributes can be assigned on top of existing ones.

**5.4.6.2   Resource Class Attributes and Traffic Trunks.** The resource class affinity attributes associated with a traffic trunk are adapted directly from the resource class attributes. Similarly, these policy attributes indicate the class of resources to be explicitly included or excluded from the path of the traffic trunk at the ingress to an MPLS network. In other words, resource class affinity attributes can be used to impose additional constraints on the path traversed by a particular traffic trunk. The resource class affinity attributes for a traffic trunk are represented as a sequence of tuples:

<resource class, affinity>; <resource class, affinity>; ...

The resource class parameter denotes a resource class that is associated with the traffic trunk. The affinity parameter signifies the affinity relationship that determines whether to include or exclude members of the resource class or administrative group from the path of the traffic trunk. By default, there is no requirement to explicitly include or exclude any resources from the traffic trunk's path.

As mentioned earlier (in section 5.2), a CBR protocol is used to dynamically calculate an explicit route for a traffic trunk. This explicit route can be subjected to resource class affinity constraints with the following conditions:

- For explicit inclusion policies, prune all resources (network links) not belonging to the specified classes before performing path calculation.
- For explicit exclusion policies, prune all resources (network links) belonging to the specified classes prior to performing path calculation.

**5.4.6.3 Implementing Policies with Resource Class Attributes.** As discussed in the previous sections, the resource class attributes supports the capability to include or exclude certain links for specific traffic trunks based on a user-defined policy. The path of the traffic trunk or label switched path (LSP) (also known as TE tunnel) is characterized by a 32-bit resource class affinity string along with a 32-bit resource class mask (0 = do not care, 1 = care), and the network link is characterized by a 32-bit resource class attribute string. The default value of tunnel/link bits is $0 \times 00000000$ and the default value of the tunnel mask is $0 \times FFFFFFFF$. The affinity value and mask are used to determine to which resource class(es) a network link belongs.

Figure 5.1 illustrates a simple user-defined policy (policy 1) whereby all network links have a resource class value of $0 \times 00$ (binary value of 00000000). The LSP of the traffic trunk from R1 (ingress LSR) to R5 (egress LSR) has a resource class affinity of $0 \times 00$ and a resource class mask of $0 \times 11$ (binary value of 00010001). This policy will only match network links that have a resource class attribute binary value of nnn0nnn0 (where n can be zero or one). Therefore, the possible paths that correspond with policy 1 are R1–R2–R4–R3–R5 and R1–R2–R3–R5.

Figure 5.2 illustrates policy 2, which is adapted directly from policy 1. The LSP of the traffic trunk from R1 (ingress LSR) to R5 (egress LSR) still has a resource class affinity of $0 \times 00$ and a resource class mask of $0 \times 11$. All network links have a resource class value of $0 \times 00$ except link R2–R3, which has a resource class value of $0 \times 01$ (binary value of 00000001). Similar to policy 1, this policy will only match network links that have a resource class attribute binary value of nnn0nnn0 (where n can be zero or one). This time there is only one available path: R1–R2–R4–R3–R5. Path R1–R2–R3–R5 is not selected because link R2–R3 does not concur with policy 2.

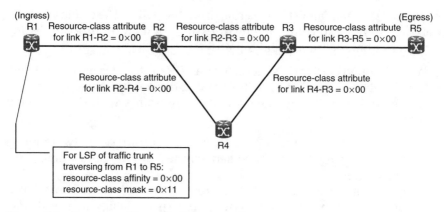

**Figure 5.1    Resource Class Attributes—Policy 1**

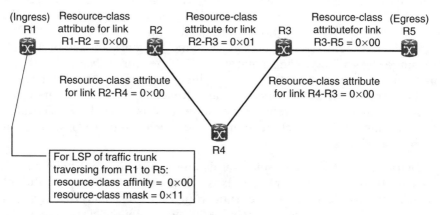

**Figure 5.2    Resource Class Attributes—Policy 2**

Policy 3 is shown in Figure 5.3. It is identical to policy 2. The only modification is on the resource class mask, which has a value of $0 \times 10$ (binary value of 00010000). This policy will only match network links that have a resource class attribute binary value of nnn0nnnn (where n can be zero or one). As in policy 1, the possible paths that concur with policy 3 are R1–R2–R4–R3–R5 and R1–R2–R3–R5.

Figure 5.4 illustrates policy 4, which again closely resembles policy 2, although the value of the resource class affinity has been changed to $0 \times 01$ (binary value of 00000001) instead. This policy will only match network links that have a resource class attribute binary value of nnn0nnn1 (where n can be zero or one). Even though the resource class attribute of link R2–R3 matches the criteria defined in policy 4, link R2–R3 is only the partial path between R1 and R5. The full paths R1–R2–R4–R3–R5 and R1–R2–R3–R5

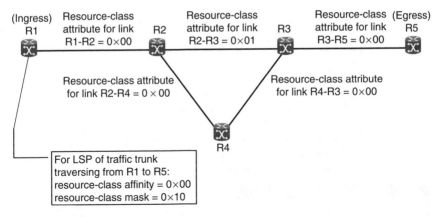

**Figure 5.3    Resource Class Attributes—Policy 3**

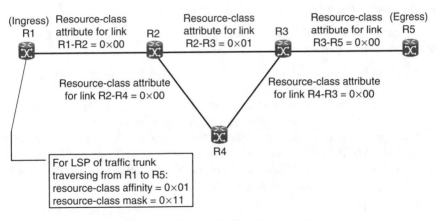

**Figure 5.4    Resource Class Attributes—Policy 4**

do not correspond with policy 4 if they are taken into consideration. Hence, there are no available paths in this scenario.

## 5.5    OSPF TRAFFIC ENGINEERING

It is a good time now to study a specific constraint-based IGP and correlate the generic resource attributes described in section 5.4 with this IGP. The constraint-based IGP of interest here is OSPF with TE extension (OSPF-TE). OSPF-TE specifies an additional link-state advertisement (LSA) type known as the traffic engineering LSA (TE LSA), which in turn makes use of the OSPF opaque LSA defined in [RFC2370]. The header and payload of the TE LSA are examined in the following subsections.

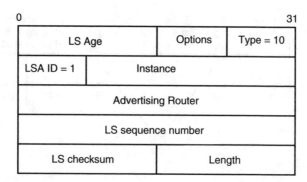

**Figure 5.5   OSPF-TE Header Format**

### 5.5.1   TE LSA Header

The TE LSA uses the standard OSPF LSA header with some slight variations in the header fields, as illustrated in Figure 5.5.

Put another way, TE LSAs are really opaque LSAs. The opaque LSA link-state type field defines the flooding scope:

- Type 9 signifies a link-local scope. Type 9 opaque LSAs are not flooded beyond the local subnet.
- Type 10 signifies an area-local scope. Type 10 opaque LSAs are not flooded beyond the borders of their associated area.
- Type 11 opaque LSAs are flooded throughout an autonomous system (AS). Their flooding scope is similar to that of AS-external (or type 5) LSAs (see [RFC2328] and [RFC2370]). Opaque LSAs with AS-wide scope are not flooded into OSPF stub areas.

The LSA ID of an opaque LSA is composed of an 8-bit type field and a 24-bit type-specific data field. In the case of the TE LSA, the 8-bit type field is assigned a value of 1 and the remaining 24 bits denote the instance field, which is an arbitrary value used to maintain multiple TE LSAs. The LSA ID has no topological significance.

### 5.5.2   TE LSA Payload

The TE LSA payload encompasses one or more nested type/length/value (TLV) triplets for extension purposes. Figure 5.6 shows the format of each TLV.

The length field defines the number of octets in the value field. The value field is padded to a four-octet alignment (32-bit aligned), but the padding is not included in the length field. The type field specifies two types of TLV:

- *Router address TLV*: The router address TLV is of type 1. It is composed of a single TLV and specifies the router ID of the advertising

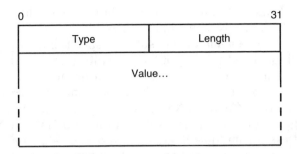

**Figure 5.6    Format of OSPF-TE TLV**

router. This TLV has a length of four octets and the value is the router ID.

- *Link TLV*: The link TLV is of type 2. It describes a single link and its value field is constructed with a set of nested TLVs or sub-TLVs. In this instance, the length field is variable because it is subjected to different sets of sub-TLVs, defined in Table 5.1.

Table 5.1 lists the link sub-TLVs required for OSPF-TE. The details of these sub-TLVs are described as follows:

- *Link type*: The link type sub-TLV is of type 1 and is one octet in length. This sub-TLV defines the type of network link. A value of 1 denotes a point-to-point link and a value of 2 signifies a multi-access link.
- *Link ID*: The link ID sub-TLV is of type 2 and is four octets in length. This sub-TLV identifies the other end of the link. For point-to-point links, this is the router ID of the neighbor, whereas for multi-access links, this is the interface address of the designated router (DR). For these link types, the link ID sub-TLV shares the same value as the link ID field in the OSPF router LSA (or type 1 LSA).

**Table 5.1    OSPF-TE Link Sub-TLVs**

| Link Sub-TLVs | Type | Length |
|---|---|---|
| Link type | 1 | 1 octet |
| Link ID | 2 | 4 octets |
| Local interface IP address (1 to N local addresses) | 3 | 4N octets |
| Remote interface IP address (1 to N neighbor addresses) | 4 | 4N octets |
| TE metric | 5 | 4 octets |
| Maximum bandwidth | 6 | 4 octets |
| Maximum reservable bandwidth | 7 | 4 octets |
| Unreserved bandwidth | 8 | 32 octets |
| Administrative group (resource class/link color) | 9 | 4 octets |

- *Local interface IP address*: The local interface IP address sub-TLV is of type 3 and is 4N octets in length, where N is the number of local addresses. The value of this sub-TLV specifies a single IP address (or multiple IP addresses) of the local interface(s) corresponding to a particular network link (point-to-point or multi-access).
- *Remote interface IP address*: The remote interface IP address sub-TLV is of type 4 and is 4N octets in length, where N is the number of neighbor addresses. The value of this sub-TLV specifies a single IP address (or multiple IP addresses) of the neighbor's interface(s) corresponding to a point-to-point link. For a multi-access link, the value of the remote interface IP address sub-TLV is set to 0.0.0.0. The local and remote interface IP addresses can be used to differentiate multiple parallel links between routers.
- *TE metric*: The TE metric sub-TLV is of type 5 and is four octets in length. This sub-TLV specifies the link metric for TE purposes. By default, the TE metric is the standard OSPF cost metric. However, this metric can be different than the standard OSPF link metric and it can be administratively assigned.
- *Maximum bandwidth*: The maximum bandwidth sub-TLV is of type 6 and is four octets in length. The maximum bandwidth sub-TLV specifies the maximum bandwidth that can be used on a particular link in the direction from the router originating the LSA to its neighbor (see Figure 5.7). By default, this is the physical capacity of the link. The units of measurement are in bytes per second.

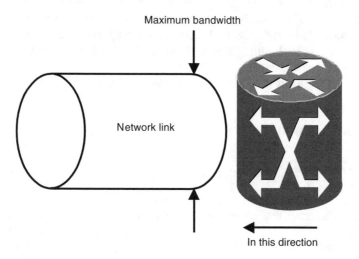

**Figure 5.7  Maximum Bandwidth Illustration**

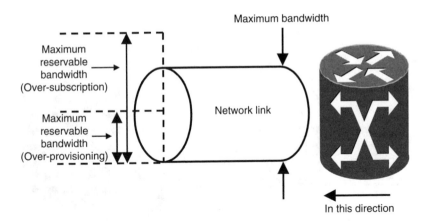

**Figure 5.8  Maximum Reservable Bandwidth Illustration**

- *Maximum reservable bandwidth*: The maximum reservable band-width sub-TLV is of type 7 and is four octets in length. This user-configurable sub-TLV specifies the maximum bandwidth that can be reserved on a particular link in the direction from the router originating the LSA to its neighbor (see Figure 5.8). By default, the value of this sub-TLV is equivalent to the maximum bandwidth. The units of measurement are in bytes per second. Overprovisioning occurs when the maximum reservable bandwidth is used up. In the case of oversubscription, the maximum reservable bandwidth can be greater than the maximum bandwidth.
- *Unreserved bandwidth*: The unreserved bandwidth sub-TLV is of type 8 and is 32 octets in length. This sub-TLV specifies the bandwidth that can be reserved with a setup priority of 0 through 7, arranged in ascending order, with priority 0 occurring at the start of the sub-TLV and priority 7 at the end of the sub-TLV. The initial values (before any bandwidth is reserved) are all set to the maximum reservable bandwidth. Each value is less than or equal to the maximum reservable bandwidth (see Figure 5.9). The units of measurement are in bytes per second.
- *Administrative group*: The administrative group sub-TLV is of type 9 and is four octets in length. This sub-TLV is composed of an administratively assigned four-octet bit mask. Each set bit corresponds to one administrative group assigned to the interface, and a link can be associated with multiple groups (see section 5.4.6). By convention, the least significant bit is referred to as group 0, and the most significant bit is referred to as group 31. The administrative group is also known as resource class or link color.

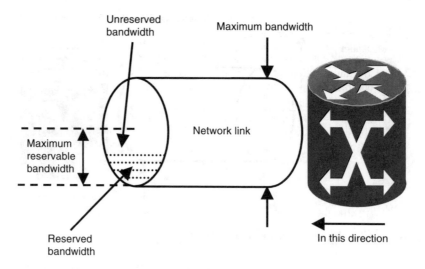

**Figure 5.9    Reserved and Unreserved Bandwidth Illustration**

## 5.6    ISIS-TE

Another alternative to OSPF-TE is ISIS with TE extension (ISIS-TE), and just like OSPF-TE, the TE capability of ISIS-TE relies on newly defined TLVs/sub-TLVs. TLVs are used to carry extra information to ISIS packets, while sub-TLVs are used to add supplement information to specific TLVs. In ISIS-TE, TLVs are located directly inside ISIS packets, whereas sub-TLVs are nested within TLVs. Each sub-TLV is composed of three fields:

- *Type*: The type field is one octet in length and is used to indicate the type of items in the value field.
- *Length*: The length field is also one octet in length and is used to denote the length of the value field in octets.
- *Value*: The value field has zero or more octets and it contains actual values of particular sub-TLVs. Typically, these sub-TLVs, as listed in Table 5.2, are part of extended intermediate system (IS) reachability TLV 22 and have the same functionalities as the link sub-TLVs of OSPF-TE.

## 5.7    CONSTRAINT-BASED ROUTING OPERATION

With the TE extensions of OSPF-TE and ISIS-TE in place, we are ready to focus on the actual operation of CBR. As illustrated in Figure 5.10, there are basically three different ways to select a path:

- Conventional topology-driven IGPs are used to select a path on a hop-by-hop basis if the traffic trunk does not have any resource requirements or policy restrictions.

**Table 5.2 Sub-TLVs of Extended IS Reachability TLV 22**

| Sub-TLVs | Type | Length |
|---|---|---|
| Administrative group (resource class/link color) | 3 | 4 octets |
| IPv4 interface address (1 to N local addresses) | 6 | 4N octets |
| IPv4 neighbor address (1 to N neighbor addresses) | 8 | 4N octets |
| Maximum link bandwidth | 9 | 4 octets |
| Maximum reservable link bandwidth | 10 | 4 octets |
| Unreserved bandwidth | 11 | 32 octets |
| TE default metric | 18 | 4 octets |

- Administratively defined explicit paths (see chapter 6) are used if automatic path selection is not needed but resource requirements or policy restrictions are required.
- Constraint-based IGPs such as OSPF-TE and ISIS-TE are used if dynamic path selection (for scalability purposes) and resource requirements or policy restrictions are required.

Regardless of whether the explicit paths are statically defined or dynamically selected, they are represented as explicit routes, which in turn are

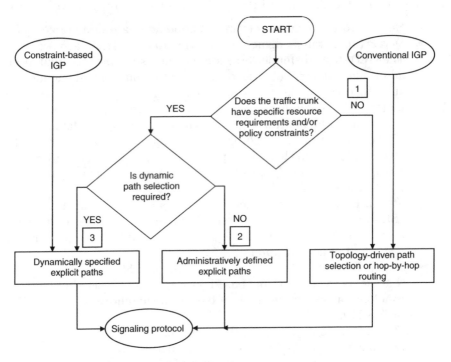

**Figure 5.10 Path Selection Flowchart**

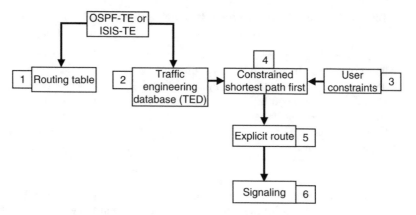

**Figure 5.11   CBR Operations Performed by the Ingress LSR**

passed to a signaling protocol for path setup (see chapter 6). In the next section, we shall discuss the detailed operations performed by the constraint-based IGP.

### 5.7.1   Operational Model of Constraint-Based Routing

As shown in Figure 5.11, the CBR operations consist of the following steps:

- *Step 1*: Store information from IGP flooding in a routing table.
- *Step 2*: Store traffic engineering information in a TE database (TED). The TED is an extended link-state database, which is built from additional link attributes associated with resources and encoded as sub-TLV objects. These link attributes include TE metric, maximum bandwidth, maximum reservable bandwidth, unreserved bandwidth, and administrative group, which are made available by the TE extensions of IGP such as OSPF-TE (see section 5.5) or ISIS-TE (see section 5.6). These new parameters are piggybacked on type 10 (opaque area-local) LSAs for OSPF-TE and are distributed by type 22 TLVs for ISIS-TE.
- *Step 3*: Examine user-defined constraints configured at the ingress LSR such as bandwidth requirement, inclusion or exclusion policies (link colors), and QoS or CoS parameters. (QoS/CoS parameters are discussed in later chapters.)
- *Step 4*: Determine the physical path for the LSP (see section 5.7.2) that meets the constraints based on steps 2 and 3 together with the topology link-state information based on the chosen IGP (OSPF-TE or ISIS-TE).
- *Step 5*: Denote computed path as an explicit route (ER). The ER can be manually specified if dynamic information distribution and automatic path calculation are not available (see chapter 6).
- *Step 6*: Pass ER to RSVP-TE for signaling (see chapter 6).

### 5.7.2 Constrained Shortest-Path-First Algorithm

CBR is a form of local constraint-based source routing and it uses the constrained shortest-path-first (CSPF) algorithm to calculate the best path for a TE tunnel to traverse. The CSPF process is instantiated by the ingress LSR (tunnel head end) when one or more of the following conditions become effective:

- A new traffic trunk requests a new tunnel.
- The existing TE tunnel of a current traffic trunk has failed.
- A current traffic trunk reoptimizes.

When one or more of the above conditions are met, the following information is taken into account during path selection:

- Attributes of traffic trunks (user-defined constraints) manually configured at ingress LSR
- Attributes associated with resources gathered from the TED of the selected IGP (OSPF-TE or ISIS-TE)
- Topology link-state information derived from the chosen IGP (OSPF-TE or ISIS-TE)

The series of tasks performed by the CSPF algorithm are listed as follows:

- *Task 1*: Prune off links that do not have sufficient resources (bandwidth) and do not comply with defined policy constraints.
- *Task 2*: Calculate the shortest-distance path on the remaining topology using standard IGP metrics or TE metrics (if specified).
- *Task 3*: Find all paths with the lowest IGP cost and choose the path with the highest minimum bandwidth. If a tie, then:
  — Choose the path with the lowest hop count (not IGP cost, just hop count). If another tie, then:
  — Just select one path at random.

Finally, when the path has been determined, it is represented as an explicit route (ER) with a sequence of router IP addresses, which are either interface addresses for numbered links or loopback addresses for unnumbered links. The derived ER is used as an input to the path setup component—a signaling protocol such as RSVP-TE (see chapter 6).

### 5.7.3 Path Selection Examples

Let us further consolidate the path selection concepts discussed in the previous section with some examples.

**5.7.3.1 Example 1.** In example 1, as shown in Figure 5.12, there are five available paths (paths 1 to 5) that a traffic trunk from R11 (the ingress) can traverse to R23 (the egress). This example does not have any imposed policy constraints, but the bandwidth requirement for all the network links

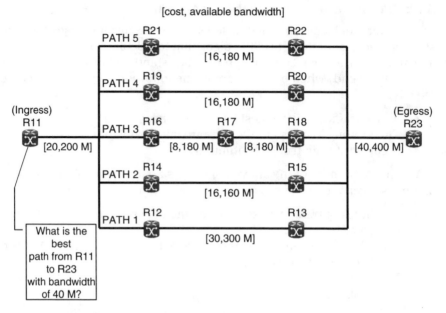

**Figure 5.12   Path Calculation Example 1**

along the path from R11 to R23 must be 40M (M stands for megabytes per second). All the network links complied with this bandwidth constraint.

Next, the cost aggregates of the five paths are as follows:

- Path 1 has a total cost of 20 + 30 + 40 = 90.
- Path 2 has a total cost of 20 + 16 + 40 = 76.
- Path 3 has a total cost of 20 + 8 + 8 + 40 = 76.
- Path 4 has a total cost of 20 + 16 + 40 = 76.
- Path 5 has a total cost of 20 + 16 + 40 = 76.

The highest IGP cost is 90 and the lowest IGP cost is 76. Based on the CSPF algorithm, path 1 is not selected (see Figure 5.13) because it holds the highest cost. The remaining four paths (paths 2 to 5) all share the same lowest cost, resulting in a tie. The breakdowns of the bandwidth allocated to the respective network links for each of these four paths are as follows:

- For path 2 (R11–R14–R15–R23):
  — Link R11–R14 has a bandwidth of 200M.
  — Link R14–R15 has a bandwidth of 160M.
  — Link R15–R23 has a bandwidth of 400M.
- For path 3 (R11–R16–R17–R18–R23):
  — Link R11–R16 has a bandwidth of 200M.
  — Link R16–R17 has a bandwidth of 180M.
  — Link R17–R18 has a bandwidth of 180M.

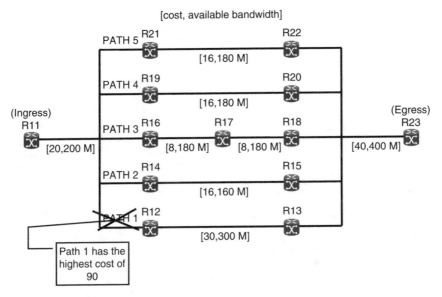

**Figure 5.13  Path Selection Criteria 1**

— Link R18–R23 has a bandwidth of 400M.
- For path 4 (R11–R19–R20–R23):
  — Link R11–R19 has a bandwidth of 200M.
  — Link R19–R20 has a bandwidth of 180M.
  — Link R20–R23 has a bandwidth of 400M.
- For path 5 (R11–R21–R22–R23):
  — Link R11–R21 has a bandwidth of 200M.
  — Link R21–R22 has a bandwidth of 180M.
  — Link R22–R23 has a bandwidth of 400M.

Because the bandwidth (200M) of the network links from R11 to its next-hops and the bandwidth (400M) of the network links from the penultimate hops to R23 remain constant for all four paths, the deciding factor lies at the transit links: R14–R15 of path 2, R16–R17 along with R17–R18 of path 3, R19–R20 of path 4, and R21–R22 of path 5. In this case, the highest minimum bandwidth is 180M. Path 2 is eliminated (see Figure 5.14) from the list of contending paths as the bandwidth of its transit link R14–R15 is only 160M, which is not the highest minimum bandwidth among the four paths.

The remaining three paths (paths 3 to 5) tie again, sharing the same highest minimum bandwidth with a value of 180M. The next tiebreaker is to select the path with the lowest hop count. Path 3 is dropped (see Figure 5.15) from the path list, as it has a hop count of 5, higher than the hop count of 4 shared by the other two paths (paths 4 and 5).

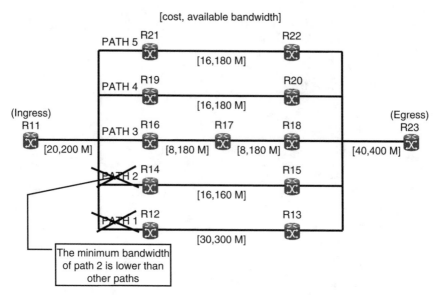

**Figure 5.14    Path Selection Criteria 2**

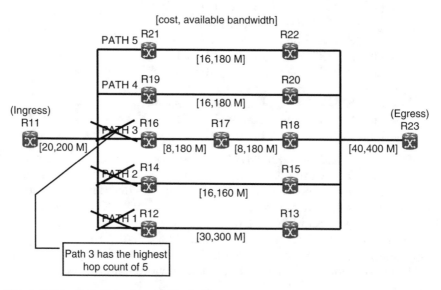

**Figure 5.15    Path Selection Criteria 3**

Because paths 4 and 5 tie yet again, the final tiebreaker is to simply choose one path at random. In this example, the final selected path is path 4 (see Figure 5.16).

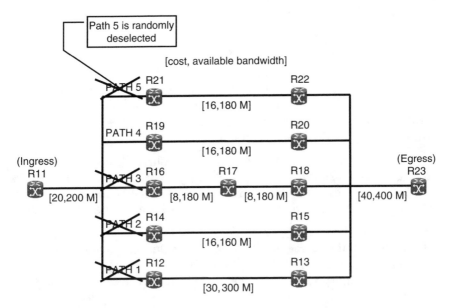

**Figure 5.16   Path Selection Criteria 4**

**5.7.3.2   Example 2.** In example 2, as shown in Figure 5.17, there are three available paths that a traffic trunk from R31 (the ingress) can traverse to R36 (the egress):

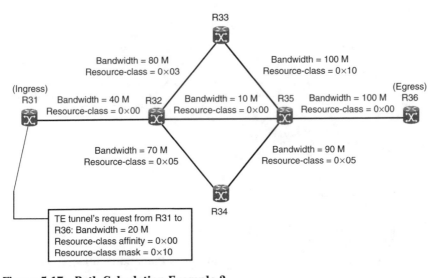

**Figure 5.17   Path Calculation Example 2**

- Path R31–R32–R35–R36
- Path R31–R32–R33–R35–R36
- Path R31–R32–R34–R35–R36

The bandwidth requirement for all the network links along the path (TE tunnel) from R31 to R36 must be 20M. All the network links except link R32–R35, which only has a bandwidth of 10M, complied with this bandwidth constraint. Therefore, link R32–R35 is pruned (see Figure 5.18) and path R31–R32–R35–R36 is not selected.

The imposed policy constraints include a resource class affinity of $0 \times 00$ and a resource class mask of $0 \times 10$. This policy will only match network links that have a resource class attribute binary value of nnn0nnnn (where n can be zero or one). The resource class attributes of the network links include:

- $0 \times 00$ = binary value of 00000000. This value is assigned to network links R31–R32, R32–R35, and R35–R36.
- $0 \times 03$ = binary value of 00000011. This value is assigned to network link R32–R33.
- $0 \times 05$ = binary value of 00000101. This value is assigned to network links R32–R34 and R34–R35.
- $0 \times 10$ = binary value of 00010000. This value is assigned to network link R33–R35.

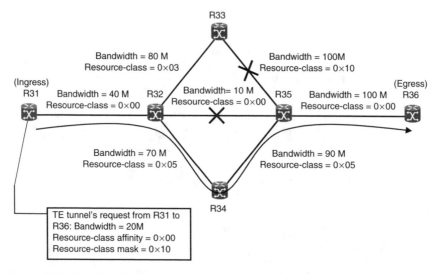

**Figure 5.18  Bandwidth Requirement and Policy Constraints Illustration**

The above resource class attribute values all match with the defined policy except the resource class attribute value of $0 \times 10$, which is assigned to link R33–R35. Therefore, link R33–R35 is also pruned (see Figure 5.18) and path R31–R32–R33–R35–R36 is omitted. The final selected path that fulfills both the specified bandwidth requirement and policy constraints is path R31–R32–R34–R35–R36.

## 5.8  SUMMARY

CBR is the key to MPLS-TE. In the MPLS-TE operation, information distribution and path calculation are taken care of dynamically with a CBR protocol. Traditional topology-driven IGPs such as OSPF and ISIS do not take into account constraints such as resource (bandwidth) availability and traffic characteristics during path calculation. Hence, there is a need to convert these conventional routing protocols into CBR protocols: OSPF-TE and ISIS-TE. This is achieved through TE extensions that incorporate link resource attributes such as maximum bandwidth, maximum reservable bandwidth, unreserved bandwidth, and administrative group (resource class/link color). These new parameters are piggybacked on type 10 (opaque area-local) LSAs for OSPF-TE and are distributed by type 22 TLVs for ISIS-TE.

The TE extensions for these CBR protocols not only provide routers a dynamic as well as a more precise perspective on network capacity, load, congestion state, and other link attributes, but they also facilitate intelligent path calculation and explicit route (ER) determination. Once the path is calculated, it is expressed as an ER. The ER is handed over to a signaling protocol (for example, RSVP-TE), which then takes care of the path setup (see chapter 6).

Put simply, CBR helps in the performance optimization of operational networks by dynamically discovering feasible paths that comply with a set of constraints defined for traffic trunks. It also considerably reduces the explicit path administrative configuration burden and the manual intervention required for the proper operation of MPLS-TE.

# Chapter 6

# Resource Reservation Protocol and Traffic Engineering

## 6.1 INTRODUCTION

Recall from chapter 5 that the Multi-Protocol Label Switching with traffic engineering extensions (MPLS-TE) operation is composed of four main processes:

- *Information distribution*: This process requires the awareness of available network resources. This "resource-aware" mechanism is implemented by defining relatively simple TE extensions to the traditional topology-driven interior gateway protocols (IGPs) (for example, open shortest path first (OSPF) and integrated intermediate system to intermediate system (ISIS)) so that additional network link attributes such as TE metric, maximum bandwidth, maximum reservable bandwidth, unreserved bandwidth, and administrative group (also known as resource class or link color) are included as part of each router's link-state advertisement (LSA).
- *Path selection*: This process uses information such as link attributes and topology-state information distributed by the IGP LSAs together with the manually configured constraints (such as path bandwidth and link color) defined at the ingress to select a constrained shortest path that fulfills the specific requirements of traffic trunks. This constraint-based path can be defined administratively offline by a network operator, or it can be computed automatically online by a constraint-based path selection algorithm, for instance, using constrained shortest path first (CSPF).
- *Path setup*: This process requires a signaling protocol to establish a label switched path (LSP) for traffic trunks along the constraint-based path or explicit route that was dynamically determined or

manually configured earlier in the path selection process. The signaling protocol of interest is the Resource Reservation Protocol with TE extensions (RSVP-TE). These extensions are the primary focus of this chapter. RSVP-TE is the most commonly used signaling protocol in MPLS-TE applications. The other alternative is CR-LDP (constraint-based LSP setup using Label Distribution Protocol [RFC3212]), which is beyond the scope of this book. At present there is no consensus on which protocol is technically better. Service providers should therefore make a choice between the two based on their requirements and networking environments. The traffic trunk attributes discussed previously in chapter 4 are mostly found in RSVP-TE.

- *Forwarding* of traffic across the established LSP.

## 6.2 TERMINOLOGY

This section gives a general conceptual overview of the terms you will come across in this chapter. Some of these terms are more specifically defined in the later sections.

- *Abstract node*: A group of nodes whose internal topology is opaque to the ingress label switch router (LSR) of the LSP. An abstract node is said to be simple if it contains only one physical node.
- *Class-Num*: A one-octet field that identifies the RSVP object class.
- *C-Type*: A one-octet field that describes an RSVP object type unique within a particular Class-Num.
- *Explicitly routed LSP*: An LSP whose path is established by means other than standard IP routing.
- *LSP (label switched path)*: The path created by the concatenation of one or more label switched hops, allowing a packet to be forwarded by swapping labels from an MPLS node to another MPLS node (see section 2.2.2).
- *LSP tunnel*: An LSP that is used to tunnel below normal IP routing or filtering mechanisms. Note that the terms *LSP* and *LSP tunnel* are used interchangeably throughout this chapter.
- *MPLS domain*: A contiguous set of nodes that operate MPLS routing as well as forwarding and are also in a single routing or administrative domain.
- *NHOP (next-hop)*: The RSVP_HOP object is referred to as a NHOP object for upstream messages.
- *PHOP (previous-hop)*: The RSVP_HOP object is referred to as a PHOP object for downstream messages.
- *Traffic-engineered tunnel (TE tunnel)*: A set of one or more LSP tunnels that carries a traffic trunk.
- *Traffic trunk*: A set of flows aggregated by their service classes and then placed on an LSP or TE tunnel (see sections 3.8 and 3.9).

- *TSPEC (traffic specification)*: Describes the characteristics of the traffic that an RSVP sender expects to generate. The TSPEC can take the form of a token bucket filter or an upper bound on the peak rate.

## 6.3 EVOLUTION OF RSVP

The standard RSVP [RFC2205] was originally defined for the integrated services (IntServ) model ([RFC1633] and [RFC2210]) in which it is used for the resource reservation of individual traffic flows between hosts to ensure that the end-to-end quality of service (QoS) requested by each of these traffic flows is satisfied. Put another way, RSVP was developed to tackle network congestion by allowing routers to decide beforehand whether they could meet the requirements of an application flow (host-to-host flow), and then reserve the desired resources if they were available. The actual (physical) path of the flow across a network was determined by destination-based hop-by-hop routing, a typical trait of conventional IGPs.

Today, RSVP is extended to RSVP-TE [RFC3209] to support the establishment and maintenance of LSPs in MPLS domains. RSVP-TE has become the unified label distribution and signaling protocol in MPLS-TE applications. It plays a crucial role in automating the TE process. The RSVP-TE specification extends the standard RSVP with new capabilities that support the following functions in an MPLS-TE environment:

- Downstream-on-demand label distribution (see section 2.2.5).
- Instantiation of explicitly routed LSPs that are no longer constrained by conventional destination-based hop-by-hop routing. Because the traffic that traverses an established LSP is identified by the label applied at the ingress LSR, the LSP can also be regarded as an LSP tunnel acting below normal IP routing and filtering mechanisms. This is because the traffic flowing through the LSP is opaque to each of the transit LSRs along the path.
- Transportation of traffic trunks. The LSPs created with RSVP-TE are used to carry traffic trunks (see sections 3.8 and 3.9). Two or more LSPs between identical source and destination can be load shared to carry a single traffic trunk. On the other hand, a single LSP can carry several traffic trunks belonging to different service classes. Either one of these approaches establishes the foundation for classes of service (CoS) (see section 2.3.5).
- Allocation of network resources (bandwidth) to explicit LSPs. The resource reservation or signaling takes place between pairs of edge LSRs, which serve as the ingress and egress points of a traffic trunk. In RSVP-TE, the resource reservation for an LSP is not mandatory. LSPs without resource reservations are used to carry best-effort traffic.

- Preemption of established LSP tunnels under administrative policy control.
- Dynamic rerouting of established LSP tunnels.
- Loop detection.
- Tracking the actual route traversed by an LSP tunnel.
- Diagnosis of LSP tunnels.
- The notion of abstract nodes. An abstract node is a group of nodes whose internal topology is opaque to the ingress node of the LSP. Using this concept of abstraction, an explicitly routed LSP can be specified as a sequence of IP prefixes or a series of autonomous systems.

There are two distinct aspects that differentiate RSVP-TE from the standard RSVP:

- RSVP-TE is implemented on LSRs (or hosts) to establish and maintain LSP tunnels as well as to reserve network resources for these LSP tunnels. Conversely, the standard RSVP was implemented only on hosts to request and reserve network resources on a per-flow basis.
- RSVP-TE allows an RSVP session to aggregate multiple traffic flows (based on local policies) between pairs of LSRs that serve as the ingress and egress points of an LSP tunnel. In other words, the resource reservation in RSVP-TE applies to a collection of flows or a traffic trunk sharing the same path and the same pool of shared network resources. This is known as aggregate reservation. Because in standard RSVP a session was defined as a single data flow with a particular destination and transport-layer protocol, the resource reservation would only be applicable to that individual data flow. This is known as per-flow reservation. Figure 6.1 differentiates per-flow reservation by standard RSVP with aggregate reservation by RSVP-TE. Because traffic flows are aggregated in RSVP-TE, the number of LSP tunnels (hence the number of RSVP sessions) does not increase proportionally with the number of flows in the network. Therefore, RSVP-TE alleviates a key scaling issue with standard RSVP, which is the large amount of system resources that would otherwise be required to support large numbers of RSVP sessions at per-flow granularity.

Even with all the above-stated enhancements, RSVP-TE is not without limitations. The RSVP-TE specification supports only unicast and unidirectional LSP tunnels. At the moment, multicast and bidirectional LSP tunnels are not supported.

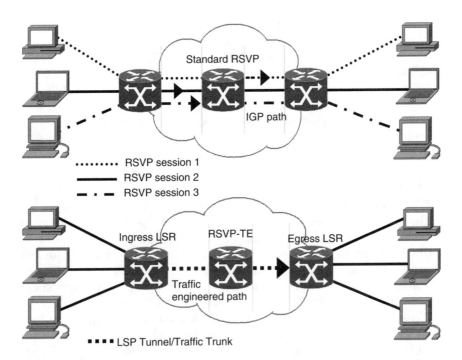

**Figure 6.1   Per-Flow Reservation versus Aggregate Reservation**

## 6.4   RSVP-TE MESSAGES AND OBJECTS

Once the path is dynamically selected or manually defined in the path selection phase (see chapter 5), it is handed over to RSVP-TE. RSVP-TE then uses PATH and RESV messages to set up an LSP tunnel with the required bandwidth reservation along the selected path. The ingress LSR sends a PATH message downstream to the egress LSR. The egress LSR responds to the PATH message with a RESV message by sending it upstream back to the ingress LSR. The LSP tunnel is established when the ingress LSR receives the RESV message. The ingress LSR then uses the LSP tunnel to forward traffic trunks to the egress LSR.

The RSVP-TE specification introduces five new objects to be included in the PATH or RESV messages. The functionalities of RSVP-TE (see previous section) are made possible through the introduction of these new RSVP objects:

- LABEL_REQUEST (found in PATH messages)
- LABEL (found in RESV messages)
- EXPLICIT_ROUTE (found in PATH messages)
- RECORD_ROUTE (found in PATH and RESV messages)
- SESSION_ATTRIBUTE (found in PATH messages)

The above-mentioned new objects are used to hold label binding and explicit routing information to support essential MPLS-TE requirements such as label distribution and explicit routing. These new object extensions do not make an RSVP-TE implementation incompatible with standard RSVP implementations. They are optional except for the LABEL-REQUEST and LABEL objects, which are both mandatory for the establishment of LSP tunnels. An RSVP implementation can easily be distinguished between LSP signaling and standard RSVP reservations by examining the objects that are encompassed within the message.

### 6.4.1 PATH Message

PATH messages are sent downstream. Specifically, a PATH message is sent by the ingress LSR toward the egress LSR when it needs to establish an LSP tunnel. In other words, the ingress LSR initiates the path setup and bandwidth reservation (optional) for this LSP tunnel. Furthermore, the PATH message addressed to the egress LSR contains the router alert IP option [RFC2113] in its IP header, indicating that the IP packet requires special processing by intermediate LSRs. When a PATH message establishes an LSP tunnel between the ingress and egress LSRs, a path state is stored in each LSR along the way using a path state block (PSB). Each PSB holds the path state for a particular (session, sender) pair, defined by the SESSION and SENDER_TEMPLATE objects found in the PATH message. The format of a PATH message is as follows:

- <Common Header>
- [<INTEGRITY>] ← optional
- <SESSION> ← new C-Type
- <RSVP_HOP>
- <TIME_VALUES>
- [<EXPLICIT_ROUTE>] ← new object, optional
- <LABEL_REQUEST> ← new object, mandatory
- [<SESSION_ATTRIBUTE>] ← new object, optional
- [<POLICY_DATA>] ← optional
- <Sender Descriptor>
  — <SENDER_TEMPLATE> ← new C-Type
  — <SENDER_TSPEC>
  — [<ADSPEC>] ← optional
  — [<RECORD_ROUTE>] ← new object, optional, sender specific

An RSVP message is composed of a common header, followed by a body consisting of variable-length objects. The format of the common header is illustrated in Figure 6.2.

The fields found in the common header are described as follows:

| Vers | Flags | Message-Type | RSVP Checksum |
|------|-------|--------------|---------------|
| Send_TTL | | (Reserved) | RSVP Length |

**Figure 6.2  RSVP Common Header Format**

- *Version field*: This field is 4 bits in length and it describes the protocol version number. The current version is 1.
- *Flags field*: This field is 4 bits in length. The values 0 × 01 to 0 × 08 are reserved.
- *Message type field*: This field is one octet in length and has the following values:
  — 0 × 01 = PATH
  — 0 × 02 = RESV
  — 0 × 03 = PATHERR
  — 0 × 04 = RESVERR
  — 0 × 05 = PATHTEAR
  — 0 × 06 = RESVTEAR
  — 0 × 07 = RESVCONF
- *RSVP checksum field*: This field is two octets in length and it denotes the checksum of an RSVP message.
- *SEND_TTL field*: This field is one octet in length. It takes its value from the IP TTL (time-to-live) value with which the message was sent.
- *RSVP length field*: This field is two octets in length and it represents the total length of an RSVP message in bytes, including the common header and the consecutive variable-length objects.

Every object contains one or more four-octet word with a one-word or four-octet header. Figure 6.3 illustrates the format of a typical RSVP object.

The object header has the following common fields:

- *Length field*: This two-octet field measures the total object length in bytes. The value is a multiple of 4 and must be at least 4.

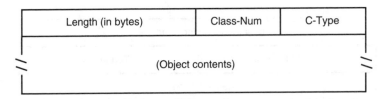

**Figure 6.3  RSVP Object Format**

- *Class-Num field*: This one-octet field identifies the object class and each object class is given a name.
- *C-Type field*: This one-octet field describes the object type, which is unique within a particular Class-Num.

The maximum object content length is 65,528 bytes. The Class-Num and C-Type fields are used together as a two-octet number to define a unique type for each object. The high-order 2 bits of the Class-Num are used to determine what action an LSR should take when it does not recognize the Class-Num of an object (see [RFC2205]). The various objects that appear in an RSVP-TE PATH message are discussed in the following subsections.

**6.4.1.1 INTEGRITY Object (RSVP Authentication).** RSVP-TE messages can be authenticated to prevent unauthorized LSRs from setting up reservations. RSVP-TE sessions must rely on other security mechanisms other than the connection-oriented level of security that comes with a reliable transport protocol such as Transmission Control Protocol (TCP). The MD5 (message-digest 5 algorithm [RFC1321]) signature authentication is used to protect RSVP-TE sessions. Only LSRs with identical secret keys running the RSVP-TE will be able to authenticate and participate in the RSVP-TE process. In other words, all LSRs connected to the same IP subnet must use the same secret key. The INTEGRITY object (Class-Num = 4) is used to carry the cryptographic data (or hashed value) that authenticates the originating LSR and verifies the contents of this RSVP message. The detail application of the INTEGRITY object is described in [RFC2747].

**6.4.1.2 SESSION Object.** The SESSION object has a Class-Num of 1 and two new C-Types are defined for RSVP-TE: LSP_TUNNEL_IPv4, C-Type = 7, and LSP_TUNNEL_IPv6, C-Type = 8. The SESSION object uniquely identifies the LSP tunnel being established. The LSP_TUNNEL_IPv4 session object carries the IPv4 address (or IPv6 address for the LSP_TUNNEL_IPv6 session object) for the egress LSR (endpoint) of an LSP tunnel, a unique 16-bit tunnel ID, and a unique 32-bit (or 16 bytes for the LSP_TUNNEL_IPv6 session object) extended tunnel ID (ingress LSR places its IPv4 address in this field). Both tunnel IDs remain constant over the life of the LSP tunnel, even when the tunnel is rerouted. Figure 6.4 illustrates the format of the SESSION object.

| IPv4 tunnel endpoint address | |
|---|---|
| MUST be zero | Tunnel ID |
| Extended Tunnel ID | |

**Figure 6.4    SESSION Object Format**

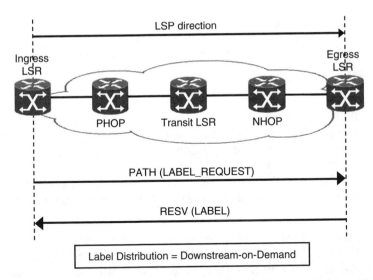

**Figure 6.5   PHOP and NHOP with Reference to the Flow of PATH and RESV Messages**

**6.4.1.3   RSVP_HOP Object.** The RSVP_HOP object has a Class-Num of 3 and two C-Types are defined: IPv4, C-Type = 1, and IPv6, C-Type = 2. The IPv4 RSVP_HOP object carries the IPv4 address (or IPv6 address for the IPv6 RSVP_HOP object) of the RSVP-capable LSR that sent this message and an interface handle that provides the necessary information to attach the reservation to the correct interface. The RSVP_HOP object is referred to as a previous-hop (PHOP) object for downstream messages and a next-hop (NHOP) object for upstream messages (see Figure 6.5).

**6.4.1.4   TIME_VALUES Object.** The TIME_VALUES object (Class-Num = 5, C-Type = 1) contains the value for the refresh period used by the origina-tor of the message. This object is required in every PATH and RESV mes-sage. Each LSR chooses the refresh time locally when originating these messages.

**6.4.1.5   EXPLICIT_ROUTE Object.** The EXPLICIT_ROUTE object (ERO) has a Class-Num of 20 and a C-Type of 1. It defines a predetermined path for the LSP tunnel across the MPLS domain. When the ERO is present, the RSVP PATH message is forwarded toward the egress LSR along the path specified by the ERO, independent of the conventional IP routing. In other words, ERO is the representation of the path selected in the constraint-based routing (CBR) process and it explicitly specifies the route that the RSVP PATH message must take. The ERO is intended to be used only for unicast applications and only when all routers along the explicit route sup-port RSVP and the ERO.

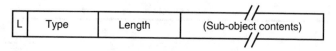

**Figure 6.6    ERO Subobject Format**

An explicit route is encoded as a series of variable-length subobjects contained in the ERO. Each subobject can identify a group of LSRs in the explicit route or can specify an operation to be performed along the path. An ERO is therefore a specification of which LSRs are to be traversed along the path. Each group of LSRs is referred to as an abstract node. When an abstract node is composed of only one node, it is called a simple abstract node.

For instance, an explicit route might contain only autonomous system (AS) number subobjects, in which each subobject is associated with an AS in the global topology. In this case, each AS is an abstract node, and the explicit route is a path that composes each of these specified ASs. Although there might be multiple hops within each AS, they are opaque to the source node of the explicit route. Figure 6.6 illustrates how each subobject is encoded in the ERO.

The ERO subobject has the following common fields:

- *L-bit*: The L-bit is an attribute of the subobject. When the L-bit is set, the subobject represents a loose hop in which the explicit route from the previous LSR to this hop can include other LSRs found by IGPs. When the L-bit is cleared, it represents a strict hop in which the explicit route from the previous LSR to this hop is directly connected and cannot include other LSRs.
- *Type field*: The type field indicates the type of contents for the subobject. Three types of ERO subobjects are currently defined with the following values:
  — 0 × 01 = IPv4 prefix: This value indicates an abstract node composed of a set of LSRs that have an IPv4 prefix. A prefix with a 32-bit length represents a single IPv4 LSR.
  — 0 × 02 = IPv6 prefix: This value indicates an abstract node consisting of a set of LSRs that have an IPv6 prefix. A prefix with a 128-bit length represents a single IPv6 LSR.
  — 0 × 20 = autonomous system number: This value identifies an abstract node composed of a set of LSRs belonging to this autonomous system.
- *Length field*: The length field contains the total length of the subobject in bytes, including the L-bit, type field, and length field. The length value is a multiple of 4 bytes and the minimum is 4 bytes.

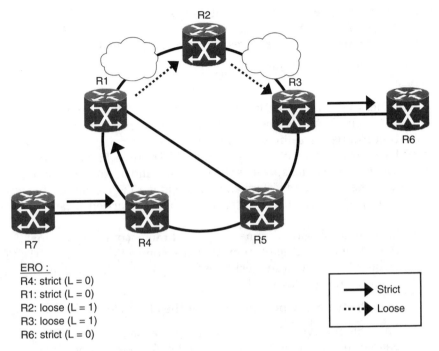

ERO :
R4: strict (L = 0)
R1: strict (L = 0)
R2: loose (L = 1)
R3: loose (L = 1)
R6: strict (L = 0)

→ Strict
·····▶ Loose

**Figure 6.7   Explicit Route Example**

An example of a strict/loose explicit route defined using 32-bit IPv4 prefixes is illustrated in Figure 6.7.

The reader should take note that the presence of loose hops in an explicit route can result in forwarding loops formed by the underlying routing protocol during transients. Nevertheless, loops in an LSP tunnel can be detected using the RECORD_ROUTE object (see section 6.4.1.12).

**6.4.1.6   LABEL_REQUEST Object.**   The LABEL_REQUEST object (LRO) is a new mandatory object in a PATH message added at the ingress LSR requesting each LSR to provide an MPLS label to its upstream LSR. Put another way, when the ingress LSR establishes an LSP tunnel, a PATH message containing an LRO is generated. The presence of an LRO signifies that a label binding is requested for this LSP. The LSR that accepts the LRO will assign a label and place it in a LABEL object on the corresponding RESV message. If a label range is specified, the label allocation process must assign a label from that range.

The LRO also contains the layer-3 protocol ID (L3PID) that identifies the L3 protocol that will traverse the LSP tunnel. The L3PID is required because the L3 protocol cannot be inferred from the L2 header, which simply identifies the higher-layer protocol as MPLS. When an LSR receives a PATH message, it stores the LRO in the local PSB (see section 6.4.1) of the LSP so that PATH refresh messages will also contain the LRO.

The LRO has a Class-Num of 19 and there are currently three possible C-Types:

- Label request without label range (type 1). This is the typical case in which the MPLS label is carried in a standard MPLS shim header that sits between the L2 and L3 headers.
- Label request with an Asynchronous Transfer Mode (ATM) label range (type 2) that specifies the minimum and maximum virtual path identifier (VPI) and virtual channel identifier (VCI) values. This type of request is used when the MPLS label is carried in an L2 ATM header.
- Label request with a frame relay label range (type 3) that specifies the minimum and maximum data-link connection identifier (DLCI) values. This type of request is used when the MPLS label is carried in an L2 frame relay header.

Some of the typical error conditions for the LRO follow:

- The LSR receiving the PATH message recognizes the LRO but is unable to support it; for instance, a label range has been specified but a label cannot be allocated from that range. When this happens, the LSR will send a PATHERR message with error code "routing problem" and error value "MPLS label allocation failure" to the ingress LSR.
- The LSR receiving the message does not recognize the LRO. When this happens, the LSR will send a PATHERR message "unknown object class" to the ingress LSR. This error will cause the LSP tunnel setup to fail.
- The LSR cannot support the L3PID. When this happens, the LSR will send a PATHERR message with error code "routing problem" and error value "unsupported L3PID" to the ingress LSR. This error will cause the LSP tunnel setup session to fail.

**6.4.1.7  SESSION_ATTRIBUTE Object.** The SESSION_ATTRIBUTE object (SAO) can be included in the RSVP PATH message for session identification and diagnosis purposes. The SAO has a Class-Num of 207 and two new C-Types are currently defined: LSP_TUNNEL, C-Type = 7, and LSP_TUNNEL_RA, C-Type = 1. The LSP_TUNNEL_RA C-Type carries resource affinity information (see section 5.4.6) in addition to all the fields found in the LSP_TUNNEL C-Type.

The SAO also controls LSP tunnel priority, preemption, and local rerouting (or fast reroute) features. Preemption is implemented by two priorities:

- Setup priority, which indicates the priority of this LSP tunnel with respect to taking resources from other existing LSP tunnels, in the range of 0 to 7. The value 0 is the highest priority. The setup priority is used to determine whether a particular LSP tunnel can preempt another. It should never be higher than the holding priority for a given session.
- Holding priority, which indicates the priority of this LSP tunnel with respect to holding resources that can be consumed by other LSP tunnels, in the range of 0 to 7. The value 0 is the highest priority. The holding priority is used to determine whether a particular LSP tunnel can be preempted by another LSP tunnel. Specifically, the holding priority is the priority at which resources assigned to this LSP tunnel will be reserved.

Local repair (or fast reroute) is signified by a specific 8-bit flag value. The SAO has an 8-bit flag field with the following values:

- $0 \times 01$ = Local protection desired: This flag permits transit LSRs to use a local repair mechanism that may result in violation of the ERO. With this flag on, a transit LSR can reroute traffic for fast service restoration when a fault is detected on an adjacent downstream link or node. The ingress LSR notifies the transit LSRs that they can implement the MPLS fast reroute feature (see chapter 8 for more details).
- $0 \times 02$ = Label recording desired: This flag indicates that label information should be included in the RRO (see section 6.4.1.12).
- $0 \times 04$ = Shared-explicit (SE) style desired (or ingress LSR may reroute bit): This flag indicates that the tunnel ingress LSR can reroute the LSP tunnel without tearing it down, thus fulfilling the concept of make-before-break (see section 6.8). With this flag on, the ingress LSR informs the egress LSR to use the SE reservation style (see section 6.4.2.2) when responding with the corresponding RESV message.

The SAO also includes fields for resource affinities (applicable to LSP_TUNNEL_RA only):

- *Exclude-any field*: A 32-bit vector representing a set of attribute filters associated with an LSP tunnel, any of which renders a link unacceptable.
- *Include-any field*: A 32-bit vector representing a set of attribute filters associated with an LSP tunnel, any of which renders a link acceptable. A null set, with all bits set to zero, automatically passes.
- *Include-all*: A 32-bit vector representing a set of attribute filters associated with a tunnel, all of which must be present for a link to be acceptable. A null set, with all bits set to zero, automatically passes.

**6.4.1.8 POLICY_DATA Object.** The POLICY_DATA object (Class-Num = 14, C-Type = 1) carries information allowing a local policy module to decide whether an associated reservation is administratively permitted. This object can appear in a PATH, RESV, PATHERR, or RESVERR message.

**6.4.1.9 SENDER_TEMPLATE Object.** RSVP PATH messages are required to carry a SENDER_TEMPLATE object (STO), which describes the type of data packets (or traffic trunk) for which this specific sender will originate. This template is used to distinguish this sender's packets from the rest in the same session on the same link. The STO is similar to the FILTER_SPEC object (see section 6.4.3.5).

The STO has a Class-Num of 11 and two new C-Types are defined for RSVP-TE: LSP_TUNNEL_IPv4, C-Type = 7, and LSP_TUNNEL_IPv6, C-Type = 8. The LSP_TUNNEL_IPv4 STO contains the IPv4 address (or IPv6 address for the LSP_TUNNEL_IPv6 STO) of the sending LSR and a unique 16-bit LSP ID that can be modified to allow a sender to share resources with itself. This LSP ID is used when an established LSP tunnel with SE reservation style is rerouted (see section 6.8). Figure 6.8 illustrates the format of the STO.

**6.4.1.10 SENDER_TSPEC Object.** The SENDER_TSPEC object (Class-Num = 12, C-Type = 2) carries link management configuration information such as the requested bandwidth reservation along with the minimum and maximum packet sizes supported by this LSP tunnel. The SENDER_TSPEC object is required in a PATH message.

**6.4.1.11 ADSPEC Object.** The ADSPEC object (Class-Num = 13, C-Type = 2) carries one-pass-with-advertising (OPWA) data in a PATH message. OPWA describes a reservation setup model in which PATH messages sent downstream gather information that the receiver can use to predict the end-to-end QoS. The information collected is referred to as an advertisement. The receiver uses the information provided by the advertisement to construct or modulate a suitable reservation request.

**6.4.1.12 RECORD_ROUTE Object.** The RECORD_ROUTE object (RRO) has a Class-Num of 21 and a C-Type of 1. The primary role of RRO is to record routes, which allows the ingress LSR to receive a listing of the LSRs

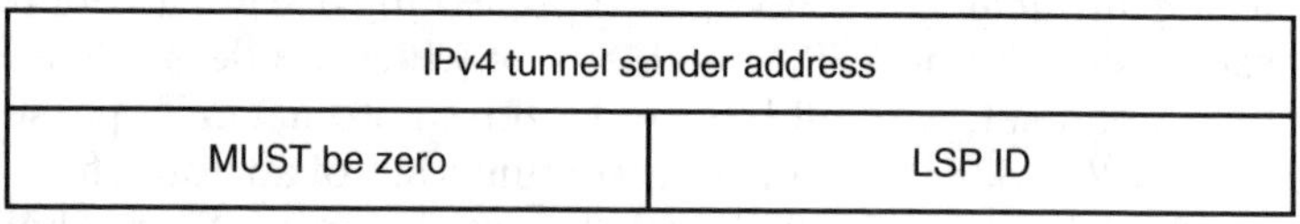

| IPv4 tunnel sender address | |
|---|---|
| MUST be zero | LSP ID |

**Figure 6.8 SENDER_TEMPLATE Object Format**

that the LSP tunnel traverses in the MPLS domain. Optionally, labels can also be recorded.

The RRO is applicable to both RSVP PATH and RESV messages. When the ingress LSR attempts to establish an LSP tunnel, it creates a PATH message that contains an LRO. The PATH message can also contain an RRO object, which initially contains the ingress LSR's IP address. After the PATH message containing the RRO is received by an intermediate LSR, the LSR stores a copy of the RRO in its PSB (see section 6.4.1) and adds its own IP address to the RRO. When the egress LSR receives a PATH message with an RRO, it adds an RRO to its subsequent RESV message. After the exchanging of PATH and RESV messages, each LSR along the path will have the complete route of the LSP tunnel from ingress to egress.

Similar to the ERO, the content of an RRO contains a series of subobjects. There are currently three types of subobjects defined: IPv4 addresses (type 1), IPv6 addresses (type 2), and MPLS labels (type 3). Furthermore, the RRO can be used to:

- Discover L3 routing loops or loops inherent in the explicit route when the ERO uses loose hops. Put another way, RRO can function as a loop detection mechanism because it is analogous to a path vector. An intermediate LSR examines all subobjects contained within an incoming RRO. If the LSR determines that it is already in the list, a forwarding loop exists. The action performed by an LSR on receipt of an RRO depends on which RSVP message type is received. For PATH messages containing a forwarding loop, the LSR sends a "routing problem" PATHERR message with the error value "loop detected" and drops the PATH message. For RESV messages containing a forwarding loop, the LSR simply drops the message. The rule of thumb is that RESV messages should not loop if PATH messages do not loop. An RSVP session is considered to be loop-free if downstream LSRs receive PATH messages or upstream LSRs receive RESV messages with no routing loops detected from the embedded RRO.
- Collect the latest detailed hop-by-hop path information on the LSP tunnel setup session, including internal AS hops.
- Support route pinning. With minor changes, the RRO can be input to the ERO to pin down a particular path session. To achieve this, the sender, after receiving the RRO from the receiver in a RESV message, applies it to the ERO in the next PATH message.

### 6.4.2 Reservation Styles

Each LSP tunnel is established with either the fixed-filter (FF) or the shared-explicit (SE) reservation styles selected exclusively by the egress LSR. Nevertheless, the ingress LSR can indicate to the egress LSR the type of reservation style it requires by setting or clearing the "SE style desired" flag in the SESSION_ATTRIBUTE object of the PATH message (see section 6.4.1.7).

The wildcard filter (WF) reservation style is not used for explicit routing because of its merging rules and lack of applicability for TE purposes [RFC3209]. The choice between the FF and SE reservation styles will be the deciding factor when determining how RSVP-TE supports the rerouting of an existing LSP tunnel. The following subsections describe the operational characteristics of these alternative reservation styles.

**6.4.2.1 FF Style.** In general, the fixed-filter (FF) reservation style specifies an explicit list of senders and creates a distinct reservation for each of them. Each sender, identified by an IP address and an LSP ID, has a dedicated reservation that is not shared with other senders. Because each sender has its own reservation, a unique label and a separate LSP tunnel can be created for each sender/receiver (or ingress/egress) pair.

In RSVP-TE, the FF reservation style allows the establishment of multiple parallel unicast point-to-point LSP tunnels. If the LSP tunnels traverse a common link, the total amount of reserved bandwidth on the shared link is the sum of the reservations for each of the individual senders (see Figure 6.9). The FF reservation style is applicable for concurrent and independent traffic originating from different senders.

**6.4.2.2 SE Style.** The shared-explicit (SE) reservation style creates a single reservation over a common link shared by an explicit list of senders. In other words, this reservation style allows a receiver to explicitly specify

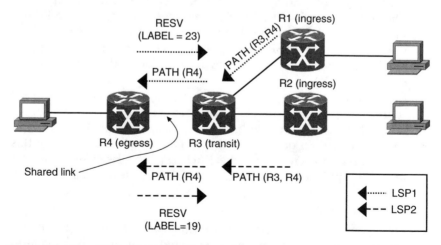

**Figure 6.9  Fixed-Filter Reservation Style Illustration**

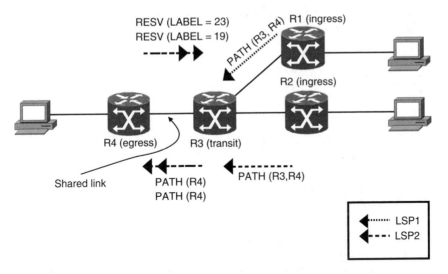

**Figure 6.10    Shared-Explicit Reservation Style Illustration**

the senders to be included in a single reservation. Because each sender is explicitly listed in the RESV message, different labels can be assigned to different sender/receiver pairs, thus creating distinct LSP tunnels.

In addition, SE style reservations can also be implemented with multi-point-to-point LSP tunnels in which PATH messages do not carry the ERO, or have the same EROs. In either of these cases, a common label can be assigned. Figure 6.10 illustrates the operation of the SE reservation style across a shared link. Although each LSP has its own unique identity across the shared link, both LSPs share the bandwidth of the largest request. As such, the SE reservation style facilitates RSVP signaling to support make-before-break when an established LSP tunnel is rerouted (see section 6.8).

### 6.4.3   RESV Message

RESV messages are sent upstream. Specifically, a RESV message is sent by the egress LSR toward the ingress LSR in response to a PATH message. The LSP tunnel is established when the ingress LSR receives the RESV message. RESV messages use the path state information in each LSR to follow exactly the reverse path in which the PATH message has traversed, thus creating and maintaining a reservation state in each LSR along the reverse path by distributing label bindings, allocating resources along the path, and specifying the reservation style (FF or SE). All of this information is stored in a reservation state block (RSB). Each RSB holds a reservation request that arrived in a particular RESV message, corresponding to the triple: SESSION, NHOP, FILTER_SPEC_LIST. In this case, FILTER_SPEC_LIST can be a single FILTER_SPEC for the FF reservation style or a list of FILTER_SPEC for the SE reservation style.

The format of a RESV message with FF reservation style is as follows:

- <Common Header>
- [<INTEGRITY>] ← optional
- <SESSION> ← new C-Type
- <RSVP_HOP>
- <TIME_VALUES>
- [<RESV_CONFIRM>] ← optional
- [<SCOPE>] ← optional
- [<POLICY_DATA>] ← optional
- <STYLE>
- <FF Flow Descriptor List>
  — <FLOWSPEC>
  — <FILTER_SPEC> ← new C-Type
  — <LABEL> ← new object, mandatory, sender specific, must appear after associated FILTER_SPEC and prior to any subsequent FILTER_SPEC
  — [<RECORD_ROUTE>] ← new object, optional, sender specific, must appear after associated FILTER_SPEC and prior to any subsequent FILTER_SPEC

Note that no more than one LABEL or RECORD_ROUTE should follow each FILTER_SPEC. The format of a RESV message with SE reservation style is as follows:

- <Common Header>
- [<INTEGRITY>] ← optional
- <SESSION> ← new C-Type
- <RSVP_HOP>
- <TIME_VALUES>
- [<RESV_CONFIRM>] ← optional
- [<SCOPE>] ← optional
- [<POLICY_DATA>] ← optional
- <STYLE>
- <SE Flow Descriptor>
  — <FLOWSPEC>
  — <SE Filter Spec>
    - <FILTER_SPEC> ← new C-Type
    - <LABEL> ← new object, mandatory, sender specific, must appear after associated FILTER_SPEC and prior to any subsequent FILTER_SPEC
    - [<RECORD_ROUTE>] ← new object, optional, sender specific, must appear after associated FILTER_SPEC and prior to any subsequent FILTER_SPEC

The various objects associated with an RSVP-TE RESV message are discussed in the following subsections.

**6.4.3.1  RESV_CONFIRM Object.** The RESV_CONFIRM object has a Class-Num of 15 and two C-Types are defined: IPv4, C-Type = 1, and IPv6, C-Type = 2. The IPv4 RESV_CONFIRM object carries the IPv4 address (or IPv6 address for the IPv6 RESV_CONFIRM object) of a receiver that has requested a reservation confirmation. The RESV_CONFIRM object can appear in a RESV or RESVCONF message.

**6.4.3.2  SCOPE Object.** The SCOPE object has a Class-Num of 7 and two C-Types are defined: IPv4, C-Type = 1, and IPv6, C-Type = 2. The SCOPE object can appear in a RESV, RESVERR, or RESVTEAR message. It carries an explicit list of senders (IPv4 addresses for the IPv4 SCOPE object or IPv6 addresses for the IPv6 SCOPE object) toward which the information in the message is to be forwarded.

**6.4.3.3  STYLE Object.** The STYLE object has a Class-Num of 8 and a C-Type of 1. It is used to define the reservation style (fixed filter or shared explicit) and also style-specific information that is not found in the FLOW-SPEC or FILTER_SPEC objects (see sections 6.4.3.4 and 6.4.3.5). The STYLE object is required in every RESV message.

**6.4.3.4  FLOWSPEC Object.** The FLOWSPEC object has a Class-Num of 9 and a C-Type of 2. It is used to define a desired QoS in a RESV message, which includes an indication of which QoS control service is being requested, and the parameters required for that service. The service number in the per-service header of the FLOWSPEC object indicates the requested QoS control service (see [RFC2210]). The desired QoS is derived from the SENDER_TSPEC object corresponding to a previous PATH message.

**6.4.3.5  FILTER_SPEC Object.** The FILTER_SPEC object, in conjunction with the SESSION object, defines the set of data packets or traffic trunk that receives the desired QoS defined by the FLOWSPEC object (see section 6.4.3.4) in a RESV message. The FILTER_SPEC object has a Class-Num of 10 and two new C-Types are defined for RSVP-TE: LSP_TUNNEL_IPv4, C-Type = 7, and LSP_TUNNEL_IPv6, C-Type = 8. The format and functionality of the new C-Types of the FILTER_SPEC object are identical to those of the SENDER_TEMPLATE object (see section 6.4.1.9).

**6.4.3.6  LABEL Object.** The LABEL object (Class-Num = 16, C-Type = 1) is carried in the RESV message. This object can be composed of either a single or a stack of MPLS labels. For the FF and SE reservation styles, a label is associated with each sender. The label for a sender must immediately follow the FILTER_SPEC object for that sender in the RESV message. Labels

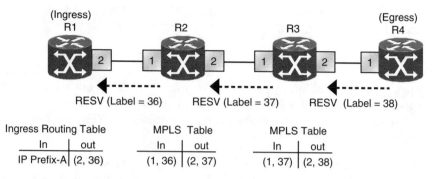

**Figure 6.11    The Processing of the LABEL Object**

in RESV messages received on different interfaces are always considered to be different, even if the label value is the same, because each label has a link-local (per-interface) scope, and not a global (per-platform) scope (see section 2.2.4). When an LSR receives a RESV message, it stores the LABEL object in its RSB (see section 6.4.3), which is then used for formatting the RESV message in the next RESV refresh event. The downstream-on-demand label distribution technique is used by RSVP-TE, in which a downstream LSR will assign a label to a forwarding equivalence class (FEC) only when the corresponding upstream LSR requests the label.

Figure 6.11 illustrates how various LSRs along the LSP tunnel process the LABEL object carried within a RESV message. When R4 (egress LSR) receives a label request derived from the LRO in the PATH message, it generates a local label (38), formats it into a LABEL object, and sends the new LABEL object as part of the RESV message to its upstream neighbor R3.

R3 binds the incoming label (also known as outbound label) of value 38 to outbound interface-2, where the RESV message is received. It also locally generates a new label (also known as inbound label) of value 37, binds it to inbound interface-1, where the PATH message is received, replaces the top label (38) in the received RESV message with this new label (37), and forwards the RESV message upstream to R2.

This process continues likewise for R2 until the RESV message reaches R1 (ingress LSR), which binds the incoming label to destination IP prefix A (the FEC in this example). The LSP tunnel is established when R1 receives the RESV message.

## 6.5    RSVP-TE PATH SETUP OPERATION

So far we have only covered the bits and pieces of RSVP-TE messages and its respective objects. This section provides a detailed example of how RSVP-TE establishes an LSP tunnel from the perspective of the PATH/RESV

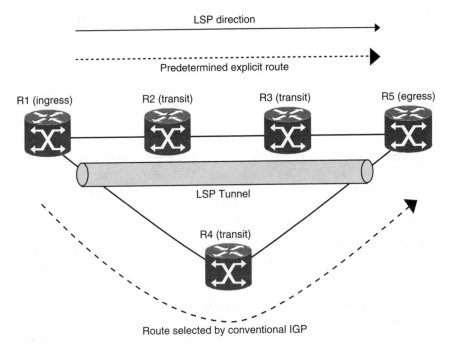

**Figure 6.12 Network Topology 1**

messages and objects as a whole. Figure 6.12 illustrates the network topology that is used in this example.

The following assumptions were made for this example:

- MPLS and RSVP-TE have been configured and enabled on R1, R2, R3, R4, and R5.
- The path selection process either dynamically or manually predetermines the physical path to be traversed by the LSP tunnel. Specifically, R1 (ingress LSR) knows the LSP tunnel should follow the explicit route R1–R2–R3–R5 determined previously by the path selection process (see chapter 5).
- Each abstract node in the ERO has the L-bit cleared (L = 0), indicating a strict hop in the explicit route, and is a simple abstract node composed of only a single LSR identified by a 32-bit IPv4 prefix.

Put another way, we want to set up an LSP tunnel for traffic trunks entering the MPLS network at R1 and exiting at R5. This is where TE comes into play. The traffic trunks should be forwarded along the physical path (R1–R2–R3–R5) taken by the LSP tunnel rather than the least-cost route (R1–R4–R5) calculated by the conventional IGP through the network. The rationale behind selecting explicit path R1–R2–R3–R5 is to:

- Reduce the amount of traffic flowing along the IGP route
- Optimize the overall bandwidth utilization of both paths
- Enhance the traffic-oriented performance characteristics for the traffic trunk and the entire MPLS network

The details of the path setup operation with the PATH and RESV messages are discussed in the following subsections.

### 6.5.1 PATH Message Analysis

R1 (ingress) is responsible for initiation of the LSP tunnel setup. It sends a PATH message to R5. In this instance, the PATH message traverses through R2 and R3 on its way to R5. The PATH message that is addressed to R5 also contains the router alert IP option [RFC2113], indicating that the IP packet requires special processing by the transit LSRs (R2 and R3).

**6.5.1.1  At R1 (Ingress LSR).** Figure 6.13 illustrates the processing of the PATH message at R1. To establish the LSP tunnel, R1 constructs a PATH message composed of:

- SESSION object that uniquely identifies this LSP tunnel with the IPv4 address of the egress LSR (loopback address of R5 denoted by R5-lo0), a unique 16-bit tunnel ID (01), and a unique 32-bit extended tunnel ID (loopback address of R1 denoted by R1-lo0).

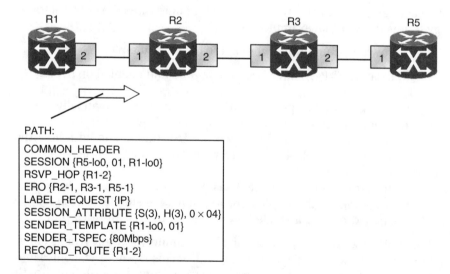

PATH:

```
COMMON_HEADER
SESSION {R5-lo0, 01, R1-lo0}
RSVP_HOP {R1-2}
ERO {R2-1, R3-1, R5-1}
LABEL_REQUEST {IP}
SESSION_ATTRIBUTE {S(3), H(3), 0 × 04}
SENDER_TEMPLATE {R1-lo0, 01}
SENDER_TSPEC {80Mbps}
RECORD_ROUTE {R1-2}
```

**Figure 6.13   PATH Message Processing at R1 (Ingress LSR)**

- RSVP_HOP object (PHOP) that carries the source IPv4 address (R1) and the source interface (2) in which this message is sent out. In this case, the PHOP is denoted by R1-2.
- EXPLICIT_ROUTE object (ERO) that specifies the physical path {R2-1, R3-1, R5-1} that the PATH message should follow to set up the LSP tunnel between R1 and R5.
- LABEL_REQUEST object, indicating that a label binding is requested for the LSP tunnel from all LSRs in the path and that the L3 protocol (IP) identified by the L3PID is to be carried over the LSP tunnel.
- SESSION_ATTRIBUTE object that controls LSP tunnel priority (setup priority S = 3, holding priority H = 3), preemption, and rerouting (flag = 0 × 04).
- SENDER_TEMPLATE object that contains the sender's IP address (denoted by R1-lo0) and the LSP ID (01).
- SENDER_TSPEC object specifying the traffic characteristics (bandwidth reservation request of 80 Mbps) for the traffic trunk that will traverse the established LSP tunnel. R5 will in turn use this information to define the desired QoS with an appropriate FLOWSPEC object (see section 6.4.3.4).
- RECORD_ROUTE object (RRO) that allows the ingress LSR (R1) to gather information regarding the exact routing path traversed by the LSP tunnel. This information is useful for loop detection as well as diagnosis purposes. The initial RRO contains the IP address of R1 denoted by R1-2.

**6.5.1.2   At R2 (Transit LSR).** When R2 receives the PATH message from R1, it stores the information found in the PATH message in its path state block (PSB), as shown in Figure 6.14. This information is used to route the corresponding RESV message back to R1. Before R2 forwards the PATH message toward R5 along the path specified in the ERO, the following updates are made to some of the fields in the PATH message:

- The RSVP_HOP is modified from R1-2 to R2-2.
- The ERO reflects the physical path {R3-1, R5-1}.
- R2 appends its own IP address (denoted by R2-2) to the RRO, which has become {R1-2, R2-2}.

If R2 cannot allocate a label for the LSP tunnel, it responds by sending a PATHERR message with an "unknown object class" error back to R1.

**6.5.1.3   At R3 (Transit LSR).** When R3 receives the PATH message from R2, it stores the information found in the PATH message in its PSB, as shown in Figure 6.15. This information is used to route the corresponding RESV message back to R2. Before R3 forwards the PATH message toward R5 along the path specified in the ERO, the following updates are made to some of the fields in the PATH message:

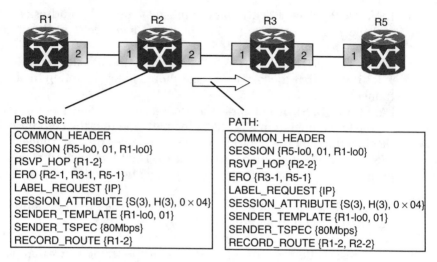

**Figure 6.14   PATH Message Processing at R2 (Transit LSR)**

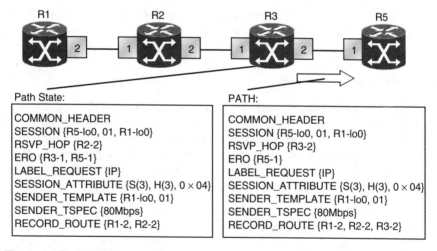

**Figure 6.15   PATH Message Processing at R3 (Transit LSR)**

- The RSVP_HOP is modified from R2-2 to R3-2.
- The ERO reflects the physical path {R5-1}.
- R3 appends its own IP address (denoted by R3-2) to the RRO, which has become {R1-2, R2-2, R3-2}.

If R3 cannot allocate a label for the LSP tunnel, it responds by sending a PATHERR message with an "unknown object class" error back to R1.

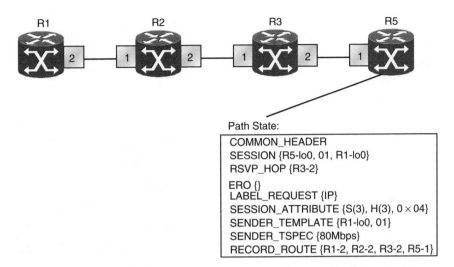

Path State:

| |
|---|
| COMMON_HEADER |
| SESSION {R5-lo0, 01, R1-lo0} |
| RSVP_HOP {R3-2} |
| ERO {} |
| LABEL_REQUEST {IP} |
| SESSION_ATTRIBUTE {S(3), H(3), 0 × 04} |
| SENDER_TEMPLATE {R1-lo0, 01} |
| SENDER_TSPEC {80Mbps} |
| RECORD_ROUTE {R1-2, R2-2, R3-2, R5-1} |

**Figure 6.16   PATH Message Processing at R5 (Egress LSR)**

**6.5.1.4   At R5 (Egress LSR).** When R5 receives the PATH message from R3, the PATH request operation is completed. R3 stores the information found in the PATH message in its PSB, as shown in Figure 6.16. Prior to storing the information, the following adjustments are made to some of the fields in the PATH message:

- The ERO is reflected as null.
- R5 appends its own IP address (denoted by R5-1) to the RRO, which has become {R1-2, R2-2, R3-2, R5-1}.

The information stored in the PSB is used to route the corresponding RESV message back to R3. From the SESSION object, R5 knows that it is the egress LSR for the LSP tunnel. It then proceeds to process the LABEL_REQUEST and constructs the corresponding RESV message (see section 6.5.2). If R5 cannot allocate a label for the LSP tunnel, it responds by sending a PATHERR message with an "unknown object class" error back to R1.

### 6.5.2   RESV Message Analysis

The RESV message is used to distribute labels and establish forwarding state for the LSP tunnel. Each LSR processes the RESV message and uses the received label for outgoing traffic associated with this LSP tunnel. The RESV message also allocates the requested resources at each LSR.

In this example, R5 generates a corresponding RESV message in response to the PATH message received from R3. The RESV message is sent

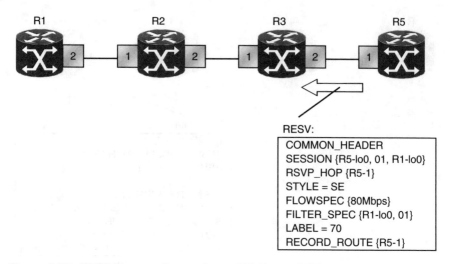

RESV:

| |
|---|
| COMMON_HEADER |
| SESSION {R5-lo0, 01, R1-lo0} |
| RSVP_HOP {R5-1} |
| STYLE = SE |
| FLOWSPEC {80Mbps} |
| FILTER_SPEC {R1-lo0, 01} |
| LABEL = 70 |
| RECORD_ROUTE {R5-1} |

**Figure 6.17    RESV Message Processing at R5 (Egress LSR)**

back toward R1 through R3, but it does not carry a reverse ERO to find its way back to R1. Instead, the RESV message follows the reverse path that is set up earlier in the PSB by the PATH message. In other words, the destination IP address of the RESV message is the unicast address of the RSVP_HOP (PHOP) obtained from each LSR's local PSB.

**6.5.2.1   At R5 (Egress LSR).** Figure 6.17 illustrates the processing of the RESV message at R5. The RESV message is composed of the following:

- SESSION object that holds the same values as those found in the PATH message.
- RSVP_HOP object (NHOP) that carries the source IPv4 address (R5) and the source interface (1) in which this message is sent out. In this case, the NHOP is denoted by R5-1.
- STYLE object that defines the shared-explicit (SE) reservation style in response to the SE flag (0 × 04) set earlier in the SESSION_ ATTRIBUTE object of the PATH message. There is a single reservation on this link for the sender R1.
- FLOWSPEC object that is constructed based on information gathered from the SENDER_TSPEC object inserted previously into the PATH message by R1. In this case, the 80-Mbps bandwidth request is granted.
- FILTER_SPEC object that takes on the same values as the SENDER_TEMPLATE object in the PATH message.
- R5 allocates a new label (InLabel) with a value of 70, places it in the LABEL object of the RESV message, and sends it upstream to R3 through interface R5-1. R5 also allocates a corresponding OutLabel

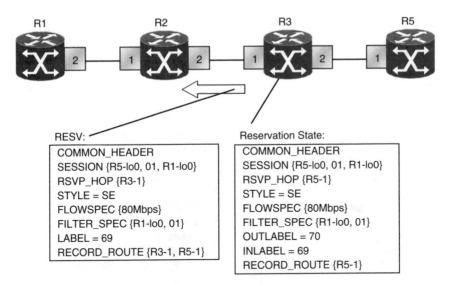

**Figure 6.18  RESV Message Processing at R3 (Transit LSR)**

with a value of 0 (IPv4 explicit null label). The value of 0 has a special meaning to R5. It indicates that the current label must be popped (see Table 2.2). When R5 receives an MPLS frame with a label value of 70 from interface R5-1, it performs a pop operation on the label and forwards the remaining packet based on the destination address in the IPv4 header.

- When the egress LSR (R5) receives a PATH message with an RRO, it adds an RRO to the corresponding RESV message. The initial RRO contains the IP address of R5 denoted by R5-1. Subsequent transit LSRs (R3, R2, and R1) will prepend their respective IP addresses (R3-1, R2-1, and R1-2) to this RRO.

**6.5.2.2  At R3 (Transit LSR).** When R3 receives the RESV message from R5, it creates a reservation state by distributing label bindings, allocating resources, and specifying the reservation style. All of this information is stored in a local reservation state block (RSB), as shown in Figure 6.18.

As for the label distribution:

- R3 stores the label (70) assigned by R5 as part of the reservation state for the LSP tunnel and uses this label (OutLabel) when forwarding outbound MPLS frames to R5 through interface R3-2.
- R3 allocates a new label (69) and places it in the LABEL object of the RESV message, replacing the previously received label. This is the label (InLabel = 69) that R3 uses to identify inbound MPLS frames

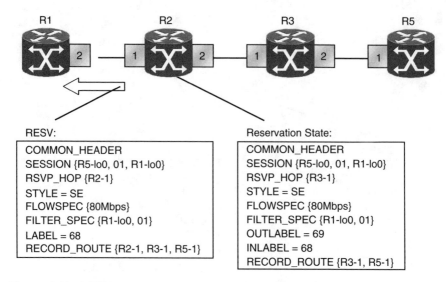

**Figure 6.19   RESV Message Processing at R2 (Transit LSR)**

from R2 through interface R3-1. The modified RESV message is then sent upstream to R2 (see Figure 6.18).

- The InLabel (69) is swapped with the OutLabel (70) for outgoing MPLS frames.

**6.5.2.3   At R2 (Transit LSR).** When R2 receives the RESV message from R3, it creates a reservation state by distributing label bindings, allocating resources, and specifying the reservation style. All of this information is stored in a local RSB, as shown in Figure 6.19.

As for the label distribution:

- R2 stores the label (69) assigned by R3 as part of the reservation state for the LSP tunnel and uses this label (OutLabel) when forwarding outbound MPLS frames to R3 through interface R2-2.
- R2 allocates a new label (68) and places it in the LABEL object of the RESV message, replacing the previously received label. This is the label (InLabel = 68) that R2 uses to identify inbound MPLS frames from R1 through interface R2-1. The modified RESV message is then sent upstream to R1 (see Figure 6.19).
- The InLabel (68) is swapped with the OutLabel (69) for outgoing MPLS frames.

**6.5.2.4   At R1 (Ingress LSR).** Once R1 receives the RESV message from R2, the LSP tunnel is established. The information gathered from the RESV message is stored in a local RSB, as shown in Figure 6.20.

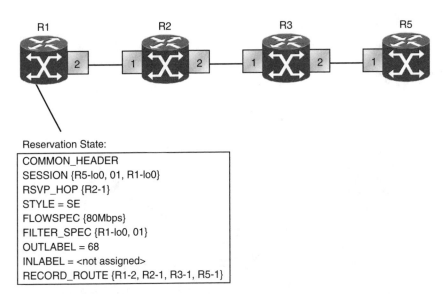

Reservation State:

```
COMMON_HEADER
SESSION {R5-lo0, 01, R1-lo0}
RSVP_HOP {R2-1}
STYLE = SE
FLOWSPEC {80Mbps}
FILTER_SPEC {R1-lo0, 01}
OUTLABEL = 68
INLABEL = <not assigned>
RECORD_ROUTE {R1-2, R2-1, R3-1, R5-1}
```

**Figure 6.20   RESV Message Processing at R1 (Ingress LSR)**

The OutLabel (68) assigned from R2 is tagged (with the push operation) by R1 for all outgoing traffic flows (traffic trunks) on interface R1-2. Because R1 is the ingress LSR, the InLabel is not assigned. The success of these operations will result in an LSP tunnel being set up from R1 to R5, following the explicitly routed path specified in the ERO.

## 6.6   ADMISSION CONTROL AND PREEMPTION

From the TE perspective, admission control is generally used to verify whether an LSR has sufficient resources to accept a new traffic trunk during path setup. Specifically, two types of admission control are involved during the path setup: trunk admission control and link admission control.

Trunk admission control determines if the requested bandwidth is available along an LSP tunnel during path setup. It is also responsible for tearing down (or preempting) existing lower-priority LSP tunnels when necessary (see sections 4.4, 4.5, and 6.4.1.7). As illustrated in Figure 6.21, LSP1 is a high-priority constraint-routed LSP, whereas LSP2 is a low-priority best-effort routed LSP. LSP1 uses path R4–R1 and LSP2 uses path R4–R3–R1.

When the network link R4–R1 fails, the high-priority LSP1 has to be rerouted and share links with LSP2, which leads to the overutilization of resources on network links R4–R3 and R3–R1. To ensure that the resource requirements of LSP1 are still guaranteed, trunk admission control preempts LSP2, which in turn is rerouted to unused path R4–R5–R3–R2–R1, as shown in Figure 6.22.

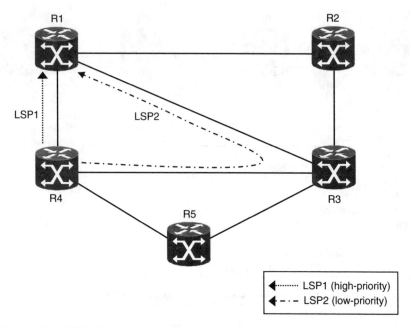

Figure 6.21 LSP Tunnel Preemption Part 1

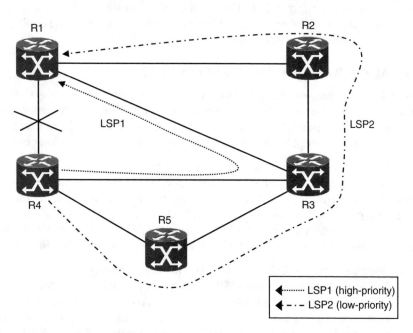

Figure 6.22 LSP Tunnel Preemption Part 2

Furthermore, trunk admission control triggers IGP information distribution (see chapter 5) whenever there is a change in the current available network resources.

Link admission control is used within the PATH message. When each LSR along an LSP tunnel receives the PATH message, the link admission control process is invoked:

- If the requested bandwidth is available, it is reserved in a waiting pool on the port outbound to the destination until a RESV message is received.
- If the required bandwidth is unavailable, the link admission control process notifies RSVP and a PATHERR message is sent upon receipt of the RESV message.

## 6.7   FORWARDING TRAFFIC ACROSS AN LSP TUNNEL

Once the LSP tunnel is set up, the next step is to forward traffic across this tunnel. The LSP tunnel established from the ingress LSR to the egress LSR is not announced to the link-state IGP (OSPF or ISIS). Therefore, any prefixes/networks announced by the egress LSR and its downstream routers would not be visible through the LSP tunnel. The LSP tunnel and all the destination prefixes/networks behind it will have to be installed in the IGP routing table (or forwarding table) of the ingress router so that traffic can be forwarded across this LSP tunnel to reach these destinations. This can be achieved:

- Administratively via static routes or policy-based routing (PBR)
- Dynamically via the Cisco internetwork operating system (IOS) autoroute announce feature

In Figure 6.23, an LSP tunnel is established from R1 (ingress LSR) to R4 (egress LSR). At R1, we can administratively define static routes or routing policies (for PBR) pointing to the LSP tunnel for any destination behind R4, in this case, R5 to R7. Alternatively, when the LSP tunnel is configured with autoroute announce, the link-state IGP will automatically install the routes announced by R4 and its downstream routers (R5 to R7) into the routing table of R1 as routes reachable via the LSP tunnel. As a result, all traffic directed to the prefixes/networks behind R4 is forwarded across the LSP tunnel. Note that the routing table at R1 remains intact with PBR but is modified accordingly by the static routes and autoroute announce feature.

## 6.8   REOPTIMIZATION AND REROUTING

The capability to reoptimize or reroute an established LSP tunnel is a mandatory requirement in TE. Some of the reasons that require the reoptimization or rerouting of an LSP tunnel include:

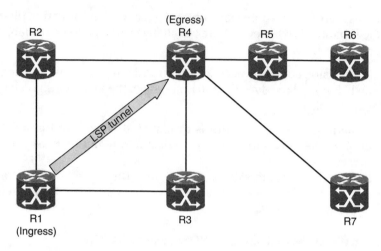

**Figure 6.23    Network Topology 2**

- The use of an administrative policy that requires an LSP tunnel to be rerouted whenever a more optimal route becomes available
- The failure of a link or node (LSR) along the established path of the LSP tunnel, which requires the LSP tunnel to be globally or locally rerouted
- The use of an administrative policy that requires a rerouted LSP tunnel to be returned to its original path when a failed link or node (LSR) becomes available again

During the reoptimization or rerouting process, it is crucial that the flow of subscriber traffic is not disrupted. This adaptive and smooth transition requires support for the make-before-break concept, whereby the new LSP tunnel must be established with traffic from the old LSP tunnel being transferred onto it before the old LSP tunnel is torn down. A problem can arise because the old and new LSP tunnels might compete with each other for resources on network segments that they have in common. Depending on the availability of resources, this contention can cause admission control to prevent the new LSP tunnel from being established. RSVP easily alleviates this challenging issue with its SE reservation style.

On links that are shared by both the old and new LSP tunnels, it is important that resources used by the old LSP tunnel are not released before traffic is transitioned to the new LSP tunnel. Reservations are not counted twice (double counting) across a common link, as this could cause admission control to reject the new LSP tunnel due to insufficient resources.

The SESSION object and the SE reservation style together provide a smooth transition to rerouting and bandwidth increase. The SE reservation style allows both the old and new LSP tunnels to share a single reservation along links that they have in common so that the new LSP tunnel does not have to compete with the old LSP tunnel for limited link resources. The SESSION object narrows the scope of the RSVP session to the particular LSP tunnel that is involved. The combination of the IP address for the egress LSR of the LSP tunnel, a tunnel ID, and the IP address for the ingress LSR of the tunnel (extended tunnel ID) is used to uniquely identify an LSP tunnel. During the reroute or bandwidth increase operation, the ingress LSR needs to appear as two different senders to the RSVP session, which is achieved by the LSP ID carried in the SENDER_TEMPLATE and FILTER_SPEC objects.

When performing a rerouting or bandwidth increase operation, in the new PATH message the ingress LSR:

- Uses the existing SESSION object to identify the LSP tunnel that will be rerouted
- Creates a new ERO for the new LSP tunnel
- Selects a new LSP ID and constructs a new SENDER_TEMPLATE object so that the ingress LSR appears as a different sender to the RSVP session

Figure 6.24 illustrates the deployment of these PATH objects. The ingress LSR (R1) sends the new PATH message toward the egress LSR (R5), as shown in Figures 6.25 and 6.26. Meanwhile, it continues to use the old LSP tunnel to forward traffic and refreshes the original PATH message.

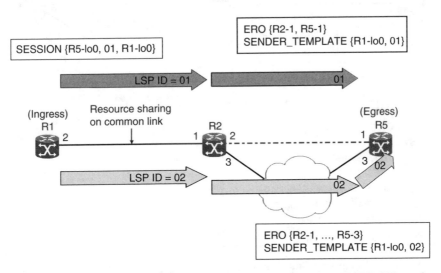

**Figure 6.24    Deployment of SESSION, ERO, and SENDER_TEMPLATE Objects for Rerouting**

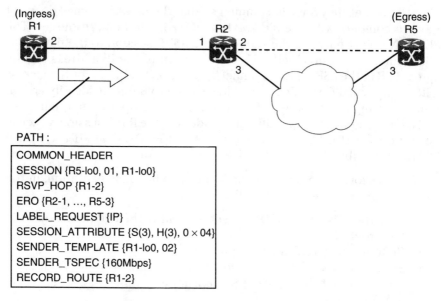

**Figure 6.25 Rerouting Details Part 1**

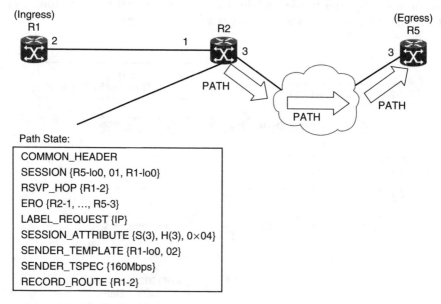

**Figure 6.26 Rerouting Details Part 2**

The egress LSR (R5) upon receipt of the new PATH message sends a RESV message that includes a number of RSVP objects, such as:

- An SE reservation style object in which bandwidth is reserved once per link
- FILTER_SPEC objects identifying an explicit list of senders over a shared link
- A LABEL object supporting downstream-on-demand label distribution. Labels are allocated for both old and new paths.

Figures 6.27 to 6.29 illustrate the contents of these RESV objects when the RESV message flows from the egress LSR (R5) back to the ingress LSR (R1).

On links that are not shared by the old and new LSP tunnels, the new PATH/RESV message pair is regarded as a typical new LSP tunnel setup. However, on links common to both the old and new LSP tunnels, the shared SESSION object and SE reservation style permit the new LSP tunnel to be established so that it shares resources with the old LSP tunnel. This takes care of the double-counting problem on shared links. Once the ingress node receives a RESV message for the new LSP tunnel, it begins to forward all traffic to the new LSP tunnel. The ingress LSR then proceeds to tear down the old LSP tunnel by sending a PATHTEAR message to all the transit LSRs to remove the old path state and free up resources.

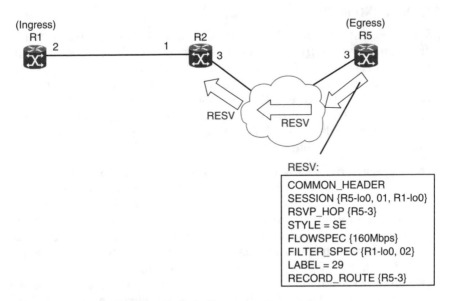

**Figure 6.27  Rerouting Details Part 3**

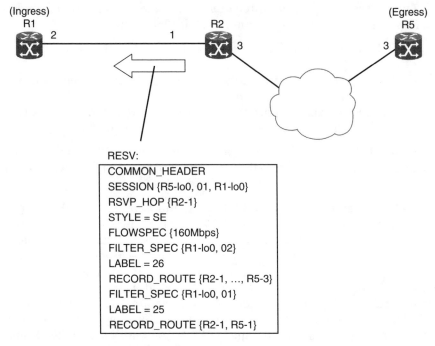

RESV:

| |
|---|
| COMMON_HEADER |
| SESSION {R5-lo0, 01, R1-lo0} |
| RSVP_HOP {R2-1} |
| STYLE = SE |
| FLOWSPEC {160Mbps} |
| FILTER_SPEC {R1-lo0, 02} |
| LABEL = 26 |
| RECORD_ROUTE {R2-1, ..., R5-3} |
| FILTER_SPEC {R1-lo0, 01} |
| LABEL = 25 |
| RECORD_ROUTE {R2-1, R5-1} |

**Figure 6.28    Rerouting Details Part 4**

## 6.9   SCALING RSVP

RSVP is a soft-state protocol that maintains state through the generation of refresh messages. Refresh messages are used to both synchronize state between RSVP neighbors and recover from lost RSVP messages. Put another way, the PATH and RESV messages must be periodically refreshed to maintain the LSP state in the LSRs along the path of an LSP tunnel. If refresh messages are not transmitted, the LSP state automatically times out and is eventually deleted. However, the use of refresh messages has resulted in two main concerns:

- Excessive overheads in terms of processing and memory incurred by LSRs running RSVP. The resource requirements increase proportionally with the number of RSVP sessions. Each session requires the generation, transmission, reception, and processing of RSVP PATH/RESV messages per refresh interval. The corresponding volume of refresh messages generated when supporting a large number of sessions poses a serious scaling problem.
- RSVP does not run over a reliable transport such as TCP. Hence, latency and reliability problems can occur when a nonrefresh RSVP message (PATHERR, RESVERR, PATHTEAR, RESVTEAR, or RESV-CONF) is lost in transit. If the message is lost, the end-to-end latency

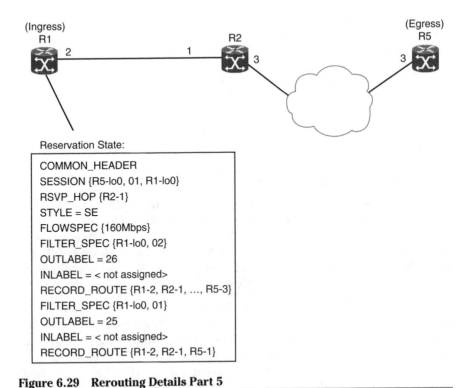

**Figure 6.29    Rerouting Details Part 5**

of RSVP signaling is based on the refresh interval of the LSRs experiencing the loss. When end-to-end response time is limited by the refresh interval, the delay incurred in the establishment or the adjustment of a reservation can exceed the range of what is acceptable for certain applications.

The scaling problem due to overheads experienced by LSRs as they attempt to process frequent refresh messages can be overcome by increasing the refresh interval, even though this considerably increases the time it takes to synchronize the LSP state. On the other hand, the reliability and latency of RSVP can be improved by decreasing the refresh interval, even though this significantly increases the volume of traffic as well as the processing overhead, and it further aggravates the refresh overhead issue.

One additional concern is how long it will take to free up resources when a tear message is lost. RSVP does not retransmit PATHTEAR or RESVTEAR messages. If the single tear message transmitted is lost, then resources will only be released after the cleanup-timer interval has passed, resulting in resources being held unnecessarily for a specific period. Even when the refresh period is adjusted, the cleanup-timer must still expire because tear messages are not retransmitted.

Therefore, to enhance the scalability, latency, and reliability of RSVP other than adjusting the refresh interval, a number of extensions have been defined:

- Bundle message extension [RFC2961]
- MESSAGE_ID extension [RFC2961]
- Summary refresh extension [RFC2961]
- Hello protocol extension [RFC3209]

The details of these extensions are discussed in the following subsections.

### 6.9.1 Bundle Message Extension

The bundle message extension helps to reduce the volume of RSVP messages that must be regularly transmitted and received at a fixed time interval. The bundle message is composed of a bundle header followed by a body consisting of a variable number of standard RSVP messages (see Figure 6.30). The bundle message is used to aggregate, or bundle, multiple RSVP messages within a single protocol data unit (PDU).

As illustrated in Figure 6.30, the bundle message header is really an RSVP common header (see also Figure 6.2) with the 4-bit flag = 0 × 01 and the message type = 12. Message type 12 denotes a bundle message, and the 4-bit flag when set to 0 × 01 indicates that the refresh-reduction-capable bit is on for this LSR. The refresh-reduction-capable bit [RFC2961] is an additional capability bit added to the common RSVP header to support refresh-overhead-reduction extensions such as bundle messages, MESSAGE_ID objects (see section 6.9.2), and summary refresh messages (see section 6.9.3). This bit is meaningful only between RSVP neighbors.

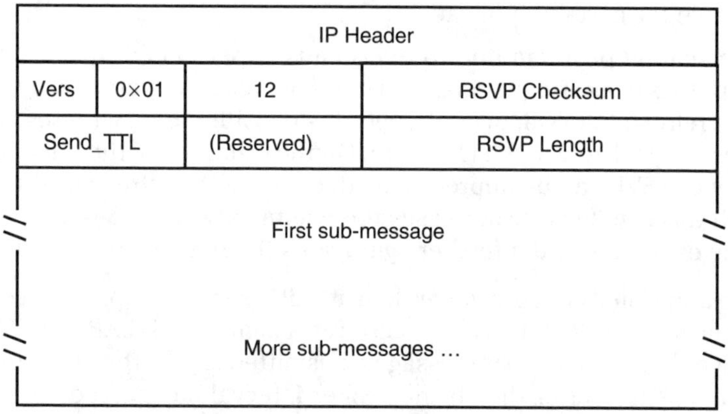

**Figure 6.30   RSVP Bundle Message**

An RSVP bundle message must contain at least one submessage. A submessage can be any standard RSVP message type except for another bundle message. Bundle messages are addressed directly to the RSVP neighbors and are sent as raw IP packets with protocol number 46. The source IP address is an address local to the LSR that originates the bundle message, and the destination IP address is the RSVP neighbor for which the submessages are intended. The bundle header immediately follows the IP header, and there is no intermediate transport header (see Figure 6.30).

When an RSVP-capable LSR receives a bundle message that is not addressed to any of its local IP addresses, it forwards the message. Non-RSVP LSRs treat RSVP bundle messages just like any other IP packet. The receiver of the bundle message processes each submessage as if it were received individually.

RSVP bundle messages can only be sent to RSVP neighbors that support bundling, which can be verified with the refresh-reduction-capable bit in the received RSVP messages. Nevertheless, the support for RSVP bundle messages is optional. Although message bundling helps in scaling RSVP, by reducing processing overhead and bandwidth utilization, it is not mandatory for an RSVP-capable LSR to transmit every standard RSVP message in a bundle message. However, this LSR must always be ready to receive standard RSVP messages.

### 6.9.2  MESSAGE_ID Extension

Two new objects are defined as part of the MESSAGE_ID extension:

- MESSAGE_ID object
- MESSAGE_ID_ACK object

These objects support the reliable delivery of RSVP messages with an acknowledgment mechanism. In addition, the MESSAGE_ID object can be used to provide an indication as to whether a received refresh message (PATH or RESV) represents a new state. The receiving LSR can use such information to reduce the amount of time it takes to process refresh messages.

**6.9.2.1  MESSAGE_ID Object.** The MESSAGE_ID object (Class-Num = 23, C-Type = 1) reduces refresh message processing by allowing the receiver to easily identify a message containing unchanged state information. When an LSR sends a refresh message with a MESSAGE_ID object, it retains the same 32-bit message identifier value used in the RSVP message that first advertised the state being refreshed. Conversely, the message identifier value is changed to a value greater than any other previously used value when an LSR sends a message representing a new or modified state. Moreover, the ACK_Desired bit [RFC2961] can be set by the LSR that sends a refresh message containing a MESSAGE_ID object if this LSR has the capability to process MESSAGE_ID_ACK objects (see section 6.9.2.2).

When an LSR receives a refresh message containing a MESSAGE_ID object, it first determines the associated RSVP session and then verifies the message identifier value it previously stored in its local state block for this particular RSVP session. The receiver must fully process the received message if the message identifier does not concur with a local state, as a new message identifier implies a new or modified state. However, the receiver will treat the inbound message (PATH or RESV) as a state refresh if this message contains the same message identifier value that was used by the most recently received message for the same RSVP session.

**6.9.2.2 MESSAGE_ID_ACK Object.** The MESSAGE_ID_ACK object (Class-Num = 24, C-Type = 1) is sent to acknowledge the receipt of messages containing MESSAGE_ID objects that were sent with the ACK_Desired bit set. ACK messages can carry one or more MESSAGE_ID_ACK objects, but they cannot contain any MESSAGE_ID objects. This mechanism can ensure the reliable delivery of error and confirm messages as well as the quicker detection of link transmission losses.

### 6.9.3 Summary Refresh Extension

A summary refresh message supports the refreshing of RSVP state along an LSP tunnel without the transmission of standard PATH and RESV messages. The main advantage of the summary refresh extension is that it reduces the amount of information that must be sent and processed to maintain RSVP state synchronization. The extension also preserves RSVP's ability to handle non-RSVP next-hops and to adjust to changes in routing.

A summary refresh message carries a set of MESSAGE_ID objects identifying the PATH and RESV states that should be refreshed. When an RSVP-capable LSR receives a summary refresh message, it matches each received MESSAGE_ID object with the locally installed PATH or RESV state. The state is updated as if a standard RSVP refresh message has been received when the MESSAGE_ID objects match the local installed state. However, the receiver informs the sender of the summary refresh message by sending a refresh NACK if a matching locally installed state cannot be found. This refresh NACK is sent via the MESSAGE_ID_NACK object (Class-Num = 24, C-Type = 2).

Upon receiving a MESSAGE_ID_NACK object, the receiver of this object performs an installed PATH or RESV state lookup based on the message identifier value contained in the object. If a matching state is found, the receiver must transmit the matching state via a standard PATH or RESV message.

The transmission of standard refresh messages should be suppressed when a summary refresh message is used to refresh the state of an RSVP session. The summary refresh extension cannot be used for a PATH or

RESV message that contains changes to a previously advertised state. Only a state that was previously advertised in PATH or RESV messages containing MESSAGE_ID objects can be refreshed by a summary refresh message.

### 6.9.4 Hello Extension

The hello extension enables RSVP LSRs to detect the loss of a neighbor LSR or when a neighbor's RSVP state information is reset. Put simply, this mechanism provides node-to-node (LSR-to-LSR) failure detection and is intended to be used when notification of link-layer failures is unavailable and unnumbered links are not used, or when the existing failure detection mechanisms provided by the link layer are not sufficient for timely node failure detection.

The hello extension consists of a hello message, a HELLO_REQUEST object (Class-Num = 22, C-Type = 1,) and a HELLO_ACK object (Class-Num = 22, C-Type = 2). Hello processing between two neighbors supports independent selection of failure detection intervals. A hello message containing a HELLO_REQUEST object is periodically generated for each neighbor whose status is being tracked. Each neighbor can independently issue HELLO_REQUEST objects, and each request is responded to with a HELLO_ACK object.

The support for hello messages is optional. Nevertheless, it is designed such that one side can still use the mechanism even if the other side is unable to do so, and neighbor failure detection can be initiated any time. A hello message can be included as a submessage within a bundle message.

### 6.10 SUMMARY

RSVP-TE provides all the functions that are required to support signaling (or path setup) for MPLS-TE applications. Succinctly, RSVP-TE allows the establishment of explicitly routed LSP tunnels, distributes label bindings, supports resource reservations along an LSP tunnel, and tracks the physical route traversed by an LSP tunnel. Moreover, RSVP-TE gracefully supports the concept of make-before-break when reoptimizing or rerouting an existing LSP tunnel, allows the reallocation of previously assigned network resources when establishing a new LSP tunnel, and performs loop detection during LSP tunnel setup. Last but not least, RSVP-TE overcomes the scalability, latency, and reliability issues posed by the soft-state model of standard RSVP. With all these enhancements, RSVP-TE is dynamic and robust enough to become the key signaling protocol deployed for MPLS-TE applications, particularly in metropolitan area network (MAN) environments.

# Chapter 7
# Traffic Engineering Metro Area Networks

## 7.1 INTRODUCTION

This chapter converts the theoretical Multi-Protocol Label Switching with traffic engineering extensions (MPLS-TE) concepts previously discussed in chapters 3 to 6 into a series of hands-on case studies so that readers can associate what they have learned thus far with their new or existing metropolitan area network (MAN) environments. These case studies are implemented using Riverstone 8600 and 38000 metro routers running Rapid OS (ROS) Version 9.3.0.1.

The scenarios presented in the case studies are based on a hypothetical MAN environment and they all demonstrate the two important aspects of MPLS-TE: traffic control and resource optimization. All names used are fictitious and all IP addresses as well as configurations are provided in the case studies for illustrative purposes only. Nevertheless, all configurations were tested and verified in a lab environment, so they can actually be deployed in the field.

## 7.2 BACKGROUND INFORMATION

The setting is based on a mythical city known as the Wind Metropolis. Wind Metropolis is in the midst of an upgrade program that seeks to overhaul or replace (if need be) its existing time division multiplexing (TDM) communication infrastructure with a high-speed broadband MAN. The main objective of the program is to propel Wind Metropolis into a commercial hub equipped with a highly advanced and efficient metro communication infrastructure. Aeolus Communications, the city's incumbent Internet service provider, has been appointed to take on this challenging task.

To date Wind Metropolis is segregated into the following four main districts:

- *Boreas District in the north*: Commercial estate 1, technological park 1, and two state universities are all situated in this northerly district.
- *Notus District in the south*: Internet data center 1 and minicity 1 are located in this southerly district.

- *Eurus District in the east*: Internet data center 2 and minicity 2 are located in this easterly district.
- *Zephir District in the west*: Commercial estate 2 and government compound 1 are situated in this westerly district.

Commercial estate 1 comprises mainly small or medium local enterprises, while technological park 1 houses world-renowned research organizations. Commercial estate 2 comprises mainly large multinational companies and financial institutions, whereas government compound 1 accommodates two ministries and one state hospital. Minicities 1 and 2 are primarily residential estates. Each of these residential areas has a world-class shopping mall and an urban supermarket. Figure 7.1 illustrates the MAN topology and the approximate distances in between the respective districts.

Internet access and Voice-over-IP (VoIP) will be the two initial services offered by Aeolus to the metro subscribers in the four different districts. The equipment pertaining to these two services will be housed at Internet data center 1. Web hosting and off-site storage are two other services that Aeolus is considering for distribution from Internet data center 2.

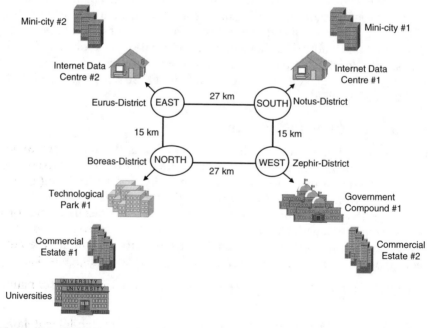

**Figure 7.1   MAN Topology**

The overall upgrade program consists of a few phases and each phase is subdivided into different implementation stages. In the first phase, Aeolus will attempt to implement the various attributes of MPLS-TE into the MAN as shown in Figure 7.1. All the implementation stages involved are documented as case studies.

## 7.2.1 Case Study 7.1: Hop-by-Hop Routed LSPs

**7.2.1.1 Case Overview and Network Topology.** Case study 7.1 describes the implementation stage for which a conventional topology-driven interior gateway protocol (IGP) is used to select a path on a hop-by-hop basis given that the traffic trunk at the ingress point does not have any resource requirements or policy constraints. This method determines a path through the network according to the IGP's perspective of the network. The IGP at each router is free to select active next-hops based on the link-state database (LSDB). In the event of a path failure (for instance, a link failure somewhere in the network), the hop-by-hop forwarding method will eventually establish a path around the failure based on updated LSDB information. As far as the deployed IGP is concerned, all routers participating in MPLS-TE are within a single area or routing domain. This applies to all the case studies throughout the chapter.

Standard open shortest path first (OSPF) is the IGP choice for the path selection. Once the path is determined, it is handed over to the Resource Reservation Protocol with TE extensions (RSVP-TE) in charge of the path setup. Figure 7.2 illustrates the network diagram for this case study. One Riverstone metro router is assigned to each point of presence (POP) of the four different districts. Specifically, Boreas-R1 is the PE router situated at the northerly POP, Zephir-R2 is the PE router located at the westerly POP, Notus-R3 is the PE router situated at the southerly POP, and Eurus-R4 is the PE router located at the easterly POP.

The setting in this case study pertains to the users in the Boreas district utilizing the Internet and VoIP services provided by Aeolus from Internet data center 1 situated at the Notus district. For simplicity, the focus is on one particular user in the Boreas district represented by host address 223.0.0.1, accessing the VoIP and Internet services from Internet data center 1, represented by a single host address 223.0.0.3. The entire MPLS-TE process is performed according to the following steps:

- *Step 1*: Mapping user traffic onto forwarding equivalence classes (FECs). The user traffic originates from 223.0.0.1, and its destination address or FEC is 223.0.0.3.
- *Step 2*: Mapping FECs onto traffic trunks with associated constraints. In this case study, there are no associated resource requirements or policy constraints with the traffic trunk.

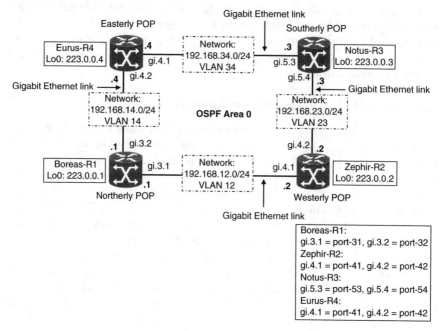

**Figure 7.2   Network Diagram for Case Study 7.1**

- *Step 3*: Mapping traffic trunks to physical network topology. The appropriate paths for the traffic trunk through the physical network are determined by standard OSPF in a hop-by-hop manner. When these paths are handed over to RSVP-TE for path setup, the established label switched path (LSP) will be hop by hop routed.

The detailed network configurations and monitoring for the above-stated procedures are discussed in the subsequent subsections.

**7.2.1.2   Network Configurations.** Listing 7.1 illustrates the detailed network configuration (described with in-line headers) for Boreas-R1. Boreas-R1 is the ingress (or POP) label switch router (LSR) serving a particular user in the Boreas district represented by loopback address 223.0.0.1, which also happens to be the router ID of Boreas-R1.

---

**Listing 7.1   Boreas-R1 Configuration**

**Define hostname**
system set name Boreas-R1

**VLAN configuration**
vlan create vlan14 id 14 ip

```
vlan create vlan12 id 12 ip
vlan add ports gi.3.2 to vlan14
vlan add ports gi.3.1 to vlan12
```

**IP address configuration**
```
interface create ip port-32 vlan vlan14 address-netmask 192.168.14.1/24
interface create ip port-31 vlan vlan12 address-netmask 192.168.12.1/24
```

**Create loopback address**
```
interface add ip lo0 address-netmask 223.0.0.1/32
```

**Set loopback address as router identifier**
```
ip-router global set router-id 223.0.0.1
```

**Single-area OSPF configuration**
```
ospf create area backbone
ospf add stub-host 223.0.0.1 to-area backbone cost 10
ospf add interface port-32 to-area backbone
ospf add interface port-31 to-area backbone
ospf set interface port-32 cost 10
ospf start
```

**MPLS configuration**
```
mpls add interface port-31
mpls add interface port-32
mpls create label-switched-path R1-to-R3 to 223.0.0.3 adaptive no-cspf
mpls create policy P1 dst-ipaddr-mask 223.0.0.3/32
mpls set label-switched-path R1-to-R3 policy P1
mpls start
```

**RSVP configuration**
```
rsvp add interface port-31
rsvp add interface port-32
rsvp set interface port-31 aggregate-enable msgid-extensions-enable
rsvp set interface port-32 aggregate-enable msgid-extensions-enable
rsvp set interface port-31 hello-enable
rsvp set interface port-32 hello-enable
rsvp start
```

To access the VoIP and Internet services distributed by Internet data center 1, there are two paths the user traffic can traverse to reach Notus-R3 from Boreas-R1. These are equal-cost multipaths determined by OSPF; thus, load sharing between the two paths is automatically enabled. To disable this feature, we deliberately set the cost of path R1–R4–R3 to be higher

119

than that of path R1–R2–R3 by overriding the default cost of both network links R1–R4 and R4–R3 (see also Listings 7.3 and 7.4) with a cost of 10. Internet data center 1 is represented by loopback address 223.0.0.3, which also happens to be the FEC, tunnel endpoint of LSP "R1–to–R3" and Notus-R3's router ID (see Listing 7.3).

A hop-by-hop routed LSP "R1–to–R3" is set up from Boreas-R1 (223.0.0.1) to Notus-R3 (223.0.0.3). The make-before-break feature is enabled on this LSP by the "adaptive" parameter, which sets the "SE style desired" flag (= 0 × 04) in the RSVP SESSION_ATTRIBUTE object (see section 6.4.1.7). The "no-cspf" parameter disables constrained shortest-path-first (CSPF) computations by the ingress LSR, but the standard shortest-path-first calculation is still applicable. Policy-based routing (PBR) is then applied to forward user traffic onto the established LSP (see Listing 7.5).

The parameters "aggregate-enable" (see section 6.9.1) and "msgid-extensions-enable" (see sections 6.9.2 and 6.9.3) help to scale RSVP by reducing the volume and processing time of RSVP refresh messages. RSVP hello extension (see section 6.9.4) is also enabled by the "hello-enable" parameter. These parameters are enabled on all the routers (see also Listings 7.2 to 7.4).

Listing 7.2 illustrates the detailed network configuration (described with in-line headers) for Zephir-R2. As Zephir-R2 is a transit (or core) LSR, the MPLS and RSVP configurations are very straightforward. MPLS and RSVP are enabled on individual ports of Zephir-R2 that participate in the MPLS-TE operation.

---

### Listing 7.2   Zephir-R2 Configuration

```
Define hostname
system set name Zephir-R2


VLAN configuration
vlan create vlan12 id 12 ip
vlan create vlan23 id 23 ip
vlan add ports gi.4.1 to vlan12
vlan add ports gi.4.2 to vlan23


IP address configuration
interface create ip port-41 vlan vlan12 address-netmask 192.168.12.2/24
interface create ip port-42 vlan vlan23 address-netmask 192.168.23.2/24


Create loopback address
interface add ip lo0 address-netmask 223.0.0.2/32
```

**Set loopback address as router identifier**
```
ip-router global set router-id 223.0.0.2
```

**Single-area OSPF configuration**
```
ospf create area backbone
ospf add stub-host 223.0.0.2 to-area backbone cost 10
ospf add interface port-41 to-area backbone
ospf add interface port-42 to-area backbone
ospf start
```

**MPLS configuration**
```
mpls add interface port-41
mpls add interface port-42
mpls start
```

**RSVP configuration**
```
rsvp add interface port-41
rsvp add interface port-42
rsvp set interface port-41 aggregate-enable msgid-extensions-enable
rsvp set interface port-42 aggregate-enable msgid-extensions-enable
rsvp set interface port-41 hello-enable
rsvp set interface port-42 hello-enable
rsvp start
```

Listing 7.3 illustrates the detailed network configuration (described with in-line headers) for Notus-R3, the egress LSR. Similar to Zephir-R2, MPLS and RSVP are enabled on individual ports of Notus-R3 that participate in the MPLS-TE operation. In addition, penultimate hop popping (PHP) is disabled on these ports by the "no-php" parameter to ensure that only Notus-R3 (the ultimate hop) performs the pop operation.

### Listing 7.3 Notus-R3 Configuration

**Define hostname**
```
system set name Notus-R3
```

**VLAN configuration**
```
vlan create vlan34 id 34 ip
vlan create vlan23 id 23 ip
vlan add ports gi.5.3 to vlan34
vlan add ports gi.5.4 to vlan23
```

**IP address configuration**
```
interface create ip port-53 vlan vlan34 address-netmask 192.168.34.3/24
interface create ip port-54 vlan vlan23 address-netmask 192.168.23.3/24
```

**Create loopback address**
```
interface add ip lo0 address-netmask 223.0.0.3/32
```

**Set loopback address as router identifier**
```
ip-router global set router-id 223.0.0.3
```

**Single-area OSPF configuration**
```
ospf create area backbone
ospf add stub-host 223.0.0.3 to-area backbone cost 10
ospf add interface port-53 to-area backbone
ospf add interface port-54 to-area backbone
ospf set interface port-53 cost 10
ospf start
```

**MPLS configuration**
```
mpls add interface port-53
mpls add interface port-54
mpls set interface port-53 no-php
mpls set interface port-54 no-php
mpls start
```

**RSVP configuration**
```
rsvp add interface port-53
rsvp add interface port-54
rsvp set interface port-53 aggregate-enable msgid-extensions-enable
rsvp set interface port-54 aggregate-enable msgid-extensions-enable
rsvp set interface port-53 hello-enable
rsvp set interface port-54 hello-enable
rsvp start
```

Listing 7.4 illustrates the detailed network configuration (described with in-line headers) for Eurus-R4, a transit (or core) LSR just like Zephir-R2. MPLS and RSVP are enabled on individual ports of Eurus-R4 that participate in the MPLS-TE operation.

### Listing 7.4   Eurus-R4 Configuration

**Define hostname**
```
system set name Eurus-R4
```

**VLAN configuration**
```
vlan create vlan14 id 14 ip
vlan create vlan34 id 34 ip
vlan add ports gi.4.2 to vlan14
vlan add ports gi.4.1 to vlan34
```

**IP address configuration**
```
interface create ip port-42 vlan vlan14 address-netmask 192.168.14.4/24
interface create ip port-41 vlan vlan34 address-netmask 192.168.34.4/24
```

**Create loopback address**
```
interface add ip lo0 address-netmask 223.0.0.4/32
```

**Set loopback address as router identifier**
```
ip-router global set router-id 223.0.0.4
```

**Single-area OSPF configuration**
```
ospf create area backbone
ospf add stub-host 223.0.0.4 to-area backbone cost 10
ospf add interface port-42 to-area backbone
ospf add interface port-41 to-area backbone
ospf set interface port-41 cost 10
ospf set interface port-42 cost 10
ospf start
```

**MPLS configuration**
```
mpls add interface port-41
mpls add interface port-42
mpls start
```

**RSVP configuration**
```
rsvp add interface port-41
rsvp add interface port-42
rsvp set interface port-41 aggregate-enable msgid-extensions-enable
rsvp set interface port-42 aggregate-enable msgid-extensions-enable
rsvp set interface port-41 hello-enable
rsvp set interface port-42 hello-enable
rsvp start
```

For brevity, the case study only illustrates the traffic flow from Boreas-R1 (ingress) to Notus-R3 (egress). Remember that LSPs are unidirectional. If MPLS-TE is to be applied to the return traffic from Notus-R3 (ingress) to Boreas-R1 (egress), an LSP will have to be established in this direction too.

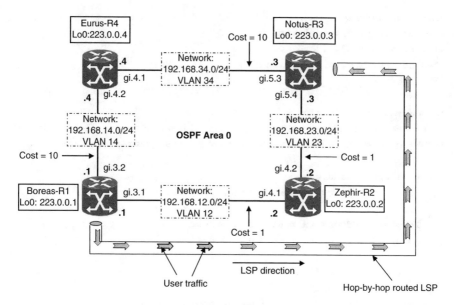

**Figure 7.3    Hop-by-Hop Routed LSP before Link R2–R3 Fails**

**7.2.1.3   Monitoring before Network Link R2–R3 Fails.** As illustrated in Figure 7.3, R1–R2–R3 is the actual physical path taken by user traffic traversing the established LSP.

Listing 7.5 illustrates the MPLS policy defined for Boreas-R1. The destination (FEC) is 223.0.0.3, which is the router ID of Notus-R3, and the source is any IP address. For testing purposes, only source IP address 223.0.0.1 (router ID of Boreas-R1) is used. The MPLS policy is applied during PBR, which forwards user traffic onto the established LSP.

---

**Listing 7.5    Boreas-R1 MPLS Policy**

```
Boreas-R1# mpls show policy


Name   Type   Destination   Port   Source    Port   TOS   Prot   Use
```

| Name | Type | Destination | Port | Source | Port | TOS | Prot | Use |
|------|------|-------------|------|--------|------|-----|------|-----|
| P1 | L3 | 223.0.0.3 | Any | 0.0.0.0 | Any | Any | IP | INUSE |

---

Listing 7.6 illustrates the LSP information of Boreas-R1. An LSP, using shared-explicit (SE) reservation style, has been established from ingress LSR Boreas-R1 (223.0.0.1) to egress LSR Notus-R3 (223.0.0.3). The assigned OutLabel received from Zephir-R2 is 4097. Because Boreas-R1 is the ingress router, it will not generate any InLabel. Instead, it will push the OutLabel onto any packets bound for 223.0.0.3 (FEC).

---

### Listing 7.6   Boreas-R1 LSP Information

```
Boreas-R1# mpls show label-switched-path brief


Ingress LSP:
```

| | | | | | Label | |
|---|---|---|---|---|---|---|
| LSP Name | To | From | State | Style | In | Out |
| **R1-to-R3** | **223.0.0.3** | **223.0.0.1** | **Up** | **SE** | **—** | **4097** |

```
Transit LSP:
```

| | | | | | Label | |
|---|---|---|---|---|---|---|
| LSP Name | To | From | State | Style | In | Out |

```
Egress LSP:
```

| | | | | | Label | |
|---|---|---|---|---|---|---|
| LSP Name | To | From | State | Style | In | Out |

---

Listing 7.7 illustrates the LSP information of Zephir-R2, the transit LSR. The locally generated InLabel that is advertised to Boreas-R1 is 4097, and the assigned OutLabel received from Notus-R3 is 16. In a transit LSR, the InLabel is swapped with the OutLabel during MPLS forwarding.

---

### Listing 7.7   Zephir-R2 LSP Information

```
Zephir-R2# mpls show label-switched-path brief


Ingress LSP:
```

| | | | | | Label | |
|---|---|---|---|---|---|---|
| LSP Name | To | From | State | Style | In | Out |

```
Transit LSP:
```

| | | | | | Label | |
|---|---|---|---|---|---|---|
| LSP Name | To | From | State | Style | In | Out |
| **R1-to-R3** | **223.0.0.3** | **223.0.0.1** | **Up** | **SE** | **4097** | **16** |

```
Egress LSP:
```

| | | | | | Label | |
|---|---|---|---|---|---|---|
| LSP Name | To | From | State | Style | In | Out |

---

Listing 7.8 illustrates the LSP information of Notus-R3, the egress LSR. The locally generated InLabel that is advertised to Zephir-R2 is 16. Label 16 is used by Riverstone to indicate the end of the TE tunnel. This label is to be popped and the lower-level label is examined for further processing. As PHP has been disabled and Notus-R3 is the egress router; it performs the pop operation, so there is no OutLabel.

---

### Listing 7.8   Notus-R3 LSP Information

```
Notus-R3# mpls show label-switched-path brief

Ingress LSP:
                                                      Label
LSP Name          To            From        State   Style   In    Out

Transit LSP:
                                                      Label
LSP Name          To            From        State   Style   In    Out

Egress LSP:
                                                      Label
LSP Name          To            From        State   Style   In    Out
R1-to-R3          223.0.0.3     223.0.0.1   Up      SE      16    -
```

---

Listing 7.9 illustrates the RSVP PATH state block (PSB) of Boreas-R1, which contains RSVP objects such as SESSION_ATTRIBUTE, SESSION, SENDER_TEMPLATE, RSVP_HOP (previous-hop or PHOP), EXPLICIT_ROUTE, and so on (see section 6.4.1). The "SE style desired" flag (= $0 \times 04$) is set in the SESSION_ATTRIBUTE object, indicating that the SE reservation style (see section 6.4.2.2) has been adopted. The PHOP is reflected as 0.0.0.0, indicating that Boreas-R1 is the ingress router. The RSVP PATH message also originates from the local RSVP application program interface (API) of Boreas-R1 and is sent out of port-31. Because path selection is hop by hop based, the EXPLICIT_ROUTE object (ERO) is not used.

---

### Listing 7.9   Boreas-R1 RSVP PSB

```
Boreas-R1# rsvp show psb

Path State Blocks:
```

```
RSVP_PSB <rsvp_1>:
session-attr: name: R1-to-R3 flags: 0x4 setup-pri: 7 holding-pri: 0
session: end-point: 223.0.0.3 tunnel-id: 6 ext-tunnel-id: 0xdf000001
send-templ: sender: 223.0.0.1 lsp-id: 1
prev-hop: 0.0.0.0
in-if: <Local-API> out-if: <port-31>
explicit-route:
```

Listing 7.10 illustrates the RSVP PSB of Zephir-R2. The PHOP is reflected as 192.168.12.1, indicating that Boreas-R1 is the previous-hop. The RSVP PATH message is received from port-41 and is sent out of port-42.

### Listing 7.10    Zephir-R2 RSVP PSB

```
Zephir-R2# rsvp show psb

Path State Blocks:


RSVP_PSB <rsvp_1>:
session-attr: name: R1-to-R3 flags: 0x4 setup-pri: 7 holding-pri: 0
session: end-point: 223.0.0.3 tunnel-id: 6 ext-tunnel-id: 0xdf000001
send-templ: sender: 223.0.0.1 lsp-id: 1
prev-hop: 192.168.12.1
in-if: <port-41> out-if: <port-42>
explicit-route:
```

Listing 7.11 illustrates the RSVP PSB of Notus-R3. The PHOP is reflected as 192.168.23.2, indicating that Zephir-R2 is the previous-hop. The RSVP PATH message is received from port-54. Because Notus-R3 is the egress router, the PATH message is sent to its local RSVP API for processing.

### Listing 7.11    Notus-R3 RSVP PSB

```
Notus-R3# rsvp show psb

Path State Blocks:


RSVP_PSB <rsvp_1>:
session-attr: name: R1-to-R3 flags: 0x4 setup-pri: 7 holding-pri: 0
session: end-point: 223.0.0.3 tunnel-id: 6 ext-tunnel-id: 0xdf000001
send-templ: sender: 223.0.0.1 lsp-id: 1
```

```
prev-hop: 192.168.23.2
in-if: <port-54> out-if: <Local-API>
explicit-route:
```

Listing 7.12 illustrates the RSVP reservation state block (RSB) of Notus-R3, which contains RSVP objects such as SESSION, STYLE, FILTER_SPEC, LABEL, RECORD_ROUTE, and so on (see section 6.4.3). The local RSVP API of Notus-R3 constructs the corresponding RESV message and generates label 16. This local label (InLabel) is placed in the LABEL object of the RESV message, which is sent upstream to Zephir-R2 based on the PHOP 192.168.23.2 found in the local RSVP PSB (see Listing 7.11). As Notus-R3 is the egress router, it performs the pop operation, so there is no remote label (OutLabel) received in this case.

### Listing 7.12   Notus-R3 RSVP RSB

```
Notus-R3# rsvp show rsb

Resv State Blocks:

RSVP_RSB <rsvp_1>:
session: end-point: 223.0.0.3 tunnel-id: 6 ext-tunnel-id: 0xdf000001
style: SE
in-if: <Local-API>
filter-spec: sender: 223.0.0.1 lsp-id: 1
remote-labels: []
local-labels: [16]
record-route:
```

Listing 7.13 illustrates the RSVP RSB of Zephir-R2. The RESV message from Notus-R3 is received at port-42. The remote label (OutLabel) of value 16 from Notus-R3 is swapped with local label (InLabel) 4097 generated by Zephir-R2. In addition, the RECORD_ROUTE object (RRO) found in the RESV message contains the IP address 192.168.23.3, which is associated with port-54 of Notus-R3, the downstream (or next-hop) neighbor of Zephir-R2. The RRO is organized in a last-in-first-out (stack) format, where the most recent LSR that has written its route information as a subobject becomes the top-level entry. The modified RESV message is sent upstream to Boreas-R1 based on the PHOP 192.168.12.1 found in the local RSVP PSB (see Listing 7.10).

---

### Listing 7.13   Zephir-R2 RSVP RSB

```
Zephir-R2# rsvp show rsb

Resv State Blocks:

RSVP_RSB <rsvp_1>:
session: end-point: 223.0.0.3 tunnel-id: 6 ext-tunnel-id: 0xdf000001
style: SE
in-if: <port-42>
filter-spec: sender: 223.0.0.1 lsp-id: 1
remote-labels: [16]
local-labels: [4097]
record-route:
192.168.23.3
```

---

Listing 7.14 illustrates the RSVP RSB of Boreas-R1. The RESV message is received at port-31. The LSP is established once Boreas-R1 receives the RESV message. The remote label (OutLabel) of value 4097 is derived from the LABEL object of the RESV message. Because Boreas-R1 is the ingress router, it does not generate any local label. Instead, it will push the remote label onto any packets bound for 223.0.0.3 (FEC). Furthermore, Zephir-R2, the downstream (or next-hop) neighbor of Boreas-R1, has prepended IP address 192.168.12.2 associated with its port-41 to the RRO.

---

### Listing 7.14   Boreas-R1 RSVP RSB

```
Boreas-R1# rsvp show rsb

Resv State Blocks:

RSVP_RSB <rsvp_1>:
session: end-point: 223.0.0.3 tunnel-id: 6 ext-tunnel-id: 0xdf000001
style: SE
in-if: <port-31>
filter-spec: sender: 223.0.0.1 lsp-id: 1
remote-labels: [4097]
local-labels: []
record-route:
192.168.12.2
192.168.23.3
```

---

Listing 7.15 illustrates the result of the traceroute test performed at Boreas-R1. The traceroute output indicates that user traffic (source 223.0.0.1) is indeed traversing path R1–R2–R3 to reach the destination (223.0.0.3).

---

### Listing 7.15　Traceroute Test at Boreas-R1

```
Boreas-R1# traceroute 223.0.0.3 source 223.0.0.1

traceroute to 223.0.0.3 (223.0.0.3) from 223.0.0.1, 30 hops max, 40 byte
packets
1 192.168.12.2 (192.168.12.2) 3 ms 1 ms 1 ms
  MPLS Label1=4097 EXP1=0 TTL=1 S=1
2 223.0.0.3 (223.0.0.3) 1 ms 0 ms 0 ms
```

---

**7.2.1.4　Monitoring after Network Link R2–R3 Fails.** As illustrated in Figure 7.4, when network link R2–R3 fails, the actual physical path taken by user traffic traversing the established LSP becomes R1–R4–R3.

Listing 7.16 illustrates the result of the traceroute test performed at Boreas-R1. The traceroute output indicates that user traffic (source 223.0.0.1) is indeed traversing path R1–R4–R3 to reach the destination (223.0.0.3).

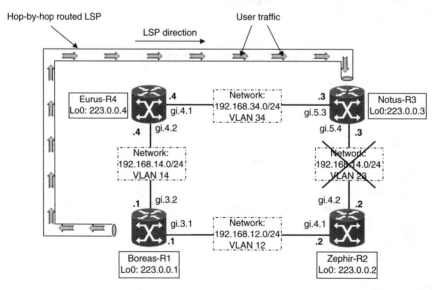

**Figure 7.4　Hop-by-Hop Routed LSP after Link R2–R3 Fails**

---

### Listing 7.16    Traceroute Test at Boreas-R1

```
Boreas-R1# traceroute 223.0.0.3 source 223.0.0.1

traceroute to 223.0.0.3 (223.0.0.3) from 223.0.0.1, 30 hops max, 40 byte
packets
1  192.168.14.4 (192.168.14.4)  3 ms  1 ms  1 ms
   MPLS Label1=4097 EXP1=0 TTL=1 S=1
2  223.0.0.3 (223.0.0.3)  3 ms  1 ms  0 ms
```

---

Listing 7.17 illustrates the LSP information of Eurus-R4, which has become the transit LSR after network link R2–R3 fails. The locally generated InLabel that is advertised to Boreas-R1 is 4097, and the assigned Out-Label received from Notus-R3 is 16. The LSP information listings for Boreas-R1 and Notus-R3 are similar to those listed in the previous subsection and are thus omitted for brevity.

---

### Listing 7.17    Eurus-R4 LSP Information

```
Eurus-R4# mpls show label-switched-path brief
```

Ingress LSP:

| | | | | | Label | |
| LSP Name | To | From | State | Style | In | Out |
| --- | --- | --- | --- | --- | --- | --- |

Transit LSP:

| | | | | | Label | |
| LSP Name | To | From | State | Style | In | Out |
| --- | --- | --- | --- | --- | --- | --- |
| R1-to-R3 | 223.0.0.3 | 223.0.0.1 | Up | SE | 4097 | 16 |

Egress LSP:

| | | | | | Label | |
| LSP Name | To | From | State | Style | In | Out |
| --- | --- | --- | --- | --- | --- | --- |

---

Listing 7.18 illustrates the RSVP PSB of Boreas-R1 after network link R2–R3 fails. The RSVP PATH message is sent out of port-32 in the direction of Eurus-R4.

---

### Listing 7.18   Boreas-R1 RSVP PSB

```
Boreas-R1# rsvp show psb

Path State Blocks:

RSVP_PSB <rsvp_1>:
session-attr: name: R1-to-R3 flags: 0x4 setup-pri: 7 holding-pri: 0
session: end-point: 223.0.0.3 tunnel-id: 6 ext-tunnel-id: 0xdf000001
send-templ: sender: 223.0.0.1 lsp-id: 2
prev-hop: 0.0.0.0
in-if: <Local-API> out-if: <port-32>
explicit-route:
```

---

Listing 7.19 illustrates the RSVP PSB of Eurus-R4. The PHOP is reflected as 192.168.14.1, indicating that Boreas-R1 is the previous-hop. The RSVP PATH message is received from port-42 and is sent out of port-41.

---

### Listing 7.19   Eurus-R4 RSVP PSB

```
Eurus-R4# rsvp show psb

Path State Blocks:

RSVP_PSB <rsvp_1>:
session-attr: name: R1-to-R3 flags: 0x4 setup-pri: 7 holding-pri: 0
session: end-point: 223.0.0.3 tunnel-id: 6 ext-tunnel-id: 0xdf000001
send-templ: sender: 223.0.0.1 lsp-id: 2
prev-hop: 192.168.14.1
in-if: <port-42> out-if: <port-41>
explicit-route:
```

---

Listing 7.20 illustrates the RSVP PSB of Notus-R3 after network link R2–R3 fails. The PHOP is reflected as 192.168.34.4, indicating that Eurus-R4 is the previous-hop. The RSVP PATH message is received from port-53.

---

### Listing 7.20   Notus-R3 RSVP PSB

```
Notus-R3# rsvp show psb
```

```
Path State Blocks:

RSVP_PSB <rsvp_1>:
session-attr: name: R1-to-R3 flags: 0x4 setup-pri: 7 holding-pri: 0
session: end-point: 223.0.0.3 tunnel-id: 6 ext-tunnel-id: 0xdf000001
send-templ: sender: 223.0.0.1 lsp-id: 2
prev-hop: 192.168.34.4
in-if: <port-53> out-if: <Local-API>
explicit-route:
```

---

The RSVP RSB listing of Notus-R3 is identical to that listed in the previous subsection before network link R2–R3 fails and is thus omitted for brevity.

Listing 7.21 illustrates the RSVP RSB of Eurus-R4. The RESV message from Notus-R3 is received at port-41. The RRO contains the IP address 192.168.34.3, which is associated with port-53 of Notus-R3, the downstream (or next-hop) neighbor of Eurus-R4.

---

### Listing 7.21   Eurus-R4 RSVP RSB

```
Eurus-R4# rsvp show rsb

Resv State Blocks:

RSVP_RSB <rsvp_1>:
session: end-point: 223.0.0.3 tunnel-id: 6 ext-tunnel-id: 0xdf000001
style: SE
in-if: <port-41>
filter-spec: sender: 223.0.0.1 lsp-id: 2
remote-labels: [16]
local-labels: [4097]
record-route:
192.168.34.3
```

---

Listing 7.22 illustrates the RSVP RSB of Boreas-R1 after network link R2–R3 fails. The RESV message is received at port-32. The new LSP is established once Boreas-R1 receives the RESV message. Eurus-R4, the downstream (or next-hop) neighbor of Boreas-R1, has prepended IP address 192.168.14.4 associated with its port-42 to the RRO.

---

**Listing 7.22   Boreas-R1 RSVP RSB**

```
Boreas-R1# rsvp show rsb

Resv State Blocks:

RSVP_RSB <rsvp_1>:
session: end-point: 223.0.0.3 tunnel-id: 6 ext-tunnel-id: 0xdf000001
style: SE
in-if: <port-32>
filter-spec: sender: 223.0.0.1 lsp-id: 2
remote-labels: [4097]
local-labels: []
record-route:
192.168.14.4
192.168.34.3
```

---

**7.2.1.5   Case Commentary.** Case study 7.1 does not really reap the full benefit of RSVP-TE without any definition of resource requirements and policy constraints. Moreover, using standard OSPF for path selection reverts the path-based nature of MPLS LSP back to hop-by-hop control exhibited by conventional IGP. Whenever there is a topology change (for instance, failure of network link R2–R3), the time taken to establish the new LSP is dependent on the convergence time of OSPF, which is on the order of several seconds (typically from 6 to 46 seconds). Because the path is determined on a hop-by-hop basis, the metro service provider (SP) can only provide best-effort service delivery in this case.

### 7.2.2   Case Study 7.2: Explicitly Routed LSPs (Loose Explicit Route Example)

**7.2.2.1   Case Overview and Network Topology.** Case study 7.2 is adapted directly from case study 7.1. It explores the setup of an explicitly routed LSP, taking into consideration that service must be able to recover without manual intervention in the event of failure. There are two types of explicit route: strict and loose. A strict explicit route with a single primary path provides precise traffic control and the flexibility to constrain routes based on policy considerations rather than always selecting the shortest-path route. However, a separate secondary path is required to back up the primary path when any of the components (links and nodes) along the primary path fail. The same outcome can be achieved with only a primary path using loose explicit routing.

In a strict explicit route, the path must only go through the specified strict hops and not any other unspecified hops. In a loose explicit route, the path can include other unspecified loose hops determined by conventional IGP. In other words, a strict hop indicates that the two LSRs must be adjacent to one another with no intermediate hops separating them, whereas a loose hop indicates that the LSRs do not have to be adjacent to each other and the IGP can be used to determine the best path to the loose hop. Therefore, through loose explicit routing, RSVP-TE is able to detour a new LSP around a failure with the help of conventional IGP but at the expense of complete control.

Standard OSPF is used once again for the selection of the transit hop to loose hop Notus-R3, which is Zephir-R2 during normal operation (see Figure 7.5) and Eurus-R4 after network link R2–R3 fails (see Figure 7.6).

The affected network configurations and the monitoring involved are discussed in the subsequent subsections.

**7.2.2.2 Network Configurations.** Listing 7.23 illustrates the portion of the modified network configuration for Boreas-R1 based on the previous configuration adapted from Listing 7.1. The unchanged portion is not shown. Instead of using OSPF for the full path selection, an explicit path is manually defined for the traffic trunk, which again does not have any associated resource requirements and policy constraints.

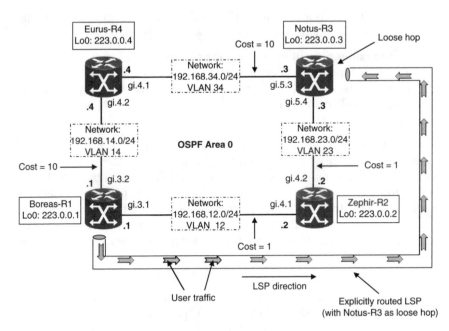

**Figure 7.5  Explicitly Routed LSP with Notus-R3 as Loose Hop before Link R2–R3 Fails**

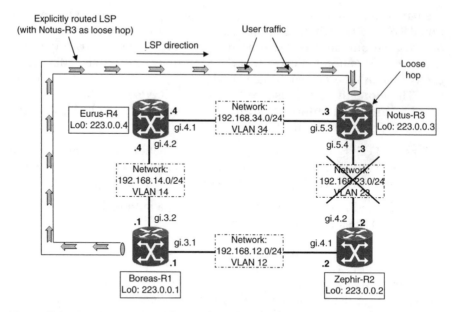

**Figure 7.6 Explicitly Routed LSP with Notus-R3 as Loose Hop after Link R2–R3 Fails**

Using the "mpls create path" and "mpls set path" commands, explicit path "R1–R2–R3" is created with the following properties:

- The maximum number of hops for the path is 1.
- The IP address of the loose hop (hop 1) in the path is 223.0.0.3, which is also the FEC. This implies that the loose hop can be reached via Zephir-R2 during normal operation or via Eurus-R4 when network link R2–R3 fails.

Explicit path "R1–R2–R3" is defined as the primary path for LSP "R1–to–R3" by specifying the "primary" parameter with the "mpls set label-switched-path" command. RSVP-TE then uses this administratively defined explicit route to set up the LSP. Note that the "no-cspf" parameter has to be specified when defining explicit paths manually. Otherwise, RSVP-TE assumes that the explicit path will be dynamically specified by a constraint-based IGP and will wait indefinitely for a valid CSPF response before setting up the LSP.

---

**Listing 7.23   Boreas-R1 Configuration**

```
system set name Boreas-R1
```

**MPLS configuration**

```
mpls add interface port-31
mpls add interface port-32
mpls create path R1-R2-R3 num-hops 1
mpls set path R1-R2-R3 hop 1 ip-addr 223.0.0.3 type loose

mpls create label-switched-path R1-to-R3 to 223.0.0.3 no-cspf
mpls set label-switched-path R1-to-R3 primary R1-R2-R3 adaptive
mpls create policy P1 dst-ipaddr-mask 223.0.0.3/32
mpls set label-switched-path R1-to-R3 policy P1
mpls start
```

The network configurations for the rest of the routers listed in section 7.2.1.2 are still valid for this case study.

**7.2.2.3  Monitoring before Network Link R2–R3 Fails.** The traceroute test shown in Listing 7.24 yields results similar to those found in Listing 7.15. Specifically, user traffic is traversing from R1 via R2 (192.168.12.2) to reach the destination (223.0.0.3).

---

### Listing 7.24   Traceroute Test at Boreas-R1

```
Boreas-R1# traceroute 223.0.0.3 source 223.0.0.1

traceroute to 223.0.0.3 (223.0.0.3) from 223.0.0.1, 30 hops max, 40 byte
packets
1  192.168.12.2 (192.168.12.2)  3 ms  1 ms  1 ms
   MPLS Label1=4097 EXP1=0 TTL=1 S=1
2  223.0.0.3 (223.0.0.3)  1 ms  0 ms  0 ms
```

---

Because a loose explicit route is derived from the ERO found in the RSVP PATH message, it would be appropriate to examine the RSVP PSB straightaway.

Listing 7.25 illustrates the RSVP PSB of Boreas-R1. To reach loose hop 223.0.0.3 from Boreas-R1, the ERO depicts the nearest next-hop found by OSPF as 192.168.12.2 (Zephir-R2).

---

### Listing 7.25   Boreas-R1 RSVP PSB

```
Boreas-R1# rsvp show psb

Path State Blocks:
```

```
RSVP_PSB <rsvp_1>:
session-attr: name: R1-to-R3_R1-R2-R3 flags: 0x4 setup-pri: 7 holding-
pri: 0
session: end-point: 223.0.0.3 tunnel-id: 16385 ext-tunnel-id:
0xdf000001
send-templ: sender: 223.0.0.1 lsp-id: 1
prev-hop: 0.0.0.0
in-if: <Local-API> out-if: <port-31>
explicit-route: 192.168.12.2=>(Loose)223.0.0.3
```

Listing 7.26 illustrates the RSVP PSB of Zephir-R2. To reach loose hop 223.0.0.3 from Zephir-R2, the ERO depicts the nearest next-hop found by OSPF as 192.168.23.3 (Notus-R3).

### Listing 7.26   Zephir-R2 RSVP PSB

```
Zephir-R2# rsvp show psb

Path State Blocks:

RSVP_PSB <rsvp_1>:
session-attr: name: R1-to-R3_R1-R2-R3 flags: 0x4 setup-pri: 7 holding-
pri: 0
session: end-point: 223.0.0.3 tunnel-id: 16385 ext-tunnel-id:
0xdf000001
send-templ: sender: 223.0.0.1 lsp-id: 1
prev-hop: 192.168.12.1
in-if: <port-41> out-if: <port-42>
explicit-route: 192.168.23.3=>(Loose)223.0.0.3
```

Listing 7.27 illustrates the RSVP PSB of Notus-R3. Loose hop 223.0.0.3 is at Notus-R3 itself, so the next-hop is no longer required to be derived from the ERO.

### Listing 7.27   Notus-R3 RSVP PSB

```
Notus-R3# rsvp show psb

Path State Blocks:

RSVP_PSB <rsvp_1>:
session-attr: name: R1-to-R3_R1-R2-R3 flags: 0x4 setup-pri: 7 holding-
pri: 0
```

```
session: end-point: 223.0.0.3 tunnel-id: 16385 ext-tunnel-id:
0xdf000001
send-templ: sender: 223.0.0.1 lsp-id: 1
prev-hop: 192.168.23.2
in-if: <port-54> out-if: <Local-API>
explicit-route:
```

**7.2.2.4  Monitoring after Network Link R2–R3 Fails.** The traceroute test shown in Listing 7.28 yields results similar to those found in Listing 7.16. Specifically, user traffic is traversing from R1 via R4 (192.168.14.4) to reach the destination (223.0.0.3) after network link R2–R3 fails.

### Listing 7.28   Traceroute Test at Boreas-R1

```
Boreas-R1# traceroute 223.0.0.3 source 223.0.0.1

traceroute to 223.0.0.3 (223.0.0.3) from 223.0.0.1, 30 hops max, 40 byte
packets
1  192.168.14.4 (192.168.14.4)  3 ms  1 ms  1 ms
   MPLS Label1=4097 EXP1=0 TTL=1 S=1
2  223.0.0.3 (223.0.0.3)  2 ms  1 ms  0 ms
```

Listing 7.29 illustrates the RSVP PSB of Boreas-R1 after network link R2–R3 fails. To reach loose hop 223.0.0.3 from Boreas-R1, the ERO depicts the next-hop found by OSPF as 192.168.14.4 (Eurus-R4).

### Listing 7.29   Boreas-R1 RSVP PSB

```
Boreas-R1# rsvp show psb

Path State Blocks:

RSVP_PSB <rsvp_1>:
session-attr: name: R1-to-R3_R1-R4-R3 flags: 0x4 setup-pri: 7 holding-
pri: 0
session: end-point: 223.0.0.3 tunnel-id: 16385 ext-tunnel-id:
0xdf000001
send-templ: sender: 223.0.0.1 lsp-id: 2
prev-hop: 0.0.0.0
in-if: <Local-API> out-if: <port-32>
explicit-route: 192.168.14.4=>(Loose)223.0.0.3
```

Listing 7.30 illustrates the RSVP PSB of Eurus-R4 after network link R2–R3 fails. To reach loose hop 223.0.0.3 from Eurus-R4, the ERO depicts the next-hop found by OSPF as 192.168.34.3 (Notus-R3).

---

**Listing 7.30   Eurus-R4 RSVP PSB**

```
Eurus-R4# rsvp show psb


Path State Blocks:


RSVP_PSB <rsvp_1>:

session-attr: name: R1-to-R3_R1-R4-R3 flags: 0x4 setup-pri: 7 holding-
pri: 0

session: end-point: 223.0.0.3 tunnel-id: 16385 ext-tunnel-id:
0xdf000001

send-templ: sender: 223.0.0.1 lsp-id: 2

prev-hop: 192.168.14.1

in-if: <port-42> out-if: <port-41>

explicit-route: 192.168.34.3=>(Loose)223.0.0.3
```

---

Listing 7.31 illustrates the RSVP PSB of Notus-R3. Loose hop 223.0.0.3 is at Notus-R3 itself, so the next-hop is no longer required to be derived from the ERO.

---

**Listing 7.31   Notus-R3 RSVP PSB**

```
Notus-R3# rsvp show psb


Path State Blocks:


RSVP_PSB <rsvp_1>:

session-attr: name: R1-to-R3_R1-R4-R3 flags: 0x4 setup-pri: 7 holding-
pri: 0

session: end-point: 223.0.0.3 tunnel-id: 16385 ext-tunnel-id:
0xdf000001

send-templ: sender: 223.0.0.1 lsp-id: 2

prev-hop: 192.168.34.4

in-if: <port-53> out-if: <Local-API>

explicit-route:
```

---

**7.2.2.5   Case Commentary.** Case study 7.2 demonstrates that RSVP-TE is able to detour a new LSP around the point of failure by specifying the LSR immediately downstream of the outage as a loose hop in an administratively defined explicit path. In the case study, there are two paths to reach a specific loose hop. During normal operation, the nearest next-hop for Boreas-R1 to reach loose hop Notus-R3 (223.0.0.3) is Zephir-R2 (192.168.12.2) because R1–R2–R3 is the least-cost path found by OSPF. When network link R2–R3 fails, Eurus-R4 (192.168.14.4) becomes the next-hop for Boreas-R1 to reach 223.0.0.3 because the alternative path found by OSPF is now R1–R4–R3. In both instances, the final selected path is handed to RSVP-TE for the actual LSP setup.

As OSPF is still required to determine the missing link between Boreas-R1 and the loose hop, the time incurred to establish the new LSP is still dependent on the convergence time of OSPF (see section 7.2.1.5). Moreover, control will not be as precise as in a strict explicit route if there are several alternative paths to the same loose hop because OSPF will only choose the least-cost path or equal-cost paths (multipath).

A better way to avert a failure outage is through a global repair technique known as path protection (see section 8.3.5), which supports the configuration of primary and secondary physical paths for an LSP to protect against link and node failures. The primary path is the preferred path, while the secondary path is used as an alternative route only when the primary path fails. Both of these paths are administratively defined strict explicit routes. Besides, the use of hot standby secondary paths improves service recovery time by eliminating the call-setup delay that is required to establish a new physical path for the LSP.

Because there are no resource requirements and policy constraints defined for the traffic trunk, disabling the CSPF attribute in OSPF will suffice for this case study. However, when the need for the specification of resource requirements and policy constraints does arise, CSPF will be the impeccable feature to have. Constraint-based routing (CBR) and CSPF are covered extensively in the next two case studies.

### 7.2.3 Case Study 7.3: Bandwidth Manipulation (CBR Example 1)

**7.2.3.1 Case Overview and Network Topology.** Case study 7.3 describes the next implementation stage for which a constraint-based IGP is used to select a path whereby the traffic trunk at the ingress point has a specific resource requirement, which happens to be a guaranteed bandwidth request from a premium user. Constraint-based OSPF is enabled for the path selection. Once the path is determined, it is handed over to RSVP-TE, in charge of the path setup as before.

Figure 7.7 illustrates the network diagram for this case study. The setting is still associated with the users in the Boreas district, but this time the services provided by Aeolus will be Web hosting and off-site storage from

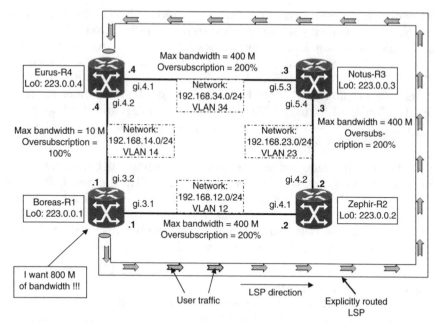

**Figure 7.7    Network Diagram for Case Study 7.3**

Internet data center 2 located at the Eurus district. Due to some renovation work occurring on the stretch of road for which the fiber link between R1 and R4 (shortest path R1–R4) is laid alongside, Aeolus has instead designated path R1–R2–R3–R4 to be traversed by user traffic bound for Internet data center 2.

As an avid advocate in the outsourcing business model, the previous user, represented by host address 223.0.0.1, further subscribes to Aeolus's Web hosting and off-site storage services, and is upgraded to premium-class user. Any user subscribed under the premium-class scheme will have the privilege of enjoying guaranteed bandwidth for the Web hosting and off-site storage services.

**7.2.3.2    Network Configurations.** Listing 7.32 illustrates the portion of the modified network configuration for Boreas-R1 (ingress LSR) based on the previous configuration adapted from Listing 7.1. The unchanged portion is not shown. Turning on the TE extensions for OSPF with the "ospf set traffic-engineering on" command enables CBR. Furthermore, do not specify the "no-cspf" parameter with the "mpls create label-switched-path" or "mpls set label-switched-path" commands. The default costs of the network links are also retained in this case study. A constrained shortest path is to be selected for the traffic trunk, which has an end-to-end bandwidth requirement of 800M. The "bps" parameter is used to configure an LSP with the bandwidth request of 800M.

Oversubscription can be used to utilize the statistical behavior of traffic for the implementation of more efficient resource allocation policies. Oversubscription is implemented in the case study because the traffic patterns for Web hosting and off-site storage services tend to be bursty in nature and Ethernet can typically facilitate oversubscription up to a ratio of 1:10 (see section 1.4). The objective is to manipulate the maximum link bandwidth value (see section 5.4.2) to be less than the actual link capacity and oversubscribe the adjusted maximum link bandwidth two times for designated path R1–R2–R3–R4.

The "bandwidth" and "subscription" parameters are used with the "mpls set interface" command to define the maximum link bandwidth and subscription ratio (or maximum allocation multiplier; see section 5.4.3).

Note that adjusting a link's bandwidth in this way will automatically result in the area or level flooding of new link-state information. For port-31, the maximum link bandwidth is 400M and the subscription ratio is 200 percent; thus, the maximum reservable link bandwidth (see section 5.4.4) is 800M. For port-32, the maximum link bandwidth is 10M and the subscription ratio is 100 percent; thus, the maximum reservable link bandwidth is 10M. A subscription ratio of 100 percent advertises the maximum reservable bandwidth equal to the maximum link bandwidth. As port-32 is part of path R1–R4, a maximum reservable link bandwidth of 10M will discourage constraint-based OSPF from selecting this path for RSVP-TE to set up the required LSP (see section 5.7.2).

Internet data center 2 is represented by loopback address 223.0.0.4 (also the FEC and router ID of Eurus-R4). Thus, the PBR policy for forwarding user traffic onto the established LSP is based on destination IP address 223.0.0.4 and any source IP address. For testing purposes, only source IP address 223.0.0.1 (router ID of Boreas-R1) is used.

---

### Listing 7.32   Boreas-R1 Configuration

```
system set name Boreas-R1

OSPF configuration
ospf create area backbone
ospf add stub-host 223.0.0.1 to-area backbone cost 10
ospf add interface port-32 to-area backbone
ospf add interface port-31 to-area backbone
ospf set traffic-engineering on
ospf start
```

**MPLS configuration**
```
mpls add interface port-31
mpls add interface port-32
mpls set interface port-31 bandwidth 400000000
mpls set interface port-31 subscription 200
mpls set interface port-32 bandwidth 10000000
mpls set interface port-32 subscription 100

mpls create label-switched-path 800M-PATH to 223.0.0.4
mpls set label-switched-path 800M-PATH bps 800000000

mpls create policy P1 dst-ipaddr-mask 223.0.0.4
mpls set label-switched-path 800M-PATH policy P1
mpls start
```

Listing 7.33 illustrates the portion of the modified network configuration for Zephir-R2 (transit LSR) based on the previous configuration adapted from Listing 7.2. The unchanged portion is not shown. The TE extensions for OSPF have been turned on. As both port-41 and port-42 are part of designated path R1–R2–R3–R4 with an end-on-end bandwidth requirement of 800M, their maximum link bandwidth and subscription ratio are set to 400M and 200 percent, respectively, so that the maximum reservable link bandwidth can be 800M.

### Listing 7.33   Zephir-R2 Configuration

```
system set name Zephir-R2
```

**OSPF configuration**
```
ospf create area backbone
ospf add stub-host 223.0.0.2 to-area backbone cost 10
ospf add interface port-41 to-area backbone
ospf add interface port-42 to-area backbone
ospf set traffic-engineering on
ospf start
```

**MPLS configuration**
```
mpls add interface port-41
mpls add interface port-42
mpls set interface port-41 bandwidth 400000000
mpls set interface port-41 subscription 200
mpls set interface port-42 bandwidth 400000000
```

```
mpls set interface port-42 subscription 200
mpls start
```

Listing 7.34 illustrates the portion of the modified network configuration for Notus-R3 (transit LSR) based on the previous configuration adapted from Listing 7.3. The unchanged portion is not shown. The configuration is identical to Zephir-R2.

### Listing 7.34   Notus-R3 Configuration

```
system set name Notus-R3

OSPF configuration
ospf create area backbone
ospf add stub-host 223.0.0.3 to-area backbone cost 10
ospf add interface port-53 to-area backbone
ospf add interface port-54 to-area backbone
ospf set traffic-engineering on
ospf start

MPLS configuration
mpls add interface port-53
mpls add interface port-54
mpls set interface port-53 bandwidth 400000000
mpls set interface port-53 subscription 200
mpls set interface port-54 bandwidth 400000000
mpls set interface port-54 subscription 200
mpls start
```

Listing 7.35 illustrates the portion of the modified network configuration for Eurus-R4 (egress LSR) based on the previous configuration adapted from Listing 7.4. The unchanged portion is not shown. Because port-41 belongs to designated path R1–R2–R3–R4, its maximum link bandwidth and subscription ratio are set to 400M and 200 percent, respectively, so that the maximum reservable link bandwidth can be 800M. Because port-42 is associated with undesirable path R1–R4, the combination of a maximum link bandwidth of 10M and a subscription ratio of 100 percent gives a maximum reservable link bandwidth of 10M, which will result in constraint-based OSPF pruning this link (see section 5.7.2). In addition, penultimate hop popping (PHP) is disabled on these ports by the "no-php" parameter to ensure that only Eurus-R4 performs the pop operation.

---

### Listing 7.35    Eurus-R4 Configuration

```
system set name Eurus-R4


OSPF configuration
ospf create area backbone
ospf add stub-host 223.0.0.4 to-area backbone cost 10
ospf add interface port-42 to-area backbone
ospf add interface port-41 to-area backbone
ospf set traffic-engineering on
ospf start


MPLS configuration
mpls add interface port-41
mpls add interface port-42
mpls set interface port-41 no-php
mpls set interface port-42 no-php

mpls set interface port-41 bandwidth 400000000
mpls set interface port-41 subscription 200
mpls set interface port-42 bandwidth 10000000
mpls set interface port-42 subscription 100
mpls start
```

---

**7.2.3.3   Traceroute Tests.** Listing 7.36 illustrates the result of the traceroute test performed at Boreas-R1. The traceroute output indicates that user traffic (source 223.0.0.1) is really traversing path R1–R2–R3–R4 to reach the destination (223.0.0.4).

---

### Listing 7.36    Traceroute Test at Boreas-R1

```
Boreas-R1# traceroute 223.0.0.4 source 223.0.0.1


traceroute to 223.0.0.4 (223.0.0.4) from 223.0.0.1, 30 hops max, 40 byte
packets
1 192.168.12.2 (192.168.12.2) 4 ms 1 ms 1 ms
   MPLS Label1=4097 EXP1=0 TTL=1 S=1
2 192.168.23.3 (192.168.23.3) 4 ms 1 ms 1 ms
   MPLS Label1=4097 EXP1=0 TTL=1 S=1
3 223.0.0.4 (223.0.0.4) 1 ms 1 ms 0 ms
```

---

**7.2.3.4 MPLS-TE and RSVP-TE Monitoring.** Listing 7.37 illustrates the new MPLS policy defined for Boreas-R1. The destination (FEC) is 223.0.0.4 and the source is any IP address.

---

### Listing 7.37  Boreas-R1 MPLS Policy

```
Boreas-R1# mpls show policy
```

| Name | Type | Destination | Port | Source | Port | TOS | Prot | Use |
|------|------|-------------|------|--------|------|-----|------|------|
| P1 | L3 | 223.0.0.4 | Any | 0.0.0.0 | Any | Any | IP | INUSE |

---

Listing 7.38 illustrates the LSP information of Boreas-R1. An LSP, using fixed-filter (FF) reservation style (see section 6.4.2.1), has been established from ingress LSR Boreas-R1 (223.0.0.1) to egress LSR Eurus-R4 (223.0.0.4). The assigned OutLabel received from Zephir-R2 is 4097. As Boreas-R1 is the ingress router, it will not generate any InLabel. Instead, it will push the Out-Label onto any packets bound for 223.0.0.4 (FEC).

---

### Listing 7.38  Boreas-R1 LSP Information

```
Boreas-R1# mpls show label-switched-path brief
```

Ingress LSP:

| | | | | | Label | |
|---------|---------|-----------|-------|-------|-------|------|
| LSP Name | To | From | State | Style | In | Out |
| 800M-PATH | 223.0.0.4 | 223.0.0.1 | Up | FF | — | 4097 |

Transit LSP:

| | | | | | Label | |
|---------|----|------|-------|-------|-------|-----|
| LSP Name | To | From | State | Style | In | Out |

Egress LSP:

| | | | | | Label | |
|---------|----|------|-------|-------|-------|-----|
| LSP Name | To | From | State | Style | In | Out |

---

Listing 7.39 illustrates the LSP information of Zephir-R2, the transit LSR. The locally generated InLabel that is advertised to Boreas-R1 is 4097, and the assigned OutLabel received from Notus-R3 is also 4097. The RSVP-TE label distribution scheme by Riverstone reuses labels on a per-interface basis.

---

### Listing 7.39   Zephir-R2 LSP Information

```
Zephir-R2# mpls show label-switched-path brief

Ingress LSP:
                                                          Label
LSP Name          To            From           State   Style   In      Out

Transit LSP:
                                                          Label
LSP Name          To            From           State   Style   In      Out
800M-PATH         223.0.0.4     223.0.0.1      Up      FF      4097    4097

Egress LSP:
                                                          Label
LSP Name          To            From           State   Style   In      Out
```

---

Listing 7.40 illustrates the LSP information of Notus-R3, the transit LSR. The locally generated InLabel that is advertised to Zephir-R2 is 4097, and the assigned OutLabel received from Eurus-R4 is 16.

---

### Listing 7.40   Notus-R3 LSP Information

```
Notus-R3# mpls show label-switched-path

Ingress LSP:
                                                          Label
LSP Name          To            From           State   Style   In      Out

Transit LSP:
                                                          Label
LSP Name          To            From           State   Style   In      Out
800M-PATH         223.0.0.4     223.0.0.1      Up      FF      4097    16

Egress LSP:
                                                          Label
LSP Name          To            From           State   Style   In      Out
```

---

Listing 7.41 illustrates the LSP information of Eurus-R4, the egress LSR. The locally generated InLabel that is advertised to Notus-R3 is 16. Because PHP has been disabled and Eurus-R4 is the egress router, it performs the pop operation, so there is no OutLabel.

---

### Listing 7.41   Eurus-R4 LSP Information

```
Eurus-R4# mpls show label-switched-path brief


Ingress LSP:
                                                        Label
LSP Name           To           From        State  Style  In   Out


Transit LSP:
                                                        Label
LSP Name           To           From        State  Style  In   Out


Egress LSP:
                                                        Label
LSP Name           To           From        State  Style  In   Out
800M-PATH          223.0.0.4    223.0.0.1    Up     FF     16   —
```

---

Listing 7.42 illustrates the RSVP PSB of Boreas-R1. The derived ERO from constraint-based OSPF is 192.168.12.2 (R2)=>192.168.23.3 (R3)=>192.168.34.4 (R4). The "FF style desired" flag (= 0 × 00) is set in the SESSION_ATTRIBUTE object, indicating that the FF reservation style has been adopted and the RSVP PATH message is sent out of port-31 to next-hop 192.168.12.2 in the direction of Zephir-R2. Normally, if the requested bandwidth is available, it is reserved in a waiting pool on the port outbound to the destination (in this case, port-31) until a corresponding RESV message is received.

---

### Listing 7.42   Boreas-R1 RSVP PSB

```
Boreas-R1# rsvp show psb

Path State Blocks:

RSVP_PSB <rsvp_1>:
session-attr: name: 800M-PATH flags: 0x0 setup-pri: 7 holding-pri: 0
session: end-point: 223.0.0.4 tunnel-id: 6 ext-tunnel-id: 0xdf000001
send-templ: sender: 223.0.0.1 lsp-id: 1
```

```
prev-hop: 0.0.0.0
in-if: <Local-API> out-if: <port-31>
explicit-route: 192.168.12.2=>192.168.23.3=>192.168.34.4
```

Listing 7.43 illustrates the RSVP PSB of Zephir-R2. The derived ERO has become 192.168.23.3 (R3)=>192.168.34.4 (R4). The RSVP PATH message is sent out of port-42 to next-hop 192.168.23.3 in the direction of Notus-R3. The requested bandwidth is reserved in a waiting pool on this port until a corresponding RESV message is received.

### Listing 7.43    Zephir-R2 RSVP PSB

```
Zephir-R2# rsvp show psb

Path State Blocks:

RSVP_PSB <rsvp_1>:
session-attr: name: 800M-PATH flags: 0x0 setup-pri: 7 holding-pri: 0
session: end-point: 223.0.0.4 tunnel-id: 6 ext-tunnel-id: 0xdf000001
send-templ: sender: 223.0.0.1 lsp-id: 1
prev-hop: 192.168.12.1
in-if: <port-41> out-if: <port-42>
explicit-route: 192.168.23.3=>192.168.34.4
```

Listing 7.44 illustrates the RSVP PSB of Notus-R3. The derived ERO has become 192.168.34.4 (R4). The RSVP PATH message is sent out of port-53 to next-hop 192.168.34.4 in the direction of Eurus-R4. The requested bandwidth is reserved in a waiting pool on this port until a corresponding RESV message is received.

### Listing 7.44    Notus-R3 RSVP PSB

```
Notus-R3# rsvp show psb

Path State Blocks:

RSVP_PSB <rsvp_1>:
session-attr: name: 800M-PATH flags: 0x0 setup-pri: 7 holding-pri: 0
session: end-point: 223.0.0.4 tunnel-id: 6 ext-tunnel-id: 0xdf000001
send-templ: sender: 223.0.0.1 lsp-id: 1
prev-hop: 192.168.23.2
```

```
in-if: <port-54> out-if: <port-53>
explicit-route: 192.168.34.4
```

Listing 7.45 illustrates the RSVP PSB of Eurus-R4. As the tunnel endpoint 223.0.0.4 is at Eurus-R4 itself, the next-hop is no longer required to be derived from the ERO.

### Listing 7.45    Eurus-R4 RSVP PSB

```
Eurus-R4# rsvp show psb

Path State Blocks:

RSVP_PSB <rsvp_1>:
session-attr: name: 800M-PATH flags: 0x0 setup-pri: 7 holding-pri: 0
session: end-point: 223.0.0.4 tunnel-id: 6 ext-tunnel-id: 0xdf000001
send-templ: sender: 223.0.0.1 lsp-id: 1
prev-hop: 192.168.34.3
in-if: <port-41> out-if: <Local-API>
explicit-route:
```

Listing 7.46 illustrates the RSVP RSB of Eurus-R4. The FF reservation style is used for the corresponding RESV message. The local RSVP API of Eurus-R4 constructs the RESV message and also generates label 16. This local label (InLabel) is placed in the LABEL object of the RESV message, which is sent upstream to Notus-R3 based on the PHOP 192.168.34.3 found in the local RSVP PSB (see Listing 7.45). Because Eurus-R4 is the egress router, it performs the pop operation, so there is no remote label (OutLabel).

### Listing 7.46    Eurus-R4 RSVP RSB

```
Eurus-R4# rsvp show rsb

Resv State Blocks:

RSVP_RSB <rsvp_1>:
session: end-point: 223.0.0.4 tunnel-id: 6 ext-tunnel-id: 0xdf000001
style: FF
in-if: <Local-API>
filter-spec: sender: 223.0.0.1 lsp-id: 1
```

```
remote-labels: []
local-labels: [16]
record-route:
```

Listing 7.47 illustrates Eurus-R4 MPLS/RSVP ports and their reserved bandwidth information for each of the eight holding priority levels (see sections 4.4 and 6.4.1.7). Because the holding priority setting is 0 (highest) in the SESSION_ATTRIBUTE object found in the local RSVP PSB (see Listing 7.45), the reserved bandwidth will be based on holding priority level 0.

As mentioned earlier, if the requested bandwidth is available, it is reserved in a waiting pool on the port outbound to the destination (FEC) until a corresponding RESV message is received. Because Eurus-R4 is the destination (223.0.0.4) itself and it also originates the RESV message, the requested bandwidth is not reserved on any of its MPLS/RSVP ports.

### Listing 7.47    Eurus-R4 MPLS Port Information

```
Eurus-R4# mpls show interface verbose


Interface: <port-42>


Subscription 100, StaticBW 10000000bps, AvailableBW 10000000bps
ReservedBW              [0]   0bps  [1]  0bps  [2]  0bps  [3]  0bps
                        [4]   0bps  [5]  0bps  [6]  0bps  [7]  0bps


Interface: <port-41>


Subscription 200, StaticBW 400000000bps, AvailableBW 800000000bps
ReservedBW              [0]   0bps  [1]  0bps  [2]  0bps  [3]  0bps
                        [4]   0bps  [5]  0bps  [6]  0bps  [7]  0bps
```

Listing 7.48 illustrates the RSVP RSB of Notus-R3. The RESV message from Eurus-R4 is received at port-53, where the requested bandwidth is granted (see Listing 7.49). The remote label (OutLabel) of value 16 from Eurus-R4 is swapped with local label (InLabel) 4097 generated by Notus-R3. In addition, the RECORD_ROUTE object (RRO) found in the RESV message contains the IP address 192.168.34.4, which is associated with port-41 of Eurus-R4, the downstream (or next-hop) neighbor of Notus-R3. The modified RESV message is sent upstream to Zephir-R2 based on the PHOP 192.168.23.2 found in the local RSVP PSB (see Listing 7.44).

## Listing 7.48    Notus-R3 RSVP RSB

```
Notus-R3# rsvp show rsb

Resv State Blocks:

RSVP_RSB <rsvp_1>:
session: end-point: 223.0.0.4 tunnel-id: 6 ext-tunnel-id: 0xdf000001
style: FF
in-if: <port-53>
filter-spec: sender: 223.0.0.1 lsp-id: 1
remote-labels: [16]
local-labels: [4097]
record-route:
192.168.34.4
```

Listing 7.49 illustrates Notus-R3 MPLS/RSVP ports and their reserved bandwidth information. The 800M requested bandwidth reserved in the waiting pool associated with holding priority level 0 is granted for port-53, which is outbound to the destination. No bandwidth is reserved for port-54 because it is an inbound port.

## Listing 7.49    Notus-R3 MPLS Port Information

```
Notus-R3# mpls show interface verbose

Interface: <port-53>

Subscription 200, StaticBW 400000000bps, AvailableBW 800000000bps
ReservedBW         [0]  800000000bps [1]  0bps [2]  0bps [3]  0bps
                   [4]  0bps         [5]  0bps [6]  0bps [7]  0bps

Interface: <port-54>

Subscription 200, StaticBW 400000000bps, AvailableBW 1600000000bps
ReservedBW         [0]  0bps         [1]  0bps [2]  0bps [3]  0bps
                   [4]  0bps         [5]  0bps [6]  0bps [7]  0bps
```

Listing 7.50 illustrates the RSVP RSB of Zephir-R2. The RESV message from Notus-R3 is received at port-42, where the requested bandwidth is granted (see Listing 7.51). The remote label (OutLabel) of value 4097 from Eurus-R4 is reused for the local label (InLabel). Furthermore, Notus-R3, the downstream (or next-hop) neighbor of Zephir-R2, has prepended IP address 192.168.23.3 associated with its port-54 to the RRO. The modified RESV message is sent upstream to Boreas-R1 based on the PHOP 192.168.12.1 found in the local RSVP PSB (see Listing 7.43).

---

### Listing 7.50   Zephir-R2 RSVP RSB

```
Zephir-R2# rsvp show rsb

Resv State Blocks:

RSVP_RSB <rsvp_1>:
session: end-point: 223.0.0.4 tunnel-id: 6 ext-tunnel-id: 0xdf000001
style: FF
in-if: <port-42>
filter-spec: sender: 223.0.0.1 lsp-id: 1
remote-labels: [4097]
local-labels: [4097]
record-route:
192.168.23.3
192.168.34.4
```

---

Listing 7.51 illustrates Zephir-R2 MPLS/RSVP ports and their reserved bandwidth information. The 800M requested bandwidth reserved in the waiting pool associated with holding priority level 0 is granted for port-42, which is outbound to the destination. No bandwidth is reserved for port-41 because it is an inbound port.

---

### Listing 7.51   Zephir-R2 MPLS Port Information

```
Zephir-R2# mpls show interface verbose

Interface: <port-41>

Subscription 200, StaticBW 400000000bps, AvailableBW 1600000000bps
ReservedBW          [0]  0bps        [1]  0bps [2]  0bps [3]  0bps
                    [4]  0bps        [5]  0bps [6]  0bps [7]  0bps

Interface: <port-42>

Subscription 200, StaticBW 400000000bps, AvailableBW 800000000bps
ReservedBW          [0]  800000000bps [1]  0bps [2]  0bps [3]  0bps
                    [4]  0bps        [5]  0bps [6]  0bps [7]  0bps
```

---

Listing 7.52 illustrates the RSVP RSB of Boreas-R1. The RESV message from Zephir-R2 is received at port-31, where the requested bandwidth is granted (see Listing 7.53). The LSP is established once Boreas-R1 receives the RESV message. The remote label (OutLabel) of value 4097 is derived from the LABEL object of the RESV message. Because Boreas-R1 is the ingress router, it does not generate any local label. Instead, it will push the remote label onto any packets bound for 223.0.0.4 (FEC). Furthermore, Zephir-R2, the downstream (or next-hop) neighbor of Boreas-R1, has prepended IP address 192.168.12.2 associated with its port-41 to the RRO.

---

### Listing 7.52   Boreas-R1 RSVP RSB

```
Boreas-R1# rsvp show rsb

Resv State Blocks:

RSVP_RSB <rsvp_1>:
session: end-point: 223.0.0.4 tunnel-id: 6 ext-tunnel-id: 0xdf000001
style: FF
in-if: <port-31>
filter-spec: sender: 223.0.0.1 lsp-id: 1
remote-labels: [4097]
local-labels: []
record-route:
192.168.12.2
192.168.23.3
192.168.34.4
```

---

Listing 7.53 illustrates Boreas-R1 MPLS/RSVP ports and their reserved bandwidth information. The 800M requested bandwidth reserved in the waiting pool associated with holding priority level 0 is granted for port-31, which is outbound to the destination. No bandwidth is reserved for port-32 because it is an inbound port.

---

### Listing 7.53   Boreas-R1 MPLS Port Information

```
Boreas-R1# mpls show interface
```

```
Interface: <port-32>

Subscription 100, StaticBW 10000000bps, AvailableBW 10000000bps
ReservedBW              [0]  0bps        [1]  0bps [2]  0bps [3]  0bps
                        [4]  0bps        [5]  0bps [6]  0bps [7]  0bps


Interface: <port-31>

Subscription 200, StaticBW 400000000bps, AvailableBW 800000000bps
ReservedBW              [0]  800000000bps [1]  0bps [2]  0bps [3]  0bps
                        [4]  0bps        [5]  0bps [6]  0bps [7]  0bps
```

**7.2.3.5  OSPF TED Monitoring.** Listing 7.54 illustrates the OSPF TED of Boreas-R1 after the LSP has been established by RSVP-TE. The OSPF link sub-TLV (type/length/value) details for the four routers are indicated clearly on the TED listing with the help of in-line headers. For brevity, only the local address, maximum bandwidth, maximum reservable bandwidth, and maximum unreserved bandwidth sub-TLVs are examined in this section. For more details on the other sub-TLVs, refer to section 5.5.2.

The local address sub-TLV specifies the IP address of a local interface:

- In Eurus-R4's case, the local address is 192.168.34.4 for port-41 and 192.168.14.4 for port-42.
- In Notus-R3's case, the local address is 192.168.34.3 for port-53 and 192.168.23.3 for port-54.
- In Zephir-R2's case, the local address is 192.168.23.2 for port-42 and 192.168.12.2 for port-41.
- In Boreas-R1's case, the local address is 192.168.14.1 for port-32 and 192.168.12.1 for port-31.

The maximum bandwidth sub-TLV specifies the maximum bandwidth that can be used on a particular link:

- In Eurus-R4's case, port-41 has a maximum bandwidth of 400M and port-42 has a maximum bandwidth of 10M.
- In Notus-R3's case, both port-53 and port-54 have a maximum bandwidth of 400M.
- In Zephir-R2's case, both port-42 and port-41 have a maximum bandwidth of 400M.
- In Boreas-R1's case, port-32 has a maximum bandwidth of 10M and port-31 has a maximum bandwidth of 400M.

The maximum reservable bandwidth sub-TLV specifies the maximum bandwidth that can be reserved on a particular link. By default, the value of this sub-TLV is equivalent to the maximum bandwidth. In the event oversubscription is implemented, the maximum reservable bandwidth can be greater than the maximum bandwidth:

- In Eurus-R4's case, port-41 has an oversubscription ratio of 200 percent, so the maximum reservable bandwidth becomes 800M. Meanwhile, port-42 has an oversubscription ratio of only 100 percent, so the maximum reservable bandwidth remains at the default value of 10M.
- In Notus-R3's case, both port-53 and port-54 have an oversubscription ratio of 200 percent, so the maximum reservable bandwidth becomes 800M.
- In Zephir-R2's case, both port-42 and port-41 have an oversubscription ratio of 200 percent, so the maximum reservable bandwidth becomes 800M.
- In Boreas-R1's case, port-31 has an oversubscription ratio of 200 percent, so the maximum reservable bandwidth becomes 800M. Meanwhile, port-32 has an oversubscription ratio of only 100 percent, so the maximum reservable bandwidth remains at the default value of 10M.

The maximum unreserved bandwidth sub-TLV specifies the bandwidth that can be reserved with a setup priority of 0 through 7, arranged in ascending order, with priority 0 occurring at the start of the sub-TLV and priority 7 at the end of the sub-TLV. In this case study, a setup priority of 7 is used (see Listings 7.42 to 7.45). The initial values (before any bandwidth is reserved) are all set to the maximum reservable bandwidth unless reservation has already taken place when RSVP-TE successfully establishes the required LSP with bandwidth constraint. If this is the case, the values will all be set to zero because the maximum reservable bandwidth (800M) is all expended:

- In Eurus-R4's case, the maximum unreserved bandwidth values based on the eight setup priority levels for both port-41 and port-42 remain the same as their respective maximum reservable bandwidth values of 800M and 10M because no reservation has taken place for these two ports yet.
- In Notus-R3's case, the bandwidth request has been granted for port-53 (see Listing 7.49), so the maximum unreserved bandwidth values for the eight setup priority levels become zero, while those for port-54 remain the same as the maximum reservable bandwidth of 800M because no reservation has taken place for this port yet.
- In Zephir-R2's case, the bandwidth request has been granted for port-42 (see Listing 7.51), so the maximum unreserved bandwidth values for the eight setup priority levels become zero, while those for port-41 remain the same as the maximum reservable bandwidth of 800M because no reservation has taken place for this port yet.

- In Boreas-R1's case, the bandwidth request has been granted for port-31 (see Listing 7.53), so the maximum unreserved bandwidth values for the eight setup priority levels become zero, while those for port-32 remain the same as the maximum reservable bandwidth of 10M because no reservation has taken place for this port yet.

The OSPF TED reflects the change in resources usage throughout the network. It is important to remember that an LSP is unidirectional, whereas the Gigabit Ethernet connections between the routers are full duplex. This implies that links (for port-31 of Boreas-R1, port-42 of Zephir-R2, and port-53 of Notus-R3) advertised in the direction of the LSP will now indicate the new resource information. However, the links (for port-41 of Eurus-R4, port-54 of Notus-R3, and port-41 of Zephir-R2) advertised in the reverse direction have not reduced the maximum unreserved bandwidth values because there is currently no corresponding LSP in the reverse direction to enable bidirectional conversations; thus, the maximum reservable bandwidth for these ports in the opposite direction remains available. Note that the maximum reservable bandwidth will be available on all ports before RSVP-TE has made the resource reservation during path setup.

---

### Listing 7.54   Boreas-R1 OSPF TED

```
Boreas-R1# ospf show ted

OSPF Router with ID (223.0.0.1)

Link Sub-TLV Info for Eurus-R4 (Router-ID=223.0.0.4)

Link Type 2
Link ID 192.168.34.4
Local Addr 192.168.34.4
Remote Addr 0.0.0.0
Metric 2
Max BW              :              400 Mbps
Max reservable BW :              800 Mbps
Max unreserved BW :
[0]              800 Mbps  [1]              800 Mbps
[2]              800 Mbps  [3]              800 Mbps
[4]              800 Mbps  [5]              800 Mbps
[6]              800 Mbps  [7]              800 Mbps
Resource class 0

Link Type 2
Link ID 192.168.14.4
```

**Local Addr 192.168.14.4**
Remote Addr 0.0.0.0
Metric 2
Max BW                 :           10 Mbps
Max reservable BW :           10 Mbps
Max unreserved BW :
[0]              10 Mbps   [1]              10 Mbps
[2]              10 Mbps   [3]              10 Mbps
[4]              10 Mbps   [5]              10 Mbps
[6]              10 Mbps   [7]              10 Mbps
Resource class 0

**Link Sub-TLV Info for Notus-R3 (Router-ID=223.0.0.3)**

Link Type 2
Link ID 192.168.34.4
**Local Addr 192.168.34.3**
Remote Addr 0.0.0.0
Metric 2
Max BW                 :           400 Mbps
**Max reservable BW :           800 Mbps**
Max unreserved BW :
[0]              0 bps   [1]              0 bps
[2]              0 bps   [3]              0 bps
[4]              0 bps   [5]              0 bps
[6]              0 bps   [7]              0 bps
Resource class 0

Link Type 2
Link ID 192.168.23.3
**Local Addr 192.168.23.3**
Remote Addr 0.0.0.0
Metric 2
Max BW                 :           400 Mbps
Max reservable BW :           800 Mbps
Max unreserved BW :
[0]              800 Mbps   [1]              800 Mbps
[2]              800 Mbps   [3]              800 Mbps
[4]              800 Mbps   [5]              800 Mbps
[6]              800 Mbps   [7]              800 Mbps
Resource class 0

**Link Sub-TLV Info for Zephir-R2 (Router-ID=223.0.0.2)**

Link Type 2

159

**Link ID 192.168.23.3**
**Local Addr 192.168.23.2**
Remote Addr 0.0.0.0
Metric 2
Max BW               :              400 Mbps
**Max reservable BW :**              **800 Mbps**
Max unreserved BW :
[0]              0 bps    [1]              0 bps
[2]              0 bps    [3]              0 bps
[4]              0 bps    [5]              0 bps
[6]              0 bps    [7]              0 bps
Resource class 0


Link Type 2
Link ID 192.168.12.2
**Local Addr 192.168.12.2**
Remote Addr 0.0.0.0
Metric 2
Max BW               :              400 Mbps
Max reservable BW :              800 Mbps
Max unreserved BW :
[0]              800 Mbps [1]              800 Mbps
[2]              800 Mbps [3]              800 Mbps
[4]              800 Mbps [5]              800 Mbps
[6]              800 Mbps [7]              800 Mbps
Resource class 0

**Link Sub-TLV Info for Boreas-R1 (Router-ID=223.0.0.1)**

Link Type 2
Link ID 192.168.12.2
**Local Addr 192.168.12.1**
Remote Addr 0.0.0.0
Metric 2
Max BW               :              400 Mbps
**Max reservable BW :**              **800 Mbps**
Max unreserved BW :
[0]              0 bps    [1]              0 bps
[2]              0 bps    [3]              0 bps
[4]              0 bps    [5]              0 bps
[6]              0 bps    [7]              0 bps
Resource class 0

Link Type 2
Link ID 192.168.14.4

```
Local Addr 192.168.14.1
Remote Addr 0.0.0.0
Metric 2
Max BW              :           10 Mbps
Max reservable BW :           10 Mbps
Max unreserved BW :
[0]          10 Mbps    [1]              10 Mbps
[2]          10 Mbps    [3]              10 Mbps
[4]          10 Mbps    [5]              10 Mbps
[6]          10 Mbps    [7]              10 Mbps
Resource class 0
```

## 7.2.4   Case Study 7.4: Link Affinity (CBR Example 2)

**7.2.4.1   Case Overview and Network Topology.**  In line with case study 7.3, case study 7.4 describes the implementation stage for which a constraint-based IGP is used to select a path whereby the traffic trunk at the ingress point has a specific administrative control policy. The policy constraint imposes the application of link colors to the network links. Link color is also known as resource class or administrative group. Because resource requirement has already been discussed in the previous case, the bandwidth manipulation portion is omitted in this case study for brevity.

Constraint-based OSPF is again used for the path selection. Once the constrained shortest path is determined, it is handed over to RSVP-TE for the actual path setup. Figure 7.8 illustrates the network diagram for this case study. The setting is identical to that of the previous case study, but this time with the original user (223.0.0.1), hereby referred to as Boreas Corp, requesting access to its branch office at commercial estate 2 (Zephir district) as well as using the Web hosting and off-site storage services from Internet data center 2 (Eurus district). A new user (223.0.0.11), hereby referred to as Kastor Corp, also from the Boreas district, has subscribed to the Web hosting and off-site storage services as well as the access service to its branch office at commercial estate 2 in the Zephir district. Besides, the renovation works affecting the fiber link between Boreas-R1 and Eurus-R4 have been completed.

As a premium subscriber, user traffic from Boreas Corp will have the privilege of accessing its desired services via the shortest paths. Specifically, Boreas's user traffic will traverse path R1–R2 (marked with green link color) to reach commercial estate 2 (223.0.0.2) and path R1–R4 (marked with green link color) to reach Internet data center 2 (223.0.0.4). As a standard subscriber, user traffic from Kastor Corp will traverse longer paths to reach similar destinations. Specifically, Kastor's user traffic will traverse path R1–R4–R3–R2 (marked with red link color) to reach commercial estate

161

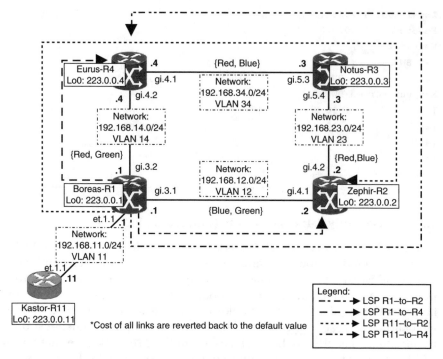

**Figure 7.8    Network Diagram for Case Study 7.4**

2 (223.0.0.2) and path R1–R2–R3–R4 (marked with blue link color) to reach Internet data center 2 (223.0.0.4).

**7.2.4.2   Network Configurations.** Listing 7.55 illustrates the network configuration for Kastor-R11 (described with in-line headers), the access router (a Riverstone 3000 metro router) to Kastor Corp that is connected to Boreas-R1 through port-11 (et.1.1).

---

### Listing 7.55    Kastor-R11 Configuration

```
system set name Kastor-R11
```

**VLAN configuration**
```
vlan create vlan11 id 11 ip
vlan add ports et.1.1 to vlan11
interface create ip port-11 vlan vlan11 address-netmask
192.168.11.11/24
```

**Create loopback address**
```
interface add ip lo0 address-netmask 223.0.0.11/32
```

**Set loopback address as router identifier**
```
ip-router global set router-id 223.0.0.11
```

**OSPF configuration**
```
ospf create area backbone
ospf add stub-host 223.0.0.11 to-area backbone cost 10
ospf add interface port-11 to-area backbone
ospf start
```

---

Listing 7.56 illustrates the portion of the modified network configuration for Boreas-R1 (ingress LSR) based on the previous configuration adapted from Listing 7.32. The unchanged portion is not shown. Boreas-R1 is interconnected to Kastor-R11 through port-11 (et.1.1).

In this case study there are four traffic trunks with four different policy constraints. These traffic trunks are hereby referred to as trunks R1–to–R4, R1–to–R2, R11–to–R4, and R11–to–R2. User traffic from Boreas Corp heading for Internet data center 2 is mapped onto trunk R1–to–R4, and user traffic from Boreas Corp heading for commercial estate 2 is mapped onto trunk R1–to–R2. Likewise, user traffic from Kastor Corp heading for Internet data center 2 is mapped onto trunk R11–to–R4, and user traffic from Kastor Corp heading for commercial estate 2 is mapped onto trunk R11–to–R2. Consequently, four corresponding LSPs need to be established separately to carry these four traffic trunks. Four different PBR policies are also required for forwarding respective user traffic onto the corresponding LSPs based on different sets of source (223.0.0.1 and 223.0.0.11) and destination (223.0.0.2 and 223.0.0.4) IP addresses.

With the help of administrative group, it is possible to "color" the network links that form the four distinct paths so that constraint-based OSPF can select the desired path for each of the traffic trunks. The "mpls create admin-group" command is used to define the following link colors:

- *Red of value 1.* Riverstone has a unique way of implementing the 32-bit administrative group value. By convention, the least significant bit is referred to as group 0 and the most significant bit is referred to as group 31. The value 1 actually turns on the group 1 bit, which gives the resource class value 0 × 00000002.
- *Blue of value 2.* The value 2 sets the group 2 bit, which gives the resource class value 0 × 00000004.
- *Green of value 3.* The value 3 sets the group 3 bit, which gives the resource class value 0 × 00000008.

The "admin-group" parameter in the "mpls set interface" command gives port-31 the link colors green and blue, and port-32 the link colors green and red. Green and blue turn on both the group 3 and group 2 bits, while green and red turn on both the group 3 and group 1 bits. Therefore, the resource class attribute value for port-31 is $0 \times 0000000C$ ($0 \times 00000008$ + $0 \times 00000004$) and the resource class attribute value for port-32 is $0 \times 0000000A$ ($0 \times 00000008$ + $0 \times 00000002$). Port-31 will only allow requests that have an affinity for green or blue network links, whereas port-32 will only allow requests that have an affinity for green or red network links.

In the same way, the green administrative group is assigned to the LSPs carrying Boreas's user traffic; thus, the link affinity for these LSPs is green. The blue administrative group is assigned to the LSP "R11–to–R4," carrying Kastor's user traffic to Internet data center 2, and the red administrative group is assigned to the LSP "R11–to–R2," carrying Kastor's user traffic to commercial estate 2. Therefore, the link affinity for LSP "R11–to–R4" is blue and the link affinity for LSP "R11–to–R2" is red. These link affinities are specified with the "include" parameter in the "mpls set label-switched-path" command.

The administrative group mask (resource class mask) for link color green is automatically inferred with a value of 1 (1 = care), corresponding to the group 3 bit, and a value of 0 (0 = don't care), corresponding to the rest of the bits, giving the mask value of $0 \times 00000008$, which happens to be the same value as link color green. Hence, the mask value is $0 \times 00000004$ for link color blue and $0 \times 00000002$ for link color red; each of these mask values corresponds to the same value as its link color.

Put simply, user traffic from Boreas Corp will only traverse green network links. On the other hand, user traffic from Kastor Corp heading for Internet data center 2 will only traverse blue network links, and user traffic from Kastor Corp heading for commercial estate 2 will only traverse red network links.

---

### Listing 7.56   Boreas-R1 Configuration

```
system set name Boreas-R1

VLAN configuration
vlan create vlan11 id 11 ip
vlan add ports et.1.1 to vlan11
interface create ip port-11 vlan vlan11 address-netmask 192.168.11.1/24

OSPF configuration
ospf add interface port-11 to-area backbone

MPLS configuration
```

```
mpls create admin-group green group-value 3
mpls create admin-group red group-value 1
mpls create admin-group blue group-value 2
mpls add interface port-31
mpls add interface port-32
mpls set interface port-32 admin-group green,red
mpls set interface port-31 admin-group green,blue


mpls create label-switched-path R1-to-R4 to 223.0.0.4
mpls create label-switched-path R11-to-R4 to 223.0.0.4
mpls create label-switched-path R1-to-R2 to 223.0.0.2
mpls create label-switched-path R11-to-R2 to 223.0.0.2


mpls set label-switched-path R1-to-R4 include green
mpls set label-switched-path R1-to-R2 include green
mpls set label-switched-path R11-to-R4 include blue
mpls set label-switched-path R11-to-R2 include red


mpls create policy R11-to-R4 src-ipaddr-mask 223.0.0.11/32 dst-ipaddr-
mask 223.0.0.4/32
mpls create policy R11-to-R2 src-ipaddr-mask 223.0.0.11/32 dst-ipaddr-
mask 223.0.0.2/32
mpls create policy R1-to-R4 src-ipaddr-mask 223.0.0.1/32 dst-ipaddr-
mask 223.0.0.4/32
mpls create policy R1-to-R2 src-ipaddr-mask 223.0.0.1/32 dst-ipaddr-
mask 223.0.0.2/32


mpls set label-switched-path R1-to-R4 policy R1-to-R4
mpls set label-switched-path R1-to-R2 policy R1-to-R2
mpls set label-switched-path R11-to-R4 policy R11-to-R4
mpls set label-switched-path R11-to-R2 policy R11-to-R2
mpls start
```

Listing 7.57 illustrates the portion of the modified network configuration for Zephir-R2 based on the previous configuration adapted from Listing 7.33. The unchanged portion is not shown. Zephir-R2 is the tunnel endpoint or egress LSR to LSPs "R1–to–R2" and "R11–to–R2" (see Listing 7.68). It is also the transit LSR to LSP "R11–to–R4."

Port-41 is given link colors green and blue, and port-42 is given link colors blue and red. Green and blue turn on both the group 3 and group 2 bits, while blue and red turn on both the group 2 and group 1 bits. Therefore, the resource class attribute value for port-41 is $0 \times 0000000C$ ($0 \times 00000008 + 0 \times 00000004$) and the resource class attribute value for port-42 is $0 \times 00000006$ ($0 \times 00000004 + 0 \times 00000002$). Port-41 will only allow requests that have an affinity for green or blue network links, whereas port-42 will only allow requests that have an affinity for blue or red network links.

---

### Listing 7.57   Zephir-R2 Configuration

```
system set name Zephir-R2

MPLS configuration
mpls create admin-group green group-value 3
mpls create admin-group blue group-value 2
mpls create admin-group red group-value 1
mpls add interface port-41
mpls add interface port-42
mpls set interface port-41 no-php
mpls set interface port-42 no-php
mpls set interface port-41 admin-group green,blue
mpls set interface port-42 admin-group blue,red
mpls start
```

---

Listing 7.58 illustrates the portion of the modified network configuration for Notus-R3 based on the previous configuration adapted from Listing 7.34. The unchanged portion is not shown. Notus-R3 is the transit LSR to LSPs "R11–to–R2" and "R11–to–R4" (see Listing 7.69).

Port-53 and port-54 are both given link colors blue and red, which turn on both the group 2 and group 1 bits. Therefore, the resource class attribute value for both ports is $0 \times 00000006$ ($0 \times 00000004 + 0 \times 00000002$). These ports will only allow requests that have an affinity for blue or red network links.

---

### Listing 7.58   Notus-R3 Configuration

```
system set name Notus-R3

MPLS configuration
mpls create admin-group blue group-value 2
mpls create admin-group red group-value 1
mpls add interface port-53
mpls add interface port-54
mpls set interface port-54 admin-group blue,red
mpls set interface port-53 admin-group blue,red
mpls start
```

---

Listing 7.59 illustrates the portion of the modified network configuration for Eurus-R4 based on the previous configuration adapted from Listing 7.35. The unchanged portion is not shown. Eurus-R4 is the tunnel endpoint or egress LSR to LSPs "R1–to–R4" and "R11–to–R4" (see Listing 7.70). It is also the transit LSR to LSP "R11–to–R2."

Port-42 is given link colors green and red, and port-41 is given link colors blue and red. Green and red turn on both the group 3 and group 1 bits, while blue and red turn on both the group 2 and group 1 bits. Therefore, the resource class attribute value for port-42 is 0 × 0000000A (0 × 00000008 + 0 × 00000002) and the resource class attribute value for port-41 is 0 × 00000006 (0 × 00000004 + 0 × 00000002). Port-42 will only allow requests that have an affinity for green or red network links, whereas port-41 will only allow requests that have an affinity for blue or red network links.

---

### Listing 7.59   Eurus-R4 Configuration

```
system set name Eurus-R4

MPLS configuration
mpls create admin-group green group-value 3
mpls create admin-group blue group-value 2
mpls create admin-group red group-value 1
mpls add interface port-41
mpls add interface port-42
mpls set interface port-41 no-php
mpls set interface port-42 no-php
mpls set interface port-42 admin-group green,red
mpls set interface port-41 admin-group blue,red
mpls start
```

---

**7.2.4.3   Traceroute Tests.** Listing 7.60 illustrates the result of the traceroute test performed at Kastor-R11. The traceroute output indicates that user traffic (source 223.0.0.11) is really traversing path R1–R2–R3–R4 to reach destination 223.0.0.4 (Internet data center 2) and path R1–R4–R3–R2 to reach destination 223.0.0.2 (commercial estate 2).

---

### Listing 7.60   Traceroute Test at Kastor-R11

```
Kastor-R11# traceroute 223.0.0.4 source 223.0.0.11
```

```
traceroute to 223.0.0.4 (223.0.0.4) from 223.0.0.11, 30 hops max, 40
byte packets
1  192.168.11.1 (192.168.11.1) 0 ms 0 ms 0 ms
2  192.168.12.2 (192.168.12.2) 2 ms 1 ms 0 ms
   MPLS Label1=4097 EXP1=0 TTL=1 S=1
3  192.168.23.3 (192.168.23.3) 2 ms 1 ms 1 ms
   MPLS Label1=4097 EXP1=0 TTL=1 S=1
4  223.0.0.4 (223.0.0.4) 1 ms 1 ms 0 ms


Kastor-R11# traceroute 223.0.0.2 source 223.0.0.11


traceroute to 223.0.0.2 (223.0.0.2) from 223.0.0.11, 30 hops max, 40
byte packets
1  192.168.11.1 (192.168.11.1) 0 ms 0 ms 0 ms
2  192.168.14.4 (192.168.14.4) 3 ms 1 ms 0 ms
   MPLS Label1=4097 EXP1=0 TTL=1 S=1
3  192.168.34.3 (192.168.34.3) 2 ms 1 ms 0 ms
   MPLS Label1=4097 EXP1=0 TTL=1 S=1
4  223.0.0.2 (223.0.0.2) 2 ms 1 ms 0 ms
```

Listing 7.61 illustrates the result of the traceroute test performed at Boreas-R1. The traceroute output indicates that user traffic (source 223.0.0.1) is really traversing path R1–R4 to reach destination 223.0.0.4 (Internet data center 2) and path R1–R2 to reach destination 223.0.0.2 (commercial estate 2).

### Listing 7.61   Traceroute Test at Boreas-R1

```
Boreas-R1# traceroute 223.0.0.4 source 223.0.0.1


traceroute to 223.0.0.4 (223.0.0.4) from 223.0.0.1, 30 hops max, 40 byte
packets
1  223.0.0.4 (223.0.0.4) 2 ms 1 ms 0 ms


Boreas-R1# traceroute 223.0.0.2 source 223.0.0.1


traceroute to 223.0.0.2 (223.0.0.2) from 223.0.0.1, 30 hops max, 40 byte
packets
1  223.0.0.2 (223.0.0.2) 2 ms 1 ms 0 ms
```

**7.2.4.4  MPLS-TE and RSVP-TE Monitoring.** Listing 7.62 illustrates the four new MPLS policies defined for Boreas-R1:

- Policy "R11–to–R4" defines the destination (FEC) as 223.0.0.4 and the source as 223.0.0.11.
- Policy "R11–to–R2" defines the destination (FEC) as 223.0.0.2 and the source as 223.0.0.11.
- Policy "R1–to–R4" defines the destination (FEC) as 223.0.0.4 and the source as 223.0.0.1.
- Policy "R1–to–R2" defines the destination (FEC) as 223.0.0.2 and the source as 223.0.0.1.

---

### Listing 7.62   Boreas-R1 MPLS Policy

```
Boreas-R1# mpls show policy

Name         Type  Destination  Port  Source       Port  TOS  Prot  Use
R11-to-R4    L3    223.0.0.4    Any   223.0.0.11   Any   Any  IP    INUSE
R11-to-R2    L3    223.0.0.2    Any   223.0.0.11   Any   Any  IP    INUSE
R1-to-R4     L3    223.0.0.4    Any   223.0.0.1    Any   Any  IP    INUSE
R1-to-R2     L3    223.0.0.2    Any   223.0.0.1    Any   Any  IP    INUSE
```

---

Listing 7.63 briefly illustrates the MPLS ports of Boreas-R1 and their respective administrative groups or link colors. Port-32 is assigned the link colors red and green, which together are equivalent to the resource class attribute value of 0 × 0000000A. Port-31 is assigned the link colors blue and green, which together are equivalent to the resource class attribute value of 0 × 0000000C.

---

### Listing 7.63   Boreas-R1 MPLS Port Information

```
Boreas-R1# mpls show interface

Interface      State      Administrative Groups
lo0            Up         <None>
port-32        Up         Red green
port-31        Up         Blue green
```

---

Listing 7.64 briefly illustrates the MPLS ports of Zephir-R2 and their respective administrative groups or link colors. Port-41 is assigned the link colors blue and green, which together are equivalent to the resource class attribute value of 0 × 0000000C. Port-42 is assigned the link colors red and blue, which together are equivalent to the resource class attribute value of $0 \times 00000006$.

---

### Listing 7.64   Zephir-R2 MPLS Port Information

```
Zephir-R2# mpls show interface

Interface        State        Administrative Groups
lo0              Up           <None>
port-41          Up           Blue green
port-42          Up           Red blue
```

---

Listing 7.65 briefly illustrates the MPLS ports of Notus-R3 and their respective administrative groups or link colors. Both port-53 and port-54 are assigned the link colors red and blue, which together are equivalent to the resource class attribute value of $0 \times 00000006$.

---

### Listing 7.65   Notus-R3 MPLS Port Information

```
Notus-R3# mpls show interface

Interface        State        Administrative Groups
lo0              Up           <None>
port-53          Up           Red blue
port-54          Up           Red blue
```

---

Listing 7.66 briefly illustrates the MPLS ports of Eurus-R4 and their respective administrative groups or link colors. Port-42 is assigned the link colors red and green, which together are equivalent to the resource class attribute value of $0 \times 0000000A$. Port-41 is assigned the link colors red and blue, which together are equivalent to the resource class attribute value of $0 \times 00000006$.

---

### Listing 7.66   Eurus-R4 MPLS Port Information

```
Eurus-R4# mpls show interface

Interface        State        Administrative Groups
lo0              Up           <None>
port-42          Up           Red green
port-41          Up           Red blue
```

---

Listing 7.67 illustrates the LSP information of Boreas-R1 (ingress LSR). Four different LSPs have been established from Boreas-R1 (223.0.0.1): two (LSPs "R11–to–R2" and "R1–to–R2") to Zephir-R2 (223.0.0.2) and the other two (LSPs "R11–to–R4" and "R1– to–R4") to Eurus-R4 (223.0.0.4). The FF reservation style is used for all four LSPs. The MPLS labels to be used for each LSP are as follows:

- For LSP "R11–to–R2," the assigned OutLabel received from Eurus-R4 is 4097. Boreas-R1 will push this OutLabel onto packets with source IP address 223.0.0.11 and destination IP address 223.0.0.2.
- For LSP "R1–to–R2," the assigned OutLabel received from Zephir-R2 is 16. Boreas-R1 will push this OutLabel onto packets with source IP address 223.0.0.1 and destination IP address 223.0.0.2.
- For LSP "R11–to–R4," the assigned OutLabel received from Zephir-R2 is 4097. Boreas-R1 will push this OutLabel onto packets with source IP address 223.0.0.11 and destination IP address 223.0.0.4.
- For LSP "R1–to–R4," the assigned OutLabel received from Eurus-R4 is 16. Boreas-R1 will push this OutLabel onto packets with source IP address 223.0.0.1 and destination IP address 223.0.0.4.

---

### Listing 7.67 Boreas-R1 LSP Information

```
Boreas-R1# mpls show label-switched-path brief


Ingress LSP:
                                                          Label
LSP Name        To              From          State   Style   In    Out
R11-to-R2       223.0.0.2       223.0.0.1     Up      FF      —     4097
R1-to-R2        223.0.0.2       223.0.0.1     Up      FF      —     16
R11-to-R4       223.0.0.4       223.0.0.1     Up      FF      —     4097
R1-to-R4        223.0.0.4       223.0.0.1     Up      FF      —     16


Transit LSP:
                                                          Label
LSP Name        To              From          State   Style   In    Out


Egress LSP:
                                                          Label
LSP Name        To              From          State   Style   In    Out
```

---

Listing 7.68 illustrates the LSP information of Zephir-R2, which serves as the transit LSR for LSP "R11–to–R4" and as the egress LSR (or tunnel end-point) for LSPs "R11–to–R2" and "R1–to–R2." The MPLS labels to be used for each LSP are as follows:

- For LSP "R11–to–R4," the assigned OutLabel received from Notus-R3 is 4097. This label is reused for the InLabel that is advertised to Boreas-R1.
- For LSP "R11–to–R2," the locally generated InLabel that is advertised to Notus-R3 is 16. Because PHP has been disabled and Zephir-R2 is the egress router for this LSP, it performs the pop operation, so there is no OutLabel.
- For LSP "R1–to–R2," the locally generated InLabel that is advertised to Boreas-R1 is 16. Because PHP has been disabled and Zephir-R2 is the egress router for this LSP, it performs the pop operation, so there is no OutLabel.

### Listing 7.68   Zephir-R2 LSP Information

```
Zephir-R2# mpls show label-switched-path brief
```

Ingress LSP:

| | | | | | Label | |
|---|---|---|---|---|---|---|
| LSP Name | To | From | State | Style | In | Out |

Transit LSP:

| | | | | | Label | |
|---|---|---|---|---|---|---|
| LSP Name | To | From | State | Style | In | Out |
| R11-to-R4 | 223.0.0.4 | 223.0.0.1 | Up | FF | 4097 | 4097 |

Egress LSP:

| | | | | | Label | |
|---|---|---|---|---|---|---|
| LSP Name | To | From | State | Style | In | Out |
| R11-to-R2 | 223.0.0.2 | 223.0.0.1 | Up | FF | 16 | — |
| R1-to-R2 | 223.0.0.2 | 223.0.0.1 | Up | FF | 16 | — |

Listing 7.69 illustrates the LSP information of Notus-R3, which serves as the transit LSR for LSPs "R11–to–R2" and "R11–to–R4." The MPLS labels to be used for each LSP are as follows:

- For LSP "R11–to–R2," the locally generated InLabel that is advertised to Eurus-R4 is 4097, and the assigned OutLabel received from Zephir-R2 is 16.
- For LSP "R11–to–R4," the locally generated InLabel that is advertised to Zephir-R2 is 4097, and the assigned OutLabel received from Eurus-R4 is 16.

### Listing 7.69   Notus-R3 LSP Information

```
Notus-R3# mpls show label-switched-path brief


Ingress LSP:
                                                          Label
LSP Name          To             From           State   Style   In      Out

Transit LSP:
                                                          Label
LSP Name          To             From           State   Style   In      Out
R11-to-R2         223.0.0.2      223.0.0.1      Up      FF      4097    16
R11-to-R4         223.0.0.4      223.0.0.1      Up      FF      4097    16

Egress LSP:
                                                          Label
LSP Name          To             From           State   Style   In      Out
```

Listing 7.70 illustrates the LSP information of Eurus-R4, which serves as the transit LSR for LSP "R11–to–R2" and as the egress LSR (or tunnel endpoint) for LSPs "R11–to–R4" and "R1–to–R4." The MPLS labels to be used for each LSP are as follows:

- For LSP "R11–to–R2," the assigned OutLabel received from Notus-R3 is 4097. This label is reused for the InLabel that is advertised to Boreas-R1.
- For LSP "R11–to–R4," the locally generated InLabel that is advertised to Notus-R3 is 16. Because PHP has been disabled and Eurus-R4 is the egress router for this LSP, it performs the pop operation, so there is no OutLabel.
- For LSP "R1–to–R4," the locally generated InLabel that is advertised to Boreas-R1 is 16. Because PHP has been disabled and Eurus-R4 is the egress router for this LSP, it performs the pop operation, so there is no OutLabel.

### Listing 7.70   Eurus-R4 LSP Information

```
Eurus-R4# mpls show label-switched-path
```

```
Ingress LSP:
                                                             Label
LSP Name        To            From          State   Style   In        Out

Transit LSP:
                                                             Label
LSP Name        To            From          State   Style   In        Out
R11-to-R2       223.0.0.2     223.0.0.1     Up      FF      4097      4097

Egress LSP:
                                                             Label
LSP Name        To            From          State   Style   In        Out
R11-to-R4       223.0.0.4     223.0.0.1     Up      FF      16        —
R1-to-R4        223.0.0.4     223.0.0.1     Up      FF      16        —
```

Listing 7.71 illustrates the RSVP PSB of Boreas-R1, identifying the four established LSP tunnels, each with a unique tunnel ID:

- LSP tunnel "R1–to–R4" is identified by tunnel ID 7.
- LSP tunnel "R11–to–R4" is identified by tunnel ID 8.
- LSP tunnel "R1–to–R2" is identified by tunnel ID 9.
- LSP tunnel "R11–to–R2" is identified by tunnel ID 10.

Note that the terms *LSP* and *LSP tunnel* are interchangeable throughout the text. In the earlier case studies, tunnel ID and LSP ID were not mentioned because there was only one established LSP involved each time. However, there are four LSP tunnels to be set up in this case study, and these tunnels need to be uniquely identified. The tunnel ID found in the SESSION object, together with the LSP ID found in the SENDER_TEMPLATE (or FILTER_SPEC) object, uniquely identifies an LSP tunnel according to the type of reservation style used. Hence, it is worthwhile to discuss further the tunnel ID and the LSP ID in association with the FF and SE reservation styles.

With the SE reservation style, bandwidth is reserved once per link regardless of the number of senders. To achieve this, a common tunnel ID is used to identify the single reservation, while each sender is represented by a unique LSP ID to facilitate make-before-break LSP rerouting or bandwidth increase (see section 6.8). The FF reservation style specifies an explicit list of senders whereby each sender has a dedicated reservation that is not shared with other senders. Because each sender has its own reservation, a separate tunnel ID is used to identify the reservation for each sender/receiver (or ingress/egress) pair. As the reservation style pertaining to this case study is FF, four unique tunnel IDs are used. Because the variation is with respect to the tunnel ID, each LSP tunnel is still uniquely identified by its tunnel ID, even if the respective LSP IDs are all identical, specifically with a constant value of 1 in this case.

The constrained shortest paths determined by constraint-based OSPF for the respective LSP tunnels to be set up by RSVP-TE are:

- For LSP tunnel "R1–to–R4," the derived ERO is 192.168.14.4 (R4). The RSVP PATH message for this LSP is sent out of port-32 to next-hop 192.168.14.4.
- For LSP tunnel "R11–to–R4," the derived ERO is 192.168.12.2 (R2)=> 192.168.23.3 (R3)=>192.168.34.4 (R4). The RSVP PATH message for this LSP is sent out of port-31 to next-hop 192.168.12.2.
- For LSP tunnel "R1–to–R2," the derived ERO is 192.168.12.2 (R2). The RSVP PATH message for this LSP is sent out of port-31 to next-hop 192.168.12.2.
- For LSP tunnel "R11–to–R2," the derived ERO is 192.168.14.4 (R4)=> 192.168.34.3 (R3)=>192.168.23.2 (R2). The RSVP PATH message for this LSP is sent out of port-32 to next-hop 192.168.14.4.

---

### Listing 7.71    Boreas-R1 RSVP PSB

```
Boreas-R1# rsvp show psb

Path State Blocks:

RSVP_PSB <rsvp_1>:
session-attr: name: R1-to-R4 flags: 0x0 setup-pri: 7 holding-pri: 0
session: end-point: 223.0.0.4 tunnel-id: 7 ext-tunnel-id: 0xdf000001
send-templ: sender: 223.0.0.1 lsp-id: 1
prev-hop: 0.0.0.0
in-if: <Local-API> out-if: <port-32>
explicit-route: 192.168.14.4

RSVP_PSB <rsvp_1>:
session-attr: name: R11-to-R4 flags: 0x0 setup-pri: 7 holding-pri: 0
session: end-point: 223.0.0.4 tunnel-id: 8 ext-tunnel-id: 0xdf000001
send-templ: sender: 223.0.0.1 lsp-id: 1
prev-hop: 0.0.0.0
in-if: <Local-API> out-if: <port-31>
explicit-route: 192.168.12.2=>192.168.23.3=>192.168.34.4

RSVP_PSB <rsvp_1>:
session-attr: name: R1-to-R2 flags: 0x0 setup-pri: 7 holding-pri: 0
session: end-point: 223.0.0.2 tunnel-id: 9 ext-tunnel-id: 0xdf000001
send-templ: sender: 223.0.0.1 lsp-id: 1
```

```
prev-hop: 0.0.0.0
in-if: <Local-API> out-if: <port-31>
explicit-route: 192.168.12.2


RSVP_PSB <rsvp_1>:
session-attr: name: R11-to-R2 flags: 0x0 setup-pri: 7 holding-pri: 0
session: end-point: 223.0.0.2 tunnel-id: 10 ext-tunnel-id: 0xdf000001
send-templ: sender: 223.0.0.1 lsp-id: 1
prev-hop: 0.0.0.0
in-if: <Local-API> out-if: <port-32>
explicit-route: 192.168.14.4=>192.168.34.3=>192.168.23.2
```

Listing 7.72 illustrates the RSVP PSB of Zephir-R2:

- For LSP tunnel "R11–to–R4," the derived ERO has become 192.168.23.3 (R3)=>192.168.34.4 (R4). The RSVP PATH message for this LSP is received via port-41 from PHOP 192.168.12.1 and is sent out of port-42 to next-hop 192.168.23.3.
- The RSVP PATH message for LSP tunnel "R1–to–R2" is received via port-41 from PHOP 192.168.12.1. Because the tunnel endpoint 223.0.0.2 is at Zephir-R2 itself, the next-hop is no longer required to be derived from the ERO.
- The RSVP PATH message for LSP tunnel "R11–to–R2" is received via port-42 from PHOP 192.168.23.3. Because the tunnel endpoint 223.0.0.2 is at Zephir-R2 itself, the next-hop is no longer required to be derived from the ERO.

**Listing 7.72   Zephir-R2 RSVP PSB**

```
Zephir-R2# rsvp show psb


Path State Blocks:


RSVP_PSB <rsvp_1>:
session-attr: name: R11-to-R4 flags: 0x0 setup-pri: 7 holding-pri: 0
session: end-point: 223.0.0.4 tunnel-id: 8 ext-tunnel-id: 0xdf000001
send-templ: sender: 223.0.0.1 lsp-id: 1
prev-hop: 192.168.12.1
in-if: <port-41> out-if: <port-42>
explicit-route: 192.168.23.3=>192.168.34.4


RSVP_PSB <rsvp_1>:
session-attr: name: R1-to-R2 flags: 0x0 setup-pri: 7 holding-pri: 0
```

```
session: end-point: 223.0.0.2 tunnel-id: 9 ext-tunnel-id: 0xdf000001
send-templ: sender: 223.0.0.1 lsp-id: 1
prev-hop: 192.168.12.1
in-if: <port-41> out-if: <Local-API>
explicit-route:


RSVP_PSB <rsvp_1>:
session-attr: name: R11-to-R2 flags: 0x0 setup-pri: 7 holding-pri: 0
session: end-point: 223.0.0.2 tunnel-id: 10 ext-tunnel-id: 0xdf000001
send-templ: sender: 223.0.0.1 lsp-id: 1
prev-hop: 192.168.23.3
in-if: <port-42> out-if: <Local-API>
explicit-route:
```

Listing 7.73 illustrates the RSVP PSB of Notus-R3:

- For LSP tunnel "R11–to–R4," the derived ERO has become 192.168.34.4 (R4). The RSVP PATH message for this LSP is received via port-54 from PHOP 192.168.23.2 and is sent out of port-53 to next-hop 192.168.34.4.
- For LSP tunnel "R11–to–R2," the derived ERO has become 192.168.23.2 (R2). The RSVP PATH message for this LSP is received via port-53 from PHOP 192.168.34.4 and is sent out of port-54 to next-hop 192.168.23.2.

**Listing 7.73   Notus-R3 RSVP PSB**

```
Notus-R3# rsvp show psb

Path State Blocks:

RSVP_PSB <rsvp_1>:
session-attr: name: R11-to-R4 flags: 0x0 setup-pri: 7 holding-pri: 0
session: end-point: 223.0.0.4 tunnel-id: 8 ext-tunnel-id: 0xdf000001
send-templ: sender: 223.0.0.1 lsp-id: 1
prev-hop: 192.168.23.2
in-if: <port-54> out-if: <port-53>
explicit-route: 192.168.34.4

RSVP_PSB <rsvp_1>:
session-attr: name: R11-to-R2 flags: 0x0 setup-pri: 7 holding-pri: 0
session: end-point: 223.0.0.2 tunnel-id: 10 ext-tunnel-id: 0xdf000001
send-templ: sender: 223.0.0.1 lsp-id: 1
```

```
prev-hop: 192.168.34.4
in-if: <port-53> out-if: <port-54>
explicit-route: 192.168.23.2
```

---

Listing 7.74 illustrates the RSVP PSB of Eurus-R4:

- The RSVP PATH message for LSP tunnel "R1–to–R4" is received via port-42 from PHOP 192.168.14.1. Because the tunnel endpoint 223.0.0.4 is at Eurus-R4 itself, the next-hop is no longer required to be derived from the ERO.
- The RSVP PATH message for LSP tunnel "R11–to–R4" is received via port-41 from PHOP 192.168.34.3. Because the tunnel endpoint 223.0.0.4 is at Eurus-R4 itself, the next-hop is no longer required to be derived from the ERO.
- For LSP tunnel "R11–to–R2," the derived ERO has become 192.168.34.3 (R3)=>192.168.23.2 (R2). The RSVP PATH message for this LSP is received via port-42 from PHOP 192.168.14.1 and is sent out of port-41 to next-hop 192.168.34.3.

---

### Listing 7.74   Eurus-R4 RSVP PSB

```
Eurus-R4# rsvp show psb

Path State Blocks:

RSVP_PSB <rsvp_1>:
session-attr: name: R1-to-R4 flags: 0x0 setup-pri: 7 holding-pri: 0
session: end-point: 223.0.0.4 tunnel-id: 7 ext-tunnel-id: 0xdf000001
send-templ: sender: 223.0.0.1 lsp-id: 1
prev-hop: 192.168.14.1
in-if: <port-42> out-if: <Local-API>
explicit-route:

RSVP_PSB <rsvp_1>:
session-attr: name: R11-to-R4 flags: 0x0 setup-pri: 7 holding-pri: 0
session: end-point: 223.0.0.4 tunnel-id: 8 ext-tunnel-id: 0xdf000001
send-templ: sender: 223.0.0.1 lsp-id: 1
prev-hop: 192.168.34.3
in-if: <port-41> out-if: <Local-API>
explicit-route:

RSVP_PSB <rsvp_1>:
session-attr: name: R11-to-R2 flags: 0x0 setup-pri: 7 holding-pri: 0
```

```
session: end-point: 223.0.0.2 tunnel-id: 10 ext-tunnel-id: 0xdf000001
send-templ: sender: 223.0.0.1 lsp-id: 1
prev-hop: 192.168.14.1
in-if: <port-42> out-if: <port-41>
explicit-route: 192.168.34.3=>192.168.23.2
```

For brevity, the RSVP RSB listings are omitted in this case study.

**7.2.4.5 OSPF TED Monitoring.** Listing 7.75 illustrates the OSPF TED of Boreas-R1. The OSPF link sub-TLV details for the four routers are indicated clearly on the TED listing with the help of in-line headers. In this section, only the resource class sub-TLVs for these routers are examined:

- For Eurus-R4 (router ID 223.0.0.4), port-41 is associated with local address 192.168.34.4 and has a resource class attribute value of 0 × 00000006 (red, blue). Port-42 is associated with local address 192.168.14.4 and has a resource class attribute value of 0 × 0000000A (red, green).
- For Notus-R3 (router ID 223.0.0.3), port-53 is associated with local address 192.168.34.3 and has a resource class attribute value of 0 × 00000006 (red, blue). Port-54 is associated with local address 192.168.23.3 and has a resource class attribute value of 0 × 00000006 (red, blue).
- For Zephir-R2 (router ID 223.0.0.2), port-42 is associated with local address 192.168.23.2 and has a resource class attribute value of 0 × 00000006 (red, blue). Port-41 is associated with local address 192.168.12.2 and has a resource class attribute value of 0 × 0000000C (blue, green).
- For Boreas-R1 (router ID 223.0.0.1), port-32 is associated with local address 192.168.14.1 and has a resource class attribute value of 0 × 0000000A (red, green). Port-31 is associated with local address 192.168.12.1 and has a resource class attribute value of 0 × 0000000C (blue, green).

**Listing 7.75   Boreas-R1 OSPF TED**

```
Boreas-R1# ospf show ted

OSPF Router with ID (223.0.0.1)


Link Sub-TLV Info for Eurus-R4 (Router-ID=223.0.0.4)


Link Type 2
Link ID 192.168.34.4
```

**Local Addr 192.168.34.4**
Remote Addr 0.0.0.0
Metric 2
Max BW                    :              1 Gbps
Max reservable BW :              1 Gbps
Max unreserved BW :
[0]              1 Gbps    [1]              1 Gbps
[2]              1 Gbps    [3]              1 Gbps
[4]              1 Gbps    [5]              1 Gbps
[6]              1 Gbps    [7]              1 Gbps
**Resource class 6**

Link Type 2
Link ID 192.168.14.4
**Local Addr 192.168.14.4**
Remote Addr 0.0.0.0
Metric 2
Max BW                    :              1 Gbps
Max reservable BW :              1 Gbps
Max unreserved BW :
[0]              1 Gbps    [1]              1 Gbps
[2]              1 Gbps    [3]              1 Gbps
[4]              1 Gbps    [5]              1 Gbps
[6]              1 Gbps    [7]              1 Gbps
**Resource class a**

**Link Sub-TLV Info for Notus-3 (Router-ID=223.0.0.3)**

Link Type 2
Link ID 192.168.34.4
**Local Addr 192.168.34.3**
Remote Addr 0.0.0.0
Metric 2
Max BW                    :              1 Gbps
Max reservable BW :              1 Gbps
Max unreserved BW :
[0]              1 Gbps    [1]              1 Gbps
[2]              1 Gbps    [3]              1 Gbps
[4]              1 Gbps    [5]              1 Gbps
[6]              1 Gbps    [7]              1 Gbps
**Resource class 6**

Link Type 2
Link ID 192.168.23.3
**Local Addr 192.168.23.3**

```
Remote Addr 0.0.0.0
Metric 2
Max BW             :          1 Gbps
Max reservable BW :          1 Gbps
Max unreserved BW :
[0]          1 Gbps   [1]          1 Gbps
[2]          1 Gbps   [3]          1 Gbps
[4]          1 Gbps   [5]          1 Gbps
[6]          1 Gbps   [7]          1 Gbps
```
**Resource class 6**

**Link Sub-TLV Info for Zephir-R2 (Router-ID=223.0.0.2)**

```
Link Type 2
Link ID 192.168.23.3
```
**Local Addr 192.168.23.2**
```
Remote Addr 0.0.0.0
Metric 2
Max BW             :          1 Gbps
Max reservable BW :          1 Gbps
Max unreserved BW :
[0]          1 Gbps   [1]          1 Gbps
[2]          1 Gbps   [3]          1 Gbps
[4]          1 Gbps   [5]          1 Gbps
[6]          1 Gbps   [7]          1 Gbps
```
**Resource class 6**

```
Link Type 2
Link ID 192.168.12.2
```
**Local Addr 192.168.12.2**
```
Remote Addr 0.0.0.0
Metric 2
Max BW             :          1 Gbps
Max reservable BW :          1 Gbps
Max unreserved BW :
[0]          1 Gbps   [1]          1 Gbps
[2]          1 Gbps   [3]          1 Gbps
[4]          1 Gbps   [5]          1 Gbps
[6]          1 Gbps   [7]          1 Gbps
```
**Resource class c**

**Link Sub-TLV Info for Boreas-R1 (Router-ID=223.0.0.1)**

```
Link Type 2
Link ID 192.168.14.4
```

```
Local Addr 192.168.14.1
Remote Addr 0.0.0.0
Metric 2
Max BW                    :              1 Gbps
Max reservable BW :              1 Gbps
Max unreserved BW :
[0]              1 Gbps    [1]              1 Gbps
[2]              1 Gbps    [3]              1 Gbps
[4]              1 Gbps    [5]              1 Gbps
[6]              1 Gbps    [7]              1 Gbps
Resource class a

Link Type 2
Link ID 192.168.12.2
Local Addr 192.168.12.1
Remote Addr 0.0.0.0
Metric 2
Max BW                    :              1 Gbps
Max reservable BW :              1 Gbps
Max unreserved BW :
[0]              1 Gbps    [1]              1 Gbps
[2]              1 Gbps    [3]              1 Gbps
[4]              1 Gbps    [5]              1 Gbps
[6]              1 Gbps    [7]              1 Gbps
Resource class c
```

## 7.3  SUMMARY

The four case studies discussed in this chapter focus on the two important aspects of TE: traffic control and resource optimization. The whole idea of MPLS-TE is to defy the limitations posed by conventional destination-based hop-by-hop forwarding. This is achieved with explicitly routed LSPs.

An explicit route can be regarded as a controlled route. In other words, the shortest path is not the primary consideration when determining a route. The priority emphasis is on the resource requirements and policy constraints imposed on the desired path.

Based on the physical topology of a specific network, an explicit route can be either administratively defined by an operator or dynamically specified with a CBR protocol such as constraint-based OSPF or integrated intermediate system to intermediate system (ISIS). A CBR protocol is enhanced with the capability to take resource availability and traffic characteristics into consideration during path determination through link attributes that are associated with network resources. Resource attributes can be defined on every link in a network and are flooded throughout the network. These resource attributes are part of the topology-state parameters that are used to constrain the routing of traffic trunks through specific resources. With a proper control policy in place at the ingress point specifying the bandwidth requirements and policy constraints such as link affinities, the CBR protocol can determine the required constrained shortest path, which in turn is handed to RSVP-TE for the actual LSP setup.

To conclude the discussion on the various aspects of MPLS-TE, a traffic-engineered MAN can:

- Route primary paths around known bottlenecks or points of congestion in the network, thus resolving the hyperaggregation phenomenon.
- Optimize available aggregate bandwidth by ensuring that resource-poor network segments do not become overutilized while resource-rich network segments along potential alternate paths do not become underutilized.
- Enhance the traffic-oriented performance characteristics of the network by reducing packet loss, purging chronic congestion, and increasing throughput.
- Improve statistically bounded performance characteristics of the network such as loss ratio, delay variation, and transfer delay that are required to support mission-critical services.
- Provide precise control over how traffic is rerouted when the primary path encounters single or multiple failures (see chapter 8).
- Give metro SPs the competitive advantage by maximizing operational efficiency and lowering operational costs.
- Offer more options, lower costs, and better services to the metro subscribers.

# Part 3
# Reliability Aspect of Metropolitan Area Networks

# Chapter 8
# Reliability in MPLS-Based Metropolitan Area Networks

## 8.1 INTRODUCTION

As defined in [RFC3272], network survivability or network reliability refers to the capability of a network to maintain service continuity in the presence of faults, which can be accomplished by promptly recovering from network impairments and maintaining the required quality of service (QoS) for existing services after recovery. Reliability has become a major concern, particularly in metro environments, due to the recent increasing demands to carry mission-critical traffic or other high-priority traffic over IP-based metropolitan area networks (MANs).

Multi-Protocol Label Switching (MPLS) is an imperative emerging technology that enhances the connectionless behavior of IP networks to become connection-oriented ones. Because MPLS is path oriented, it can potentially provide faster and more predictable protection and restoration capabilities than conventional hop-by-hop-oriented IGPs. This chapter covers the implementation aspects of the protection and restoration of IP/MPLS-based MANs.

## 8.2 TERMINOLOGY

This section gives a general conceptual overview of the terms used in this chapter. Some of these terms are more specifically defined in later sections. The reader is assumed to be familiar with the terminology in Resource Reservation Protocol (RSVP) and RSVP with traffic engineering extensions (RSVP-TE) (see chapter 6).

- *Backup LSP*: The label switched path (LSP) that is responsible for backing up the main LSP(s). A backup LSP typically refers to either a detour LSP or a bypass tunnel.
- *Bypass tunnel*: An LSP that is used to protect a set of LSPs passing over a common facility.

- *Detour LSP*: The LSP that is used to reroute traffic around a failure in one-to-one backup.
- *Facility backup*: A local repair technique where a bypass tunnel is used to protect one or more main LSPs, which traverse the point of local repair (PLR), the resource being protected, and the merge point (MP), in that order.
- *Local repair*: Techniques used to repair LSP tunnels quickly when a node or link along the main LSP fails.
- *Main LSP*: The primary LSP to be protected at a given hop if it has one or multiple associated backup LSPs originating at that hop. The main LSP is also commonly referred to as the protected LSP.
- *Merge point (MP)*: The label switch router (LSR) where one or more backup tunnels rejoin the path of the main LSP, downstream of the potential failure. The same LSR can be an MP and a PLR at the same time.
- *Next-hop (NHOP) backup tunnel*: A backup LSP that bypasses a single link of the main LSP.
- *Next-next-hop (NNHOP) backup tunnel*: A backup LSP that bypasses a single node of the main LSP.
- *One-to-one backup*: A local repair technique where a detour LSP is separately created at a PLR for each main LSP.
- *Point of local repair (PLR)*: The head-end LSR of a detour LSP or a bypass tunnel.

## 8.3 FAILURE PROTECTION TYPES

MANs are typically high-speed networks in which network reliability is mandatory. When service disruption happens, the reasons could be due to:

- Congestion at specific links or nodes along the LSP
- Link failure within the LSP
- LSR or node failure
- Administrative change that affects the LSP

With MPLS-TE, it is possible to provide nondisruptive traffic across the LSP when the above-stated conditions occur. In the case of failure outage, the upper-layer application will not notice any service disruption. Rerouting executed by IP routing protocols (or interior gateway protocols (IGPs)) is traditionally used to restore service following link or node outages. However, this only takes place after a period of routing convergence, which can range from seconds to minutes. MPLS-TE introduces the concept of protection switching or fast reroute, in which local repair mechanisms are used to achieve fast restoration in tens of milliseconds at the IP layer prior to convergence. If restoration is fast enough (typically 50 milliseconds or more), there will be no perceptible interruption of the metro subscriber's data, voice, or video applications when failure events occur, and the metro subscriber will be oblivious to what has happened.

MPLS protection switching deploys preestablished secondary LSPs that back up existing primary LSPs by bypassing failed links and nodes. In this way, the time taken for redirecting traffic onto the backup LSP during a failure outage does not include any path computation or signaling delay, including the delay to propagate failure notification between LSRs. As such, the entire LSP rerouting/restoration process is expedited.

In the event that a link or node failure occurs along a primary LSP without its corresponding secondary LSP being defined, the ingress LSR will continue to forward traffic toward its destination as native IP packets using the IGP shortest path if such a path is available. The types of protection implemented by MPLS-TE against LSP outages include:

- *Link protection*: The purpose of LSP link protection is to protect an LSP from a specific link failure. Under link protection, the path of the backup LSP (or the secondary LSP) is disjoint from the path of the primary LSP at the particular link over which protection is required. When the protected link fails, traffic on the primary LSP is switched over to the backup LSP at the head end of the failed link. This is commonly referred to as local repair (or local protection), whereby protection is against a single link or node failure.
- *Node protection*: The purpose of LSP node protection is to protect an LSP from a specific node failure. Under node protection/restoration, the path of the backup LSP is disjoint from the path of the primary LSP at the particular node to be protected. When the node fails, traffic on the primary LSP is switched over to the backup LSP at the upstream LSR that is directly connected to the failed node. Node protection is also part of the local repair scheme.
- *Path protection*: Path protection is intended to protect an LSP from failure at any point along its routed path. Under path protection, the path of the backup LSP is completely disjoint from the path of the primary LSP (with no fate sharing). The advantage of path protection is that the backup LSP protects the primary LSP from all possible link and node failures along the path, except for failures that occur at the ingress and egress LSRs, or for correlated failures that can impact both primary and backup paths at the same time. In other words, path protection is a global repair (or global protection) technique in which the protection is against any link or node failure along the LSP. Because the path selection is end to end based, path protection is more efficient from the resource usage perspective than link or node protection. Nevertheless, because restoration is localized, the failover time of link or node protection is considerably

faster than path protection, in which remote nodes must be notified of failures.

The details of these various protection schemes are discussed in the following subsections.

### 8.3.1  Link Protection

Link protection is accomplished through backup LSPs that only bypass a particular link. This is accomplished by rerouting user traffic on the main LSP to the next-hop (NHOP), which bypasses the failed link (see Figure 8.1). These backup LSPs are commonly referred to as NHOP backup tunnels because they terminate at the NHOP of the main LSP, beyond the point of failure.

In Figure 8.1, when network link R2–R4 fails, user traffic is rerouted to NHOP backup tunnel R2–R3–R4, thus bypassing the failed link.

### 8.3.2  Node Protection

In node protection, the backup LSP bypasses the next-hop node along the main LSP. These backup LSPs are commonly referred to as next-next-hop (NNHOP) backup tunnels because they terminate at the node following the next-hop node of the main LSP. During a node failure, NNHOP backup tunnels enable the node upstream of the failure to reroute user traffic on the

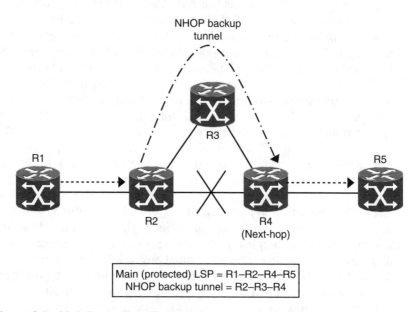

**Figure 8.1  Link Protection Illustration**

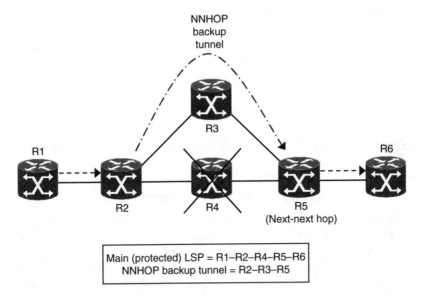

**Figure 8.2  Node Protection Illustration**

main LSP to the NNHOP, thus bypassing the failed next-hop node (see Figure 8.2). NNHOP backup tunnels also provide protection against link failures because they bypass the failed link along with the node. In addition, RSVP hello messages (see section 6.9.4) can be used in conjunction with node protection to speed up the detection of node failures.

In Figure 8.2, when node R4 fails, user traffic is rerouted to NNHOP backup tunnel R2–R3–R5, which bypasses the failed node.

### 8.3.3  Bandwidth Protection

NHOP and NNHOP backup tunnels can also be used to provide bandwidth protection for rerouted LSPs by associating a backup bandwidth with these tunnels. Bandwidth protection ensures that a particular backup tunnel has sufficient capacity to meet the bandwidth requirements of the rerouted LSPs. This backup tunnel can be configured to protect two types of backup bandwidth:

- *Limited backup bandwidth*: The backup tunnel provides bandwidth protection in which the requested bandwidth is guaranteed. Therefore, sufficient backup bandwidth must be present when rerouting LSPs to this type of backup tunnel, and the aggregated bandwidth of all LSPs using this backup tunnel must not exceed the backup bandwidth of the backup tunnel.
- *Unlimited backup bandwidth*: The backup tunnel does not provide any bandwidth protection. In this case, only best-effort (that is, no

191

guarantee) protection is present, so there is no limit to the amount of bandwidth used by the LSPs that are mapped to this backup tunnel. LSPs that did not request for any bandwidth reservation (zero bandwidth allocation) previously during path setup can only use backup tunnels that have unlimited backup bandwidth.

### 8.3.4 Local Repair Options

The head-end LSR of a NHOP or NNHOP backup tunnel is known as the point of local repair (PLR). During a link or node failure along the main LSP, the PLR repairs the broken LSP by diverting traffic around the failed link or node onto the NHOP or NNHOP backup tunnel. The PLR can deploy the NHOP or NNHOP backup tunnels in two different ways:

- *One-to-one backup*: A local repair technique where a backup tunnel is separately created for each main LSP at a PLR. One-to-one backup tunnels are referred to as detour LSPs.
- *Facility backup*: A local repair technique where a backup tunnel is used to protect one or more main LSPs that traverse between the PLR and a common LSR downstream of the potential failure. Facility backup tunnels are referred to as bypass tunnels.

To fully protect a main LSP, each component (link and node) of the main LSP has to be protected by a detour LSP or a bypass tunnel. Consequently, for a main LSP that traverses N nodes, there can be as many as (N − 1) detour LSPs or bypass tunnels. There is a clear distinction between the created detour LSPs and bypass tunnels in this respect. Each of those detour LSPs can only protect a particular main LSP, whereas each of those bypass tunnels can protect a set of main LSPs that traverse a common link or path.

As illustrated in Figure 8.3, there are five LSRs along the main LSP. Therefore, four detour LSPs or bypass tunnels are required to fully protect this main LSP. In other words, a detour LSP or bypass tunnel is created around each component (link and node) of the main LSP. Note that if the PLR is the penultimate hop, in this case R4, node protection is not possible and only the downstream link R4–R5 can be protected.

**8.3.4.1 One-to-One Backup.** For each LSP that is backed up with the one-to-one technique, a detour LSP is established. As illustrated in Figure 8.4, the main LSP runs from R21 to R25. R22 (the PLR) provides user traffic protection by creating a detour LSP that merges with the main LSP at R24. R24 is known as a merge point (MP), where the detour LSP rejoins the main LSP. It is desirable to merge a detour LSP back to its main LSP whenever feasible to minimize the number of LSPs in the network. When a detour LSP intersects its main LSP at an MP with the same outgoing interface, it will be merged. Put another way, the MP is responsible for mapping both the main

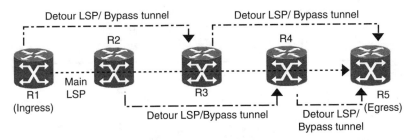

**Figure 8.3   Number of Detour LSPs or Bypass Tunnels Required for a Five-Node Main LSP**

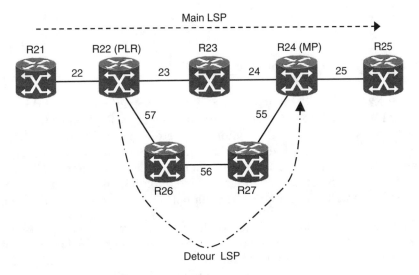

**Figure 8.4   One-to-One Backup Scenario**

LSP inbound label and the detour LSP inbound label to the same outbound label/action.

The labels for the respective routers are also indicated in Figure 8.4. When a failure occurs along the main LSP, R22 will redirect traffic from the main LSP onto the detour LSP R22–R26–R27–R24. For instance, if R23 fails (see Figure 8.5), R22 will switch any user traffic received on link R21–R22 with label 22 onto link R22–R26 with label 57 along the detour LSP, instead of link R22–R23 with label 23 along the main LSP, which has also become unavailable due to the node failure.

When R24 (the MP) receives packets containing the detoured user traffic from link R27–R24 with label 55, it will switch these packets onto link R24–R25 with label 25. Hence, the detour LSP is merged back with the main

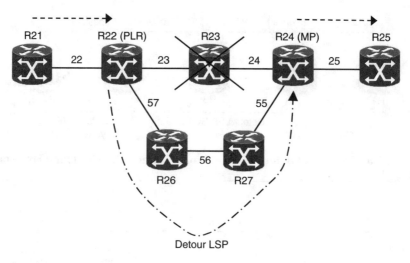

**Figure 8.5    Failover to Detour LSP**

LSP at R24. While R22 is using its detour, user traffic will take the path R21–R22–R26–R27–R24–R25. In one-to-one backup, a single label (or single-level label stack) is used throughout the detour. Instead of terminating at the MP, the endpoint of the detour LSP sustains path continuity by merging immaculately with the main LSP. In other words, when a detour LSP intersects its main LSP at an LSR with the same outgoing interface, it is merged back to the main LSP to reduce the number of LSPs in the MPLS domain.

**8.3.4.2   Facility Backup.** The facility backup creates a single bypass tunnel to back up a set of main LSPs traversing between the PLR and a common node downstream of the potential failure. This backup technique uses the MPLS label stack. As illustrated in Figure 8.6, R32 (the PLR) has constructed a bypass tunnel that protects against the failure of link R32–R34. This same bypass tunnel can be used to protect any main LSP that traverses the common link R32–R34, thus improving the scalability aspect of this backup technique. For simplicity, this example uses only one main LSP to illustrate the failover.

When a failure occurs along a main LSP, the PLR uses label stacking to redirect traffic onto the appropriate bypass tunnel. As illustrated in Figure 8.7, when link R32–R34 fails, an IP packet bounding for R35 uses a two-level label stack composed of the bypass tunnel label at the top of the stack and the main LSP label (of the MP) at the bottom of the stack as R32 redirects it from R31 on the main LSP to link R32–R36 of the bypass tunnel. R32 first swaps label 32 with label 34, which was previously used for link R32–R34 before it failed. R32 then pushes label 36 of the bypass tunnel onto the label

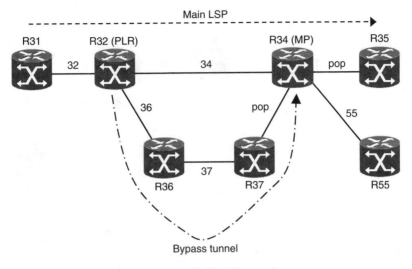

**Figure 8.6  Facility Backup Scenario 1: Link Protection**

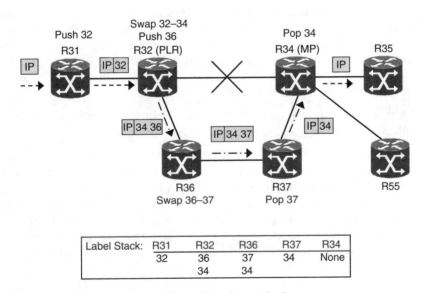

| Label Stack: | R31 | R32 | R36 | R37 | R34 |
|---|---|---|---|---|---|
| | 32 | 36 | 37 | 34 | None |
| | | 34 | 34 | | |

**Figure 8.7  Failover to Bypass Tunnel for Scenario 1**

stack and sends the dual-labeled IP packet to R36. R36 swaps top label 36 with label 37 and in turn sends the packet to R37.

Upon receipt of the packet, R37, the penultimate hop of the bypass tunnel, pops the bypass tunnel label 37 at the top of the label stack and sends the packet to R34 (the MP). R34 receives the packet with the same label

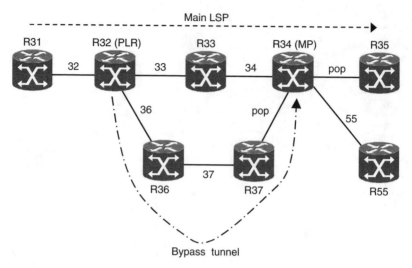

**Figure 8.8    Facility Backup Scenario 2: Node Protection**

(34) as if it were received from link R32–R34 of the main LSP before the link failure. Based on this label (34), R34 performs the same corresponding outgoing action (that is, the pop operation). When R32 is using the bypass tunnel for the main LSP, the user traffic heading for R35 takes the path R31–R32–R36–R37–R34–R35. In this case, the bypass tunnel has become the alternative connection between R32 and R34.

Figure 8.8 illustrates node protection with facility backup, which is more complicated than the previous link protection example.

In this scenario, R32 would need to know the label used on the link R33–R34 because R34 expects to receive this label through the bypass tunnel when R33 fails (see Figure 8.9). The application of an extended RECORD_ROUTE object will allow R32 to learn this label when the "label recording desired" flag (see section 8.4.3) is included in the SESSION_ATTRIBUTE object. Thus, the LSRs along the path will insert into the RECORD_ROUTE object (RRO) the label value information and then their corresponding IP addressing information above it. The RRO is organized in a last-in-first-out (stack) format, where the most recent LSR that has written its route information as a subobject becomes the top-level entry. In other words, R32 can examine the RRO in the RESV message and learn about the incoming labels that are used by all downstream nodes for this LSP. The rest of the local repair operation is similar to that in scenario 1.

### 8.3.5    Path Protection

A strict explicit route with a single primary path provides precise control over how the traffic associated with an LSP will flow. It also offers the

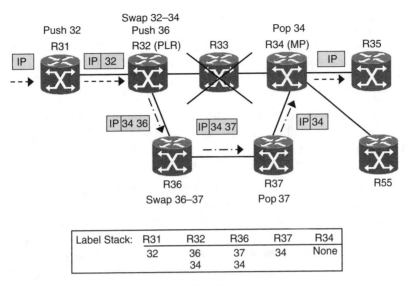

| Label Stack: | R31 | R32 | R36 | R37 | R34 |
|---|---|---|---|---|---|
| | 32 | 36 | 37 | 34 | None |
| | | 34 | 34 | | |

**Figure 8.9    Failover to Bypass Tunnel for Scenario 2**

flexibility to constrain routes based on policy considerations rather than always selecting the shortest path route. However, in the event of failure, service must be able to recover without manual intervention. One way to alleviate this is through loose explicit routing, which allows RSVP-TE to set up a new path around a failure, at the expense of control (see case study 7.2). Path protection provides a better alternative.

Path protection supports the configuration of primary and secondary physical paths for an LSP to protect against link and node failures. This process is known as global repair. The primary path is the preferred path, while the secondary path is used as an alternative route only when the primary path fails. There are two types of secondary paths: hot standby and cold standby. A hot standby secondary path is precomputed and preestablished, while a cold standby secondary path is precomputed but not preestablished. There can be more than one secondary path associated with an LSP, with a configurable order of preference. This approach has the main advantage of specifying disparate paths across the backbone, should the network have such a physical configuration. The secondary paths are selected based on preference, with the higher numerical value preferred (see section 4.7.2).

If a link or node in the primary path fails, the LSR immediately upstream of the outage notifies the ingress LSR of the failure with RSVP-TE. Upon receipt of the outage notification, the ingress LSR reroutes traffic from the failed primary path to the secondary path. The use of hot standby secondary paths improves recovery time by eliminating the call-setup delay that is required to establish a new physical path for the LSP. Because resources are reserved even when the primary path is active, hot standby secondary paths waste more resources than cold standby secondary paths. Nevertheless, the restoration time for hot standby secondary paths is much faster. When the primary path is reestablished, the ingress LSR automatically switches traffic from the secondary path back to the primary path.

This global repair approach is illustrated in Figure 8.10. If there is a link or node (LSR) failure along the primary LSP R1–R2–R6, the ingress LSR R1 detects this via RSVP-TE. The ingress LSR then begins forwarding traffic over to the secondary LSP R1–R3–R4–R5–R6.

Finally, a less preferred hop-by-hop backup path can be configured as a backup of last resort. Its main responsibility is to attempt to establish a path through a network with multiple failures that have already interrupted service on the primary and preferred secondary paths. This path is not preestablished because it would be difficult to predetermine what are the remaining paths from ingress to egress after these multiple failures have occurred.

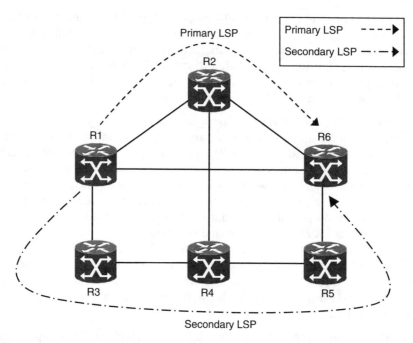

**Figure 8.10   Path Protection Scenario**

### 8.3.6 Fate Sharing

For a protection path to work optimally, it should be calculated to minimize the number of physical links shared with the primary path. This ensures that any single point of failure will not affect both the primary path and the protection path at the same time. Put another way, the ingress LSR can be configured so that if primary and secondary paths are created, the secondary path will contain none of the hops contained within the primary path. This concept is known as no fate sharing, which can be applied when calculating backup paths for link, node, or path protection.

## 8.4 RSVP EXTENSIONS FOR LOCAL REPAIR

Two new objects, FAST_REROUTE and DETOUR, are defined to extend RSVP-TE for fast reroute signaling and can only be carried in PATH messages. The SESSION_ATTRIBUTE and RECORD_ROUTE objects are also extended to support bandwidth and node protection features.

### 8.4.1 FAST_REROUTE Object

The FAST_REROUTE object is carried in the RSVP PATH message to indicate that the main LSP requires a backup path to be preestablished across the components (links and nodes). This object includes all the necessary information to allow transit routers to execute the constrained shortest-path-first (CSPF) process and select backup paths that meet the criteria of the main LSP. The object information includes:

- *Setup priority*: The priority of the backup path pertaining to the taking resources, in the range of 0 to 7. The value 0 is the highest priority. Setup priority is used in deciding whether this backup session can preempt another session.
- *Holding priority*: The priority of the backup path pertaining to the holding resources, in the range of 0 to 7. The value 0 is the highest priority. Holding priority is used in deciding whether this backup session can be preempted by another session.
- *Hop limit*: This field indicates the maximum number of extra hops the backup path is allowed to take, from the current node (a PLR) to an MP, with PLR and MP excluded in the counting. For instance, a hop limit of 0 means that only direct links between PLR and MP can be taken into account.
- The flags field that allows a specific local protection technique to be requested:
  —*Flags = 0 × 01*: This flag value indicates that protection via the one-to-one backup technique is desired.
  —*Flags = 0 × 02*: This flag value indicates that protection via the facility backup technique is desired.

- *Bandwidth*: This field indicates the bandwidth to be used for protection.
- Session attribute filters (or link affinity):
  — *Exclude-any*: A 32-bit vector representing a set of link affinity values associated with a backup path, any of which renders a link unacceptable.
  — *Include-any*: A 32-bit vector representing a set of link affinity values associated with a backup path, any of which renders a link acceptable.
  — *Include-all*: A 32-bit vector representing a set of link affinity values associated with a backup path, all of which must be present for a link to be acceptable.

The FAST_REROUTE object is only inserted into the PATH message by the ingress LSR and cannot be modified by downstream LSRs. The constraint-based information is not required to be the same on the backup path as it is on the main LSP. If this information is not specifically configured when fast reroute is configured, the constraints are inherited.

### 8.4.2  DETOUR Object

The DETOUR object is used in the one-to-one backup technique to identify the detour LSPs. The object information includes:

- *PLR ID*: This field represents the IP address of the PLR requesting the detour LSP.
- *Avoid node ID*: This field represents the IP address of the immediate downstream node that the PLR is trying to bypass in the event of failure.

The main objective of this object is to create the detour around the link for link protection and around the immediate next-hop for node protection.

### 8.4.3  SESSION_ATTRIBUTE Flags

The following SESSION_ATTRIBUTE flags are applicable to local repair:

- $0 \times 01$ = *Local protection desired*: This flag allows transit routers to use a local repair mechanism that may result in violation of the EXPLICIT_ROUTE object (ERO). Thus, the ERO of the original path is adjusted accordingly at the PLR. When a fault is detected on an adjacent downstream link or node, a transit node may reroute user traffic for fast service restoration.
- $0 \times 02$ = *Label recording desired*: This flag indicates that label information should be included when doing a route record and is particularly useful when implementing the facility backup technique during node protection (see section 8.3.4.2).

- *0 × 04 = Shared-explicit (SE) style desired*: This flag indicates that the ingress LSR of the main LSP can reroute this LSP without tearing it down. The egress LSR should use the SE style when responding with a RESV message.

To explicitly request bandwidth and node protection, two additional SESSION_ATTRIBUTE flags are defined:

- *0 × 08 = Bandwidth protection desired*: This flag indicates to the PLRs along the main LSP that a backup path with a bandwidth guarantee is desired. If no FAST_REROUTE object is included in the PATH message, the bandwidth to be guaranteed is the bandwidth requirement of the main LSP itself. If a FAST_REROUTE object is present in the PATH message, the bandwidth to be guaranteed is specified within the FAST_REROUTE object.
- *0 × 10 = Node protection desired*: This flag indicates to the PLRs along a main LSP path that a backup path bypassing at least the next node of the main LSP is desired.

### 8.4.4   RECORD_ROUTE Subobject Flags

The following RECORD_ROUTE subobject flags are applicable to local repair and are typically used in upstream RESV messages:

- *0 × 01 = Local protection available*: This flag indicates that the link downstream of this node is protected via a local repair mechanism, which can be either one-to-one or facility backup. This flag can only be set if the local protection flag is set in the SESSION_ATTRIBUTE object of the corresponding PATH message.
- *0 × 02 = Local protection in use*: This flag indicates that a local repair mechanism is in use to maintain this LSP, typically when an outage of the adjacent link or the neighboring node occurs.

To determine whether bandwidth or node protection is provided as requested, two new RECORD_ROUTE flags are defined:

- *0 × 04 = Bandwidth protection*: The PLR will set this flag when the main LSP has a backup path that is guaranteed to provide the desired bandwidth specified in the FAST_REROUTE object, or the bandwidth of the main LSP if no FAST_REROUTE object was included. The PLR sets this flag when the desired bandwidth is guaranteed and the "bandwidth protection desired" flag is set in the SESSION_ATTRIBUTE object.
- *0 × 08 = Node protection*: The PLR will set this flag when the main LSP has a backup path that provides protection against a failure of the next LSR along the main LSP. The PLR sets this flag when node protection is active and the "node protection desired" flag is set in the SESSION_ATTRIBUTE object. If a PLR can only implement a link

protection backup path, the "local protection available" flag will still be set, but the "node protection" flag will be cleared.

## 8.5 HEAD-END BEHAVIOR

The head end (or ingress LSR) of an LSP decides whether local protection is enabled and which local protection technique is preferred for this main LSP. The head end also determines what constraints are to be requested for the backup paths of a main LSP.

To indicate that a main LSP is to be locally protected, the "local protection desired" flag in the SESSION_ATTRIBUTE object is set or a FAST_REROUTE object is included in the PATH message. If a FAST_REROUTE object is signaled by the ingress LSR, it is stored in the path state block (PSB) of each LSR along the main LSP for path refreshes.

In addition, the "label recording desired" flag in the SESSION_ATTRIBUTE object is set to facilitate the use of the facility backup technique. The "node protection desired" flag in the SESSION_ATTRIBUTE object is set if node protection is required. This flag is cleared for link protection. Likewise, the "bandwidth protection desired" flag in the SESSION_ATTRIBUTE object is set if bandwidth protection is required (see section 8.3.3).

The FAST_REROUTE object is included in the PATH message if the ingress LSR needs to control the backup paths for the main LSP. The FAST_REROUTE object provides administrative control policies such as session attribute filters (or link affinities), bandwidth, hop limit, and priorities that can be used by the PLRs when determining the backup paths. Furthermore, if the main LSP is to be protected via the one-to-one backup technique, then the "one-to-one backup desired" flag is set in the FAST_REROUTE object. Similarly, the "facility backup desired" flag is set in the FAST_REROUTE object if the facility backup technique is used.

## 8.6 POINT-OF-LOCAL-REPAIR BEHAVIOR

A PLR typically supports the FAST_REROUTE object; the DETOUR object; the "local protection desired," "label recording desired," "node protection desired," and "bandwidth protection desired" flags in the SESSION_ATTRIBUTE object; and the "local protection available," "local protection in use," "bandwidth protection," and "node protection" flags in the RRO subobjects.

A PLR along a main LSP will have to provide local protection for this LSP if the "local protection desired" flag is set in the SESSION_ATTRIBUTE object or the FAST_REROUTE object is included in the received PATH message. The PLR will have to implement one-to-one protection if the "one-to-one desired" flag is set in the FAST_REROUTE object. Likewise, if the "facility backup desired" flag is set, the PLR will have to implement facility backup.

Moreover, the PLR will have to provide node protection if the "node protection desired" flag is set in the SESSION_ATTRIBUTE object. The PLR will revert to link protection if node protection is not possible. Similarly, if the "bandwidth protection desired" flag is set in the SESSION_ATTRIBUTE object, the PLR will have to provide a bandwidth guarantee. The PLR will revert to providing a backup without a guarantee of the full bandwidth of the main LSP if bandwidth protection is not possible.

Before a PLR can create a detour LSP or a bypass tunnel, the desired explicit route must be determined. This is achieved using a constraint-based routing (CBR) protocol. The PLR will attempt to set up a backup path when the explicit route is successfully computed.

## 8.7 NOTIFICATION OF LOCAL REPAIR

In most cases, the route used during a local repair is suboptimal. The main purpose of local repair is to keep high-priority and delay-sensitive traffic flowing while a more optimal rerouting of the main LSP can be initiated by the head-end (ingress) LSR of this LSP. The head-end LSR needs to be aware of the failure to resignal a new LSP that is optimal. To provide this notification, the PLR sends a PATHERR message with error code = 25 ("notify" error code) and subcode = 3 ("tunnel locally repaired" subcode) to the head-end LSR (see [RFC3209]).

## 8.8 CASE STUDY 8.1: PATH PROTECTION

This section consolidates the path protection global repair scheme discussed earlier in section 8.3.5 with case study 8.1.

### 8.8.1 Case Overview and Network Topology

Previously, case study 7.2 used loose explicit routing to restore a link or node failure at the expense of full control (see section 7.2.2.1). Case study 8.1, a variant of case study 7.2, uses the path protection alternative instead. The case study examines how a main LSP can be protected against failure (as a result of a link or node failure along its path) with a secondary LSP in hot standby. Complete control for the primary path is fulfilled with strict explicit routing, while the backup path is a completely disparate path with no common transit points that is capable of servicing the primary path when it fails. Figure 8.11 illustrates the network diagram for this case study. The setting uses R1–R4–R3 as the primary path and R1–R2–R3 as the backup path.

The primary path R1–R4–R3 is an administratively defined strict explicit route. By maintaining the cost of path R1–R4–R3 to be higher than path R1–R2–R3, the desired backup path R1–R2–R3 can be determined via standard open shortest path first (OSPF), making it a hop-by-hop routed LSP

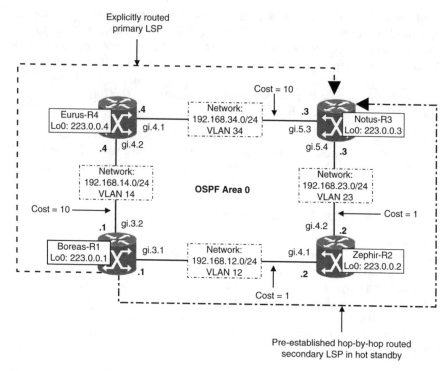

**Figure 8.11  Network Diagram for Case Study 8.1**

(see case study 7.1). The affected network configurations and the monitoring involved are discussed in the subsequent subsections.

### 8.8.2  Network Configurations

Listing 8.1 illustrates the portion of the modified network configuration for Boreas-R1 based on the configuration adapted from Listing 7.1 (see section 7.2.1.2). A strict explicit route is manually defined for the traffic trunk, which does not have any associated resource requirements or policy constraints. Using the "mpls create path" and "mpls set path" commands, explicit path "R1–R4–R3" is created with the following properties:

- The maximum number of hops for the path is 2.
- The IP address of strict hop 1 in the path is 192.168.14.4, whereas the IP address of strict hop 2 in the path is 192.168.34.3.

Explicit path "R1–R4–R3" is defined as the primary path for LSP "R1–to–R3" by specifying the "primary" parameter with the "mpls set label-switched-path" command. RSVP-TE then uses this administratively defined strict explicit route to set up the primary LSP. Note that the "no-cspf" parameter has to be specified when defining explicit paths manually.

Otherwise, RSVP-TE assumes that the explicit path will be dynamically specified by a constraint-based IGP and will wait indefinitely for a valid CSPF response before setting up the LSP. The SE reservation style is also enabled on this LSP with the "adaptive" parameter.

In addition, the command "mpls create path R1–R2–R3" creates a hop-by-hop routed LSP that is set up by RSVP-TE using the shortest path "R1–R2–R3" determined by standard OSPF. The "secondary" and "adaptive" parameters specify this LSP as a secondary path using the SE reservation style. It is also placed into a hot standby state with the "standby" parameter to allow faster cutover from the primary to the secondary path during a link or node failure. In other words, the secondary LSP is precomputed and preestablished during hot standby. Note that if a path is in hot standby state, all LSRs along the LSP must maintain this state information.

---

### Listing 8.1    Boreas-R1 Configuration

```
system set name Boreas-R1


OSPF configuration
ospf create area backbone
ospf add stub-host 223.0.0.1 to-area backbone cost 10
ospf add interface port-32 to-area backbone
ospf add interface port-31 to-area backbone
ospf set interface port-32 cost 10
ospf start


MPLS configuration
mpls add interface port-31
mpls add interface port-32


mpls create path R1-R4-R3 num-hops 2
mpls set path R1-R4-R3 hop 1 ip-addr 192.168.14.4 type strict
mpls set path R1-R4-R3 hop 2 ip-addr 192.168.34.3 type strict
mpls create label-switched-path R1-to-R3 to 223.0.0.3 no-cspf
mpls set label-switched-path R1-to-R3 primary R1-R4-R3 adaptive


mpls create path R1-R2-R3
mpls set label-switched-path R1-to-R3 secondary R1-R2-R3 adaptive
standby


mpls create policy P1 dst-ipaddr-mask 223.0.0.3/32
mpls set label-switched-path R1-to-R3 policy P1
mpls start
```

---

The network configurations for the rest of the routers listed in section 7.2.1.2 are still valid for this case study.

### 8.8.3  Monitoring before Network Link R1–R4 Fails

Listing 8.2 illustrates the result of the traceroute test performed at Boreas-R1. The traceroute output indicates that user traffic (source 223.0.0.1) is traversing primary path R1–R4–R3 to reach the destination (223.0.0.3).

---

### Listing 8.2   Traceroute Test at Boreas-R1

```
Boreas-R1# traceroute 223.0.0.3 source 223.0.0.1

traceroute to 223.0.0.3 (223.0.0.3) from 223.0.0.1, 30 hops max, 40 byte
packets
1  192.168.14.4 (192.168.14.4)  3 ms  1 ms  1 ms
   MPLS Label1=4097 EXP1=0 TTL=1 S=1
2  223.0.0.3 (223.0.0.3)  2 ms  1 ms  0 ms
```

---

Listing 8.3 illustrates the LSP information of Boreas-R1 (ingress LSR). The main LSP "R1–to–R3_R1–R4–R3" and the secondary LSP "R1–to–R3_R1–R2–R3" have been established. Both use the SE reservation style. The MPLS labels to be used for each LSP are as follows:

- For main LSP "R1–to–R3_R1–R4–R3," the assigned OutLabel received from Eurus-R4 is 4097. Boreas-R1 will push this OutLabel onto any packet that has the destination IP address 223.0.0.3 and forward the packet onto this LSP during normal operation.
- For secondary LSP "R1–to–R3_R1–R2–R3," the assigned OutLabel received from Zephir-R2 is 4097. Boreas-R1 will push this OutLabel onto any packet that has the destination IP address 223.0.0.3 and forward the packet onto this LSP when a link or node along the main LSP fails.

---

### Listing 8.3   Boreas-R1 LSP Information

```
Boreas-R1# mpls show label-switched-path brief

Ingress LSP:

                                                     Label
LSP Name              To          From        State  Style  In     Out
R1-to-R3_R1-R2-R3     223.0.0.3   223.0.0.1   Up     SE     -      4097
R1-to-R3_R1-R4-R3     223.0.0.3   223.0.0.1   Up     SE     -      4097
```

Transit LSP:

| | | | | | Label | |
|---|---|---|---|---|---|---|
| LSP Name | To | From | State | Style | In | Out |

Egress LSP:

| | | | | | Label | |
|---|---|---|---|---|---|---|
| LSP Name | To | From | State | Style | In | Out |

Listing 8.4 illustrates the LSP information of Zephir-R2, which is the transit LSR for the established secondary LSP "R1–to–R3_R1–R2–R3." The locally generated InLabel that is advertised to Boreas-R1 is 4097, and the assigned OutLabel received from Notus-R3 is 16.

### Listing 8.4   Zephir-R2 LSP Information

```
Zephir-R2# mpls show label-switched-path brief
```

Ingress LSP:

| | | | | | Label | |
|---|---|---|---|---|---|---|
| LSP Name | To | From | State | Style | In | Out |

Transit LSP:

| | | | | | Label | |
|---|---|---|---|---|---|---|
| LSP Name | To | From | State | Style | In | Out |
| R1–to–R3_R1–R2–R3 | 223.0.0.3 | 223.0.0.1 | Up | SE | 4097 | 16 |

Egress LSP:

| | | | | | Label | |
|---|---|---|---|---|---|---|
| LSP Name | To | From | State | Style | In | Out |

Listing 8.5 illustrates the LSP information of Notus-R3, which is the egress LSR (or tunnel endpoint) for main LSP "R1–to–R3_R1–R4–R3" and secondary LSP "R1–to–R3_R1–R2–R3." The MPLS labels to be used for each LSP are as follows:

- For main LSP "R1–to–R3_R1–R4–R3," the locally generated InLabel that is advertised to Eurus-R4 is 16. Because PHP has been disabled, Notus-R3 performs the pop operation, so there is no OutLabel.
- For secondary LSP "R1–to–R3_R1–R2–R3," the locally generated InLabel that is advertised to Zephir-R2 is 16. Because PHP has been disabled, Notus-R3 performs the pop operation, so there is no OutLabel.

---

### Listing 8.5    Notus-R3 LSP Information

```
Notus-R3# mpls show label-switched-path brief

Ingress LSP:
```

| LSP Name | To | From | State | Style | Label In | Out |
|----------|-----|------|-------|-------|----------|-----|

```
Transit LSP:
```

| LSP Name | To | From | State | Style | Label In | Out |
|----------|-----|------|-------|-------|----------|-----|

```
Egress LSP:
```

| LSP Name | To | From | State | Style | Label In | Out |
|----------|-----|------|-------|-------|----------|-----|
| R1-to-R3_R1-R2-R3 | 223.0.0.3 | 223.0.0.1 | Up | SE | 16 | — |
| R1-to-R3_R1-R4-R3 | 223.0.0.3 | 223.0.0.1 | Up | SE | 16 | — |

---

Listing 8.6 illustrates the LSP information of Eurus-R4, which is the transit LSR for the established main LSP "R1–to–R3_R1–R4–R3." The locally generated InLabel that is advertised to Boreas-R1 is 4097, and the assigned OutLabel received from Notus-R3 is 16.

---

### Listing 8.6    Eurus-R4 LSP Information

```
Eurus-R4# mpls show label-switched-path brief

Ingress LSP:
```

| LSP Name | To | From | State | Style | Label In | Out |
|----------|-----|------|-------|-------|----------|-----|

```
Transit LSP:
```

| LSP Name | To | From | State | Style | Label In | Out |
|----------|-----|------|-------|-------|----------|-----|
| R1-to-R3_R1-R4-R3 | 223.0.0.3 | 223.0.0.1 | Up | SE | 4097 | 16 |

```
Egress LSP:
```

| LSP Name | To | From | State | Style | Label In | Out |
|----------|-----|------|-------|-------|----------|-----|

---

Listing 8.7 illustrates the RSVP PSB of Boreas-R1. Even though the SE reservation style (flags = $0 \times 04$) is used, the two LSP tunnels are still identified by a unique tunnel ID and LSP ID pair because both of these LSP tunnels need to be precomputed and preestablished at the same time:

- Tunnel ID 16385 and LSP ID 1 identify main LSP tunnel "R1–to–R3_R1–R4–R3."
- Tunnel ID 16386 and LSP ID 9 identify secondary LSP tunnel "R1–to–R3_R1–R2–R3."

In addition:

- For main LSP tunnel "R1–to–R3_R1–R4–R3," the ERO is 192.168.14.4 (R4)=>192.168.34.3 (R3). The RSVP PATH message for this LSP is sent out of port-32 to next-hop 192.168.14.4.
- For secondary LSP tunnel "R1–to–R3_R1–R2–R3," the ERO is not used because path selection is hop by hop based. The RSVP PATH message for this LSP is sent out of port-31 to next-hop 192.168.12.2.

---

### Listing 8.7  Boreas-R1 RSVP PSB

```
Boreas-R1# rsvp show psb

Path State Blocks:

RSVP_PSB <rsvp_1>:
session-attr: name: R1-to-R3_R1-R4-R3 flags: 0x4 setup-pri: 7 holding-
pri: 0
session: end-point: 223.0.0.3 tunnel-id: 16385 ext-tunnel-id:
0xdf000001
send-templ: sender: 223.0.0.1 lsp-id: 1
prev-hop: 0.0.0.0
in-if: <Local-API> out-if: <port-32>
explicit-route: 192.168.14.4=>192.168.34.3


RSVP_PSB <rsvp_1>:
session-attr: name: R1-to-R3_R1-R2-R3 flags: 0x4 setup-pri: 7 holding-
pri: 0
session: end-point: 223.0.0.3 tunnel-id: 16386 ext-tunnel-id:
0xdf000001
send-templ: sender: 223.0.0.1 lsp-id: 9
prev-hop: 0.0.0.0
in-if: <Local-API> out-if: <port-31>
explicit-route:
```

---

Listing 8.8 illustrates the RSVP PSB of Zephir-R2. The RSVP PATH message for secondary LSP tunnel "R1–to–R3_R1–R2–R3" is received via port-41 from PHOP 192.168.12.1 and is sent out of port-42 to next-hop 192.168.23.3.

---

### Listing 8.8   Zephir-R2 RSVP PSB

```
Zephir-R2# rsvp show psb


Path State Blocks:


RSVP_PSB <rsvp_1>:
session-attr: name: R1-to-R3_R1-R2-R3 flags: 0x4 setup-pri: 7 holding-
pri: 0
session: end-point: 223.0.0.3 tunnel-id: 16386 ext-tunnel-id:
0xdf000001
send-templ: sender: 223.0.0.1 lsp-id: 9
prev-hop: 192.168.12.1
in-if: <port-41> out-if: <port-42>
explicit-route:
```

---

Listing 8.9 illustrates the RSVP PSB of Eurus-R4. The ERO for main LSP tunnel "R1–to–R3_R1–R4–R3" has become 192.168.34.3 (R3), and the RSVP PATH message received via port-42 from PHOP 192.168.14.1 is sent out of port-41 to next-hop 192.168.34.3.

---

### Listing 8.9   Eurus-R4 RSVP PSB

```
Eurus-R4# rsvp show psb


Path State Blocks:


RSVP_PSB <rsvp_1>:
session-attr: name: R1-to-R3_R1-R4-R3 flags: 0x4 setup-pri: 7 holding-
pri: 0
session: end-point: 223.0.0.3 tunnel-id: 16385 ext-tunnel-id:
0xdf000001
send-templ: sender: 223.0.0.1 lsp-id: 1
prev-hop: 192.168.14.1
in-if: <port-42> out-if: <port-41>
explicit-route: 192.168.34.3
```

---

Listing 8.10 illustrates the RSVP PSB of Notus-R3:

- The RSVP PATH message for main LSP tunnel "R1–to–R3_R1–R4–R3" is received via port-53 from PHOP 192.168.34.4. As the tunnel endpoint 223.0.0.3 is at Notus-R3 itself, the next-hop is no longer required to be derived from the ERO.
- The RSVP PATH message for secondary LSP tunnel "R1–to–R3_R1–R2–R3" is received via port-54 from PHOP 192.168.23.2.

---

### Listing 8.10 Notus-R3 RSVP PSB

```
Notus-R3# rsvp show psb

Path State Blocks:

RSVP_PSB <rsvp_1>:
session-attr: name: R1-to-R3_R1-R4-R3 flags: 0x4 setup-pri: 7 holding-
pri: 0
session: end-point: 223.0.0.3 tunnel-id: 16385 ext-tunnel-id:
0xdf000001
send-templ: sender: 223.0.0.1 lsp-id: 1
prev-hop: 192.168.34.4
in-if: <port-53> out-if: <Local-API>
explicit-route:

RSVP_PSB <rsvp_1>:
session-attr: name: R1-to-R3_R1-R2-R3 flags: 0x4 setup-pri: 7 holding-
pri: 0
session: end-point: 223.0.0.3 tunnel-id: 16386 ext-tunnel-id:
0xdf000001
send-templ: sender: 223.0.0.1 lsp-id: 9
prev-hop: 192.168.23.2
in-if: <port-54> out-if: <Local-API>
explicit-route:
```

---

For brevity, the RSVP reservation state block (RSB) listings are omitted in this case study.

### 8.8.4 Monitoring after Network Link R1–R4 Fails

As illustrated in Figure 8.12, when network link R1–R4 fails, the actual physical path taken by user traffic traversing the established LSP becomes R1–R2–R3.

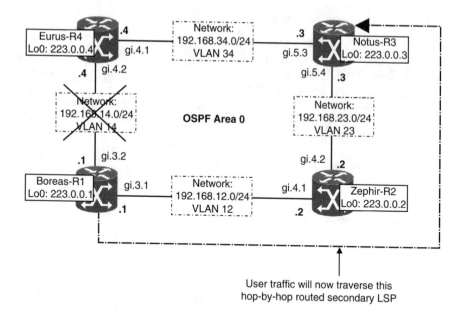

User traffic will now traverse this
hop-by-hop routed secondary LSP

**Figure 8.12   Global Repairing with Path Protection**

Listing 8.11 illustrates the output of the MPLS LSP PATHSWITCH message, indicating that the primary path has been switched over to the secondary path "R1–to–R3_R1–R2–R3" during the link failure.

---

**Listing 8.11   MPLS Message at Boreas-R1**

```
Boreas-R1#


2003-06-07 02:30:21%STP-I-POR2003-06-07 02:30:21%MPLS-I-LSPPATHSWITCH,
LSP "R1-to-R3" switching to Secondary Path "R1-to-R3_R1-R2-R3."
```

---

Listing 8.12 illustrates the LSP information of Boreas-R1 after the link
failure. Only the secondary LSP is listed.

---

**Listing 8.12   Boreas-R1 LSP Information**

```
Boreas-R1# mpls show label-switched-path brief
```

```
Ingress LSP:
                                                   Label
LSP Name            To          From      State  Style  In     Out
R1-to-R3_R1-R2-R3   223.0.0.3   223.0.0.1 Up     SE     —      4097


Transit LSP:
                                                   Label
LSP Name            To          From      State  Style  In     Out


Egress LSP:
                                                   Label
LSP Name            To          From      State  Style  In     Out
```

Listing 8.13 illustrates the result of the traceroute test performed at Boreas-R1 after the link failure. The traceroute output validates that user traffic (source 223.0.0.1) is traversing backup path R1–R2–R3 to reach the destination (223.0.0.3).

### Listing 8.13   Traceroute Test at Boreas-R1

```
Boreas-R1# traceroute 223.0.0.3 source 223.0.0.1

traceroute to 223.0.0.3 (223.0.0.3) from 223.0.0.1, 30 hops max, 40 byte
packets
1  192.168.12.2 (192.168.12.2) 3 ms 1 ms 1 ms
   MPLS Label1=4097 EXP1=0 TTL=1 S=1
2  223.0.0.3 (223.0.0.3) 2 ms 1 ms 0 ms
```

## 8.9   CASE STUDY 8.2: FAST REROUTE WITH DETOUR LSPs

Case study 8.2 reiterates the one-to-one local repair technique discussed earlier in section 8.3.4.1 using detour LSPs.

### 8.9.1   Case Overview and Network Topology

Even though case study 8.2 is adapted from case study 8.1, the setting has been changed:

- A new Gigabit Ethernet link has been provisioned between LSRs Zephir-R2 and Eurus-R4.

- Users from the Zephir district would like to access the Web hosting and off-site storage services from Internet data center 2 at the Eurus district through the new Gigabit Ethernet link.
- These top-priority users are mainly from financial institutions located at commercial estate 2 and government agencies situated in government compound 1.
- The reliability aspect of these services is a key consideration. These users are particularly interested in the level of service continuity after an ad hoc network failure outage has occurred.

The reliability aspect of the MAN can be fulfilled with MPLS fast reroute. Figure 8.13 illustrates the network diagram for this case study. Because user traffic will be traversing network link R2–R4 to reach Internet data center 2, the main LSP should also be established along this path. Link protection is implemented for network link R2–R4 to ensure fast service restoration when the link fails. During the outage, user traffic traversing the main LSP will be rerouted to detour LSPs that are precomputed and preestablished via path R2–R3–R4, which bypasses the failed link.

For completeness, this case study illustrates the bidirectional traffic flows between all three routers. To fulfill these bidirectional conversations, six LSPs have to be established because an LSP by itself is unidirectional. The associated network configurations and the monitoring involved are discussed in the subsequent subsections.

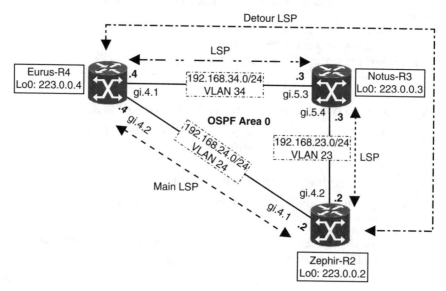

**Figure 8.13  Network Diagram for Case Study 8.2**

## 8.9.2 Network Configurations

Listing 8.14 illustrates the detailed network configuration (described with in-line headers) for Zephir-R2, which is the ingress LSR to the user traffic from the Zephir district (egress LSR for return traffic). These users are represented by loopback address 223.0.0.2, which also happens to be the router ID of Zephir-R2 and the tunnel endpoint for LSPs originating from the other two routers.

The objective of this case study is to reroute user traffic traversing the main LSP onto the detour LSP during a failure outage. Two LSPs are established from Zephir-R2: one to Eurus-R4 via network link R2–R4 with tunnel endpoint 223.0.0.4, and the other to Notus-R3 via network link R2–R3 with tunnel endpoint 223.0.0.3.

The main LSP of concern is the LSP to Eurus-R4, and the protection scheme includes only link R2–R4. Zephir-R2 is the PLR when network link R2–R4 fails. Before the PLR can set up a detour LSP with RSVP-TE, the desired explicit route for the detour path must be determined using a CBR protocol. Constraint-based OSPF is enabled for this purpose with the "ospf set traffic-engineering on" command.

The "mpls set global local-repair-enable" command specifies whether a link or node protection scheme should be used during the main LSP reroute, and the "node-prefer-protection" parameter offers the flexibility to use a node protection detour LSP starting from this PLR if it is available; otherwise, it will use a link protection detour LSP. Explicit path "R2–R4" is manually defined as the primary path for main LSP "R2-to-R4" with the following properties:

- The maximum number of hops for the path is 1.
- The IP address of the strict hop (hop 1) in the path is 192.168.24.4.

By specifying the "fast-reroute" parameter in the "mpls set label-switched-path" command, MPLS fast reroute is enabled for main LSP "R2-to-R4." Fast reroute is not enabled for the LSP to Notus-R3 (LSP "R2-to-R3"), which is hop by hop routed. User traffic on this LSP heading for Notus-R3 will be traversing the shortest path "R2–R3," determined by standard OSPF. RSVP-TE then uses the administratively defined explicit path "R2–R4" to set up LSP "R2-to-R4" and the shortest path "R2–R3" to set up LSP "R2-to-R3."

The PBR policy for forwarding user traffic onto LSP "R2-to-R4" is based on any packet with destination IP address 223.0.0.4, and the PBR policy for forwarding user traffic onto LSP "R2-to-R3" is based on any packet with destination IP address 223.0.0.3.

---

## Listing 8.14   Zephir-R2 Configuration

```
system set name Zephir-R2
```

**VLAN configuration**
```
vlan create vlan23 id 23 ip
vlan create vlan24 id 24 ip
vlan add ports gi.4.2 to vlan23
vlan add ports gi.4.1 to vlan24
```

**IP address configuration**
```
interface create ip port-42 vlan vlan23 address-netmask 192.168.23.2/24
interface create ip port-41 vlan vlan24 address-netmask 192.168.24.2/24
```

**Create loopback address**
```
interface add ip lo0 address-netmask 223.0.0.2/32
```

**Set loopback address as router identifier**
```
ip-router global set router-id 223.0.0.2
```

**OSPF configuration**
```
ospf create area backbone
ospf add stub-host 223.0.0.2 to-area backbone cost 10
ospf add interface port-41 to-area backbone
ospf add interface port-42 to-area backbone
ospf set traffic-engineering on
ospf start
```

**MPLS configuration**
```
mpls set global local-repair-enable node-prefer-protection
mpls add interface port-41
mpls add interface port-42
mpls set interface port-41 no-php
mpls set interface port-42 no-php

mpls create path R2-R4 num-hops 1
mpls set path R2-R4 hop 1 ip-addr 192.168.24.4 type strict
mpls create label-switched-path R2-to-R4 to 223.0.0.4
mpls set label-switched-path R2-to-R4 primary R2-R4
mpls set label-switched-path R2-to-R4 fast-reroute
mpls create label-switched-path R2-to-R3 to 223.0.0.3 no-cspf
mpls create policy to-R3 dst-ipaddr-mask 223.0.0.3
mpls create policy to-R4 dst-ipaddr-mask 223.0.0.4
```

```
mpls set label-switched-path R2-to-R3 policy to-R3
mpls set label-switched-path R2-to-R4 policy to-R4
mpls start
```

**RSVP configuration**
```
rsvp add interface port-41
rsvp add interface port-42
rsvp start
```

---

Listing 8.15 illustrates the detailed network configuration (described with in-line headers) for Notus-R3. Loopback address 223.0.0.3 happens to be the router ID of Notus-R3 and the tunnel endpoint for LSPs originating from the other two routers.

Notus-R3 is the transit LSR for the detour LSPs instantiated between Zephir-R2 and Eurus-R4. Therefore, constraint-based OSPF ("ospf set traffic-engineering on") and local repair function ("mpls set global local-repair-enable") need to be enabled for this transit router to execute the CSPF process and select the detour paths that meet the criteria of the main LSP.

Besides, two hop-by-hop routed LSPs are established from Notus-R3: LSP "R3–to–R4" to Eurus-R4 via the shortest path R3–R4 with tunnel endpoint 223.0.0.4, and LSP "R3–to–R2" to Zephir-R2 via the shortest path R3–R2 with tunnel endpoint 223.0.0.2. Both paths are determined by standard OSPF. RSVP-TE then uses these paths to set up the respective LSPs accordingly.

The PBR policy for forwarding user traffic onto LSP "R3–to–R4" is based on any packet with destination IP address 223.0.0.4, and the PBR policy for forwarding user traffic onto LSP "R3–to–R2" is based on any packet with destination IP address 223.0.0.2.

---

### Listing 8.15   Notus-R3 Configuration

```
system set name Notus-R3
```

**VLAN configuration**
```
vlan create vlan34 id 34 ip
vlan create vlan23 id 23 ip
vlan add ports gi.5.3 to vlan34
vlan add ports gi.5.4 to vlan23
```

**IP address configuration**
```
interface create ip port-53 vlan vlan34 address-netmask 192.168.34.3/24
interface create ip port-54 vlan vlan23 address-netmask 192.168.23.3/24
```

217

**Create loopback address**
```
interface add ip lo0 address-netmask 223.0.0.3/32
```

**Set loopback address as router identifier**
```
ip-router global set router-id 223.0.0.3
```

**OSPF configuration**
```
ospf create area backbone
ospf add stub-host 223.0.0.3 to-area backbone cost 10
ospf add interface port-53 to-area backbone
ospf add interface port-54 to-area backbone
ospf set traffic-engineering on
ospf start
```

**MPLS configuration**
```
mpls set global local-repair-enable node-prefer-protection
mpls add interface port-53
mpls add interface port-54
mpls set interface port-53 no-php
mpls set interface port-54 no-php
mpls create label-switched-path R3-to-R2 to 223.0.0.2 no-cspf
mpls create label-switched-path R3-to-R4 to 223.0.0.4 no-cspf
mpls create policy to-R2 dst-ipaddr-mask 223.0.0.2
mpls create policy to-R4 dst-ipaddr-mask 223.0.0.4
mpls set label-switched-path R3-to-R2 policy to-R2
mpls set label-switched-path R3-to-R4 policy to-R4
mpls start
```

**RSVP configuration**
```
rsvp add interface port-53
rsvp add interface port-54
rsvp start
```

---

Listing 8.16 illustrates the detailed network configuration (described with in-line headers) for Eurus-R4, which is the egress LSR to the user traffic from the Zephir district (or ingress LSR for return traffic). Loopback address 223.0.0.4 happens to be the router ID of Eurus-R4 and the tunnel endpoint for LSPs originating from the other two routers. Two LSPs are established from Eurus-R4: one to Zephir-R2 via network link R2–R4 with tunnel endpoint 223.0.0.2, and the other to Notus-R3 via network link R3–R4 with tunnel endpoint 223.0.0.3.

The main LSP of concern is the LSP to Zephir-R2, and the protection scheme includes only link R2–R4. Eurus-R4 is the PLR when network link

218

R2–R4 fails. Before the PLR can set up a detour LSP with RSVP-TE, the desired explicit route for the detour path must be determined using a CBR protocol. Constraint-based OSPF is enabled for this purpose with the "ospf set traffic-engineering on" command.

The "mpls set global local-repair-enable" command specifies whether a link or node protection scheme should be used during the main LSP reroute, and the "node-prefer-protection" parameter offers the flexibility of using a node protection detour LSP starting from this PLR if it is available; otherwise, it will use a link protection detour LSP. Explicit path "R4–R2" is manually defined as the primary path for main LSP "R4–to–R2" with the following properties:

- The maximum number of hops for the path is 1.
- The IP address of the strict hop (hop 1) in the path is 192.168.24.2.

By specifying the "fast-reroute" parameter in the "mpls set label-switched-path" command, MPLS fast reroute is enabled for main LSP "R4–to–R2." Fast reroute is not enabled for the LSP to Notus-R3 (LSP "R4–to–R3"), which is hop by hop routed. User traffic on this LSP heading for Notus-R3 will be traversing the shortest path "R4–R3" determined by standard OSPF. RSVP-TE then uses the administratively defined explicit path "R4–R2" to set up LSP "R4–to–R2" and the shortest path "R4–R3" to set up LSP "R4–to–R3."

The PBR policy for forwarding user traffic onto LSP "R4–to–R2" is based on any packet with destination IP address 223.0.0.2, and the PBR policy for forwarding user traffic onto LSP "R4–to–R3" is based on any packet with destination IP address 223.0.0.3.

---

### Listing 8.16  Eurus-R4 Configuration

```
system set name Eurus-R4
```

**VLAN configuration**
```
vlan create vlan34 id 34 ip
vlan create vlan24 id 24 ip
vlan add ports gi.4.1 to vlan34
vlan add ports gi.4.2 to vlan24
```

**IP configuration**
```
interface create ip port-41 vlan vlan34 address-netmask 192.168.34.4/24
interface create ip port-42 vlan vlan24 address-netmask 192.168.24.4/24
```

**Create loopback address**
```
interface add ip lo0 address-netmask 223.0.0.4/32
```

**Set loopback address as router identifier**
```
ip-router global set router-id 223.0.0.4
```

**OSPF configuration**
```
ospf create area backbone
ospf add stub-host 223.0.0.4 to-area backbone cost 10
ospf add interface port-42 to-area backbone
ospf add interface port-41 to-area backbone
ospf set traffic-engineering on
ospf start
```

**MPLS configuration**
```
mpls set global local-repair-enable node-prefer-protection
mpls add interface port-41
mpls add interface port-42
mpls set interface port-41 no-php
mpls set interface port-42 no-php
mpls create path R4-R2 num-hops 1
mpls set path R4-R2 hop 1 ip-addr 192.168.24.2 type strict
mpls create label-switched-path R4-to-R2 to 223.0.0.2
mpls set label-switched-path R4-to-R2 primary R4-R2
mpls set label-switched-path R4-to-R2 fast-reroute
mpls create label-switched-path R4-to-R3 to 223.0.0.3 no-cspf
mpls create policy to-R2 dst-ipaddr-mask 223.0.0.2
mpls create policy to-R3 dst-ipaddr-mask 223.0.0.3
mpls set label-switched-path R4-to-R2 policy to-R2
mpls set label-switched-path R4-to-R3 policy to-R3
mpls start
```

**RSVP configuration**
```
rsvp add interface port-41
rsvp add interface port-42
rsvp start
```

### 8.9.3 Monitoring before Network Link R2–R4 Fails

Listing 8.17 validates the two MPLS policies at Zephir-R2 defined for PBR. Policy "to–R3" specifies any packet that has a destination IP address of 223.0.0.3, and policy "to–R4" specifies any packet that has a destination IP address of 223.0.0.4.

## Listing 8.17  Zephir-R2 MPLS Policy

```
Zephir-R2# mpls show policy
```

| Name | Type | Destination | Port | Source | Port | TOS | Prot | Use |
|------|------|-------------|------|--------|------|-----|------|-----|
| To-R3 | L3 | 223.0.0.3 | Any | 0.0.0.0 | Any | Any | IP | INUSE |
| To-R4 | L3 | 223.0.0.4 | Any | 0.0.0.0 | Any | Any | IP | INUSE |

Listing 8.18 validates the two MPLS policies at Notus-R3 defined for PBR. Policy "to–R2" specifies any packet that has a destination IP address of 223.0.0.2, and policy "to–R4" specifies any packet that has a destination IP address of 223.0.0.4.

## Listing 8.18  Notus-R3 MPLS Policy

```
Notus-R3# mpls show policy
```

| Name | Type | Destination | Port | Source | Port | TOS | Prot | Use |
|------|------|-------------|------|--------|------|-----|------|-----|
| To-R2 | L3 | 223.0.0.2 | Any | 0.0.0.0 | Any | Any | IP | INUSE |
| To-R4 | L3 | 223.0.0.4 | Any | 0.0.0.0 | Any | Any | IP | INUSE |

Listing 8.19 validates the two MPLS policies at Eurus-R4 defined for PBR. Policy "to–R2" specifies any packet that has a destination IP address of 223.0.0.2, and policy "to–R3" specifies any packet that has a destination IP address of 223.0.0.3.

## Listing 8.19  Eurus-R4 MPLS Policy

```
Eurus-R4# mpls show policy
```

| Name | Type | Destination | Port | Source | Port | TOS | Prot | Use |
|------|------|-------------|------|--------|------|-----|------|-----|
| To-R2 | L3 | 223.0.0.2 | Any | 0.0.0.0 | Any | Any | IP | INUSE |
| To-R3 | L3 | 223.0.0.3 | Any | 0.0.0.0 | Any | Any | IP | INUSE |

Listing 8.20 illustrates the LSP information of Zephir-R2. To fulfill the bidirectional conversations between all three routers, six LSPs have been established:

- LSP "R2–to–R3" originating from this router (223.0.0.2) and terminating at Notus-R3 (223.0.0.3)
- Main LSP "R2–to–R4_R2–R4" originating from this router (223.0.0.2) and terminating at Eurus-R4 (223.0.0.4)
- Detour LSP "R2–to–R4_R2–R4_detour" originating from this router (223.0.0.2) and terminating at Eurus-R4 (223.0.0.4)
- LSP "R3–to–R2" originating from Notus-R3 (223.0.0.3) and terminating at this router (223.0.0.2)
- Main LSP "R4–to–R2_R4–R2" originating from Eurus-R4 (223.0.0.4) and terminating at this router (223.0.0.2)
- Detour LSP "R4–to–R2_R4–R2_detour" originating from Eurus-R4 (223.0.0.4) and terminating at this router (223.0.0.2)

---

### Listing 8.20   Zephir-R2 LSP Information

```
Zephir-R2# mpls show label-switched-path brief

Ingress LSP:

                                                      Label
LSP Name                 To          From       State Style  In    Out
R2-to-R4_R2-R4_detour    223.0.0.4   223.0.0.2  Up    FF     —     4097
R2-to-R3                 223.0.0.3   223.0.0.2  Up    FF     —     16
R2-to-R4_R2-R4           223.0.0.4   223.0.0.2  Up    FF     —     16

Transit LSP:

                                                      Label
LSP Name                 To          From       State Style  In    Out

Egress LSP:

                                                      Label
LSP Name                 To          From       State Style  In    Out
R4-to-R2_R4-R2_detour    223.0.0.2   223.0.0.4  Up    FF     16    —
R4-to-R2_R4-R2           223.0.0.2   223.0.0.4  Up    FF     16    —
R3-to-R2                 223.0.0.2   223.0.0.3  Up    FF     16    —
```

---

Listing 8.21 illustrates the LSP information of Notus-R3:

- LSP "R3–to–R4" originating from this router (223.0.0.3) and terminating at Eurus-R4 (223.0.0.4)

- LSP "R4–to–R3" originating from Eurus-R4 (223.0.0.4) and terminating at this router (223.0.0.3)
- LSP "R3–to–R2" originating from this router (223.0.0.3) and terminating at Zephir-R2 (223.0.0.2)
- LSP "R2–to–R3" originating from Zephir-R2 (223.0.0.2) and terminating at this router (223.0.0.3)
- Detour LSP "R2–to–R4_R2–R4_detour" transiting through this router (223.0.0.3)
- Detour LSP "R4–to–R2_R4–R2_detour" transiting through this router (223.0.0.3)

---

### Listing 8.21   Notus-R3 LSP Information

```
Notus-R3# mpls show label-switched-path brief


Ingress LSP:
                                                      Label
LSP Name                  To         From       State Style In    Out
R3-to-R4                  223.0.0.4  223.0.0.3  Up    FF    —     16
R3-to-R2                  223.0.0.2  223.0.0.3  Up    FF    —     16


Transit LSP:
                                                      Label
LSP Name                  To         From       State Style In    Out
R2-to-R4_R2-R4_detour     223.0.0.4  223.0.0.2  Up    FF    4097  16
R4-to-R2_R4-R2_detour     223.0.0.2  223.0.0.4  Up    FF    4097  16


Egress LSP:
                                                      Label
LSP Name                  To         From       State Style In    Out
R4-to-R3                  223.0.0.3  223.0.0.4  Up    FF    16    —
R2-to-R3                  223.0.0.3  223.0.0.2  Up    FF    16    —
```

---

Listing 8.22 illustrates the LSP information of Eurus-R4:

- LSP "R4–to–R3" originating from this router (223.0.0.4) and terminating at Notus-R3 (223.0.0.3)
- Main LSP "R4–to–R2_R4–R2" originating from this router (223.0.0.4) and terminating at Zephir-R2 (223.0.0.2)
- Detour LSP "R4–to–R2_R4–R2_detour" originating from this router (223.0.0.4) and terminating at Zephir-R2 (223.0.0.2)
- LSP "R3–to–R4" originating from Notus-R3 (223.0.0.3) and terminating at this router (223.0.0.4)

- Main LSP "R2–to–R4_R2–R4" originating from Zephir-R2 (223.0.0.2) and terminating at this router (223.0.0.4)
- Detour LSP "R2–to–R4_R2–R4_detour" originating from Zephir-R2 (223.0.0.2) and terminating at this router (223.0.0.4)

---

**Listing 8.22  Eurus-R4 LSP Information**

```
Eurus-R4# mpls show label-switched-path brief
```

Ingress LSP:

| LSP Name | To | From | State | Style | Label In | Out |
|----------|-----|------|-------|-------|----------|-----|
| R4-to-R2_R4-R2_detour | 223.0.0.2 | 223.0.0.4 | Up | FF | — | 4097 |
| R4-to-R3 | 223.0.0.3 | 223.0.0.4 | Up | FF | — | 16 |
| R4-to-R2_R4-R2 | 223.0.0.2 | 223.0.0.4 | Up | FF | — | 16 |

Transit LSP:

| LSP Name | To | From | State | Style | Label In | Out |
|----------|-----|------|-------|-------|----------|-----|

Egress LSP:

| LSP Name | To | From | State | Style | Label In | Out |
|----------|-----|------|-------|-------|----------|-----|
| R2-to-R4_R2-R4_detour | 223.0.0.4 | 223.0.0.2 | Up | FF | 16 | — |
| R2-to-R4_R2-R4 | 223.0.0.4 | 223.0.0.2 | Up | FF | 16 | — |
| R3-to-R4 | 223.0.0.4 | 223.0.0.3 | Up | FF | 16 | — |

---

Listing 8.23 illustrates the RSVP PSB of Zephir-R2. The extended tunnel ID is used to uniquely identify each pair of bidirectional LSP tunnels. However, this identifier is held the same for both the main LSP tunnel pairs and the detour LSP tunnel pairs because the detour LSP is really part of the main LSP during a failure outage, where it will eventually merge back with the main LSP at the node (or MP) immediately downstream of the point of failure. In addition:

- For main LSP tunnels "R2–to–R4_R2–R4" and "R4–to–R2_R4–R2":
  — The "local protection desired" flag (= $0 \times 01$) is set in the SESSION_ATTRIBUTE object (see section 8.4.3) so that when a fault is detected on an adjacent downstream link or node, the PLR will reroute user traffic for fast service restoration.
  — The FAST_REROUTE object (see section 8.4.1) is carried in the RSVP PATH message to indicate that each of the main LSPs requires a backup path to be precomputed and preestablished

across the components (links and nodes). This object includes information such as setup priority, holding priority, hop limit, flags, bandwidth, and session attribute filters, all of which allow transit routers to execute the CSPF process and select the proper backup paths for the main LSPs. The flags = $0 \times 01$ indicate that a one-to-one backup scheme is desired using detour LSPs. Note that the FAST_REROUTE object is only inserted into the RSVP PATH message by the ingress LSR and cannot be modified by downstream LSRs.

— The ERO for main LSP tunnel "R2–to–R4_R2–R4" is 192.168.24.4 (R4). The RSVP PATH message for this LSP is sent out of port-41 to next-hop 192.168.24.4. Zephir-R2 is the ingress LSR for this LSP.

— The RSVP PATH message for main LSP tunnel "R4–to–R2_R4–R2" is received via port-41 from PHOP 192.168.24.4. Because the tunnel endpoint 223.0.0.2 is at Zephir-R2 itself for this LSP, the next-hop is no longer required to be derived from the ERO.

- For detour LSP tunnel "R2–to–R4_R2–R4_detour":
  — The DETOUR object (see section 8.4.2) is used in the one-to-one backup scheme (indicated by flags = $0 \times 01$ in the FAST_REROUTE object) to identify the detour LSPs. The object information includes the source ID (223.0.0.2) of the PLR (Zephir-R2) requesting the detour LSP and the avoid node ID (192.168.24.4) of the immediate downstream node (Eurus-R4) that the PLR will bypass in the event of failure.
  — The ERO is 192.168.23.3 (R3)=>192.168.34.4 (R4). The RSVP PATH message for this LSP is sent out of port-42 to next-hop 192.168.23.3. Zephir-R2 is the ingress LSR for this LSP.

- For detour LSP tunnel "R4–to–R2_R4–R2_detour":
  — The DETOUR object (see section 8.4.2) is used in the one-to-one backup scheme (indicated by flags = $0 \times 01$ in the FAST_REROUTE object) to identify the detour LSPs. The object information includes the source ID (223.0.0.4) of the PLR (Eurus-R4) requesting the detour LSP and the avoid node ID (192.168.24.2) of the immediate downstream node (Zephir-R2) that the PLR will bypass in the event of failure.
  — The RSVP PATH message for this LSP tunnel is received via port-42 from PHOP 192.168.23.3. Because the tunnel endpoint 223.0.0.2 is at Zephir-R2 itself, the next-hop is no longer required to be derived from the ERO.

- For LSP tunnel "R2–to–R3," the ERO is not used because path selection is hop by hop based. The RSVP PATH message for this LSP is sent out of port-42 to next-hop 192.168.23.3. Zephir-R2 is the ingress LSR for this LSP.

- The RSVP PATH message for LSP tunnel "R3–to–R2" is received via port-42 from PHOP 192.168.23.3. Zephir-R2 is the egress LSR for this LSP.

---

**Listing 8.23   Zephir-R2 RSVP PSB**

```
Zephir-R2# mpls show rsvp psb

Path State Blocks:

RSVP_PSB <rsvp_1>:
session-attr: name: R2-to-R4_R2-R4 flags: 0x1 setup-pri: 7 holding-pri: 0
session: end-point: 223.0.0.4 tunnel-id: 16385 ext-tunnel-id:
0xdf000002
send-templ: sender: 223.0.0.2 lsp-id: 1
prev-hop: 0.0.0.0
in-if: <Local-API> out-if: <port-41>
explicit-route: 192.168.24.4
fast-reroute: setup-pri: 7 hold-pri: 7 hop-limit: 5 flags: 0x1
bandwidth: 0x0 include: 0x0 exclude: 0x0


RSVP_PSB <rsvp_1>:
session-attr: name: R4-to-R2_R4-R2 flags: 0x1 setup-pri: 7 holding-pri: 0
session: end-point: 223.0.0.2 tunnel-id: 16385 ext-tunnel-id:
0xdf000004
send-templ: sender: 223.0.0.4 lsp-id: 1
prev-hop: 192.168.24.4
in-if: <port-41> out-if: <Local-API>
explicit-route:
fast-reroute: setup-pri: 7 hold-pri: 7 hop-limit: 5 flags: 0x1
bandwidth: 0x0 include: 0x0 exclude: 0x0


RSVP_PSB <rsvp_1>:
session-attr: name: R2-to-R4_R2-R4_detour flags: 0x0 setup-pri: 7
holding-pri: 7
session: end-point: 223.0.0.4 tunnel-id: 16385 ext-tunnel-id:
0xdf000002
send-templ: sender: 223.0.0.2 lsp-id: 1
prev-hop: 0.0.0.0
in-if: <Local-API> out-if: <port-42>
explicit-route: 192.168.23.3=>192.168.34.4
detour: source-id: 223.0.0.2 avoid-node-id: 192.168.24.4


RSVP_PSB <rsvp_1>:
```

```
session-attr: name: R4-to-R2_R4-R2_detour flags: 0x0 setup-pri: 7
holding-pri: 7
session: end-point: 223.0.0.2 tunnel-id: 16385 ext-tunnel-id:
0xdf000004
send-templ: sender: 223.0.0.4 lsp-id: 1
prev-hop: 192.168.23.3
in-if: <port-42> out-if: <Local-API>
explicit-route:
detour: source-id: 223.0.0.4 avoid-node-id: 192.168.24.2


RSVP_PSB <rsvp_1>:
session-attr: name: R2-to-R3 flags: 0x0 setup-pri: 7 holding-pri: 0
session: end-point: 223.0.0.3 tunnel-id: 6 ext-tunnel-id: 0xdf000002
send-templ: sender: 223.0.0.2 lsp-id: 1
prev-hop: 0.0.0.0
in-if: <Local-API> out-if: <port-42>
explicit-route:


RSVP_PSB <rsvp_1>:
session-attr: name: R3-to-R2 flags: 0x0 setup-pri: 7 holding-pri: 0
session: end-point: 223.0.0.2 tunnel-id: 6 ext-tunnel-id: 0xdf000003
send-templ: sender: 223.0.0.3 lsp-id: 1
prev-hop: 192.168.23.3
in-if: <port-42> out-if: <Local-API>
explicit-route:
```

Listing 8.24 illustrates the RSVP PSB of Notus-R3:

- For detour LSP tunnel "R2–to–R4_R2–R4_detour," the ERO is 192.168.34.4 (R4). The RSVP PATH message for this LSP is received at port-54 from PHOP 192.168.23.2 and is sent out of port-53 to next-hop 192.168.34.4. Notus-R3 is the transit LSR for this LSP. The DETOUR object attributes are the same as those stated in Listing 8.23.
- For detour LSP tunnel "R4–to–R2_R4–R2_detour," the ERO is 192.168.23.2 (R2). The RSVP PATH message for this LSP is received at port-53 from PHOP 192.168.34.4 and is sent out of port-54 to next-hop 192.168.23.2. Notus-R3 is the transit LSR for this LSP. The DETOUR object attributes are the same as those stated in Listing 8.23.
- For LSP tunnel "R3–to–R2," the ERO is not used because path selection is hop by hop based. The RSVP PATH message for this LSP is sent out of port-54 to next-hop 192.168.23.2. Notus-R3 is the ingress LSR for this LSP.

- The RSVP PATH message for LSP tunnel "R2–to–R3" is received via port-54 from PHOP 192.168.23.2. Notus-R3 is the egress LSR for this LSP.
- For LSP tunnel "R3–to–R4," the ERO is not used because path selection is hop by hop based. The RSVP PATH message for this LSP is sent out of port-53 to next-hop 192.168.34.4. Notus-R3 is the ingress LSR for this LSP.
- The RSVP PATH message for LSP tunnel "R4–to–R3" is received via port-53 from PHOP 192.168.34.4. Notus-R3 is the egress LSR for this LSP.

---

**Listing 8.24    Notus-R3 RSVP PSB**

```
Notus-R3# rsvp show psb


Path State Blocks:


RSVP_PSB <rsvp_1>:
session-attr: name: R2-to-R4_R2-R4_detour flags: 0x0 setup-pri: 7
holding-pri: 7
session: end-point: 223.0.0.4 tunnel-id: 16385 ext-tunnel-id:
0xdf000002
send-templ: sender: 223.0.0.2 lsp-id: 1
prev-hop: 192.168.23.2
in-if: <port-54> out-if: <port-53>
explicit-route: 192.168.34.4
detour: source-id: 223.0.0.2 avoid-node-id: 192.168.24.4


RSVP_PSB <rsvp_1>:
session-attr: name: R4-to-R2_R4-R2_detour flags: 0x0 setup-pri: 7
holding-pri: 7
session: end-point: 223.0.0.2 tunnel-id: 16385 ext-tunnel-id:
0xdf000004
send-templ: sender: 223.0.0.4 lsp-id: 1
prev-hop: 192.168.34.4
in-if: <port-53> out-if: <port-54>
explicit-route: 192.168.23.2
detour: source-id: 223.0.0.4 avoid-node-id: 192.168.24.2


RSVP_PSB <rsvp_1>:
session-attr: name: R3-to-R2 flags: 0x0 setup-pri: 7 holding-pri: 0
session: end-point: 223.0.0.2 tunnel-id: 6 ext-tunnel-id: 0xdf000003
send-templ: sender: 223.0.0.3 lsp-id: 1
prev-hop: 0.0.0.0
in-if: <Local-API> out-if: <port-54>
```

```
explicit-route:

RSVP_PSB <rsvp_1>:
session-attr: name: R2-to-R3 flags: 0x0 setup-pri: 7 holding-pri: 0
session: end-point: 223.0.0.3 tunnel-id: 6 ext-tunnel-id: 0xdf000002
send-templ: sender: 223.0.0.2 lsp-id: 1
prev-hop: 192.168.23.2
in-if: <port-54> out-if: <Local-API>
explicit-route:

RSVP_PSB <rsvp_1>:
session-attr: name: R3-to-R4 flags: 0x0 setup-pri: 7 holding-pri: 0
session: end-point: 223.0.0.4 tunnel-id: 7 ext-tunnel-id: 0xdf000003
send-templ: sender: 223.0.0.3 lsp-id: 1
prev-hop: 0.0.0.0
in-if: <Local-API> out-if: <port-53>
explicit-route:

RSVP_PSB <rsvp_1>:
session-attr: name: R4-to-R3 flags: 0x0 setup-pri: 7 holding-pri: 0
session: end-point: 223.0.0.3 tunnel-id: 7 ext-tunnel-id: 0xdf000004
send-templ: sender: 223.0.0.4 lsp-id: 1
prev-hop: 192.168.34.4
in-if: <port-53> out-if: <Local-API>
explicit-route:
```

Listing 8.25 illustrates the RSVP PSB of Eurus-R4:

- For main LSP tunnels "R4–to–R2_R4–R2" and "R2–to–R4_R2–R4":
  — The "local protection desired" flag (= 0 × 01) is set in the SESSION_ATTRIBUTE object (see section 8.4.3) so that when a fault is detected on an adjacent downstream link or node, the PLR will reroute user traffic for fast service restoration.
  — The FAST_REROUTE object (see section 8.4.1) is carried in the RSVP PATH message to indicate that each of the main LSPs requires a backup path to be precomputed and preestablished across the components (links and nodes). This object includes information such as setup priority, holding priority, hop limit, flags, bandwidth, and session attribute filters, all of which allow transit routers to execute the CSPF process and select the proper backup paths for the main LSPs. The flags = 0 × 01 indicate that a one-to-one backup scheme is desired using detour LSPs. Note that the FAST_REROUTE object is only inserted into the RSVP

PATH message by the ingress LSR and cannot be modified by downstream LSRs.

— The ERO for main LSP tunnel "R4–to–R2_R4–R2" is 192.168.24.2 (R2). The RSVP PATH message for this LSP is sent out of port-42 to next-hop 192.168.24.2. Eurus-R4 is the ingress LSR for this LSP.

— The RSVP PATH message for main LSP tunnel "R2–to–R4_R2–R4" is received via port-42 from PHOP 192.168.24.2. Because the tunnel endpoint 223.0.0.4 is at Eurus-R4 itself for this LSP, the next-hop is no longer required to be derived from the ERO.

- For detour LSP tunnel "R4–to–R2_R4–R2_detour," the ERO is 192.168.34.3 (R3)=>192.168.23.2 (R2). The RSVP PATH message for this LSP is sent out of port-41 to next-hop 192.168.34.3. Eurus-R4 is the ingress LSR for this LSP. The DETOUR object attributes are the same as those stated in Listing 8.23.

- The RSVP PATH message for detour LSP tunnel "R2–to–R4_R2–R4_detour" is received via port-41 from PHOP 192.168.34.3. Because the tunnel endpoint 223.0.0.4 is at Eurus-R4 itself, the next-hop is no longer required to be derived from the ERO. The DETOUR object attributes are the same as those stated in Listing 8.23.

- For LSP tunnel "R4–to–R3," the ERO is not used because path selection is hop by hop based. The RSVP PATH message for this LSP is sent out of port-41 to next-hop 192.168.34.3. Eurus-R4 is the ingress LSR for this LSP.

- The RSVP PATH message for LSP tunnel "R3–to–R4" is received via port-41 from PHOP 192.168.34.3. Eurus-R4 is the egress LSR for this LSP.

---

**Listing 8.25    Eurus-R4 RSVP PSB**

```
Eurus-R4# rsvp show psb

Path State Blocks:

RSVP_PSB <rsvp_1>:
session-attr: name: R4-to-R2_R4-R2 flags: 0x1 setup-pri: 7 holding-pri: 0
session: end-point: 223.0.0.2 tunnel-id: 16385 ext-tunnel-id:
0xdf000004
send-templ: sender: 223.0.0.4 lsp-id: 1
prev-hop: 0.0.0.0
in-if: <Local-API> out-if: <port-42>
explicit-route: 192.168.24.2
fast-reroute: setup-pri: 7 hold-pri: 7 hop-limit: 5 flags: 0x1
bandwidth: 0x0 include: 0x0 exclude: 0x0
```

```
RSVP_PSB <rsvp_1>:
session-attr: name: R2-to-R4_R2-R4 flags: 0x1 setup-pri: 7 holding-pri: 0
session: end-point: 223.0.0.4 tunnel-id: 16385 ext-tunnel-id:
0xdf000002
send-templ: sender: 223.0.0.2 lsp-id: 1
prev-hop: 192.168.24.2
in-if: <port-42> out-if: <Local-API>
explicit-route:
fast-reroute: setup-pri: 7 hold-pri: 7 hop-limit: 5 flags: 0x1
bandwidth: 0x0 include: 0x0 exclude: 0x0


RSVP_PSB <rsvp_1>:
session-attr: name: R4-to-R2_R4-R2_detour flags: 0x0 setup-pri: 7
holding-pri: 7
session: end-point: 223.0.0.2 tunnel-id: 16385 ext-tunnel-id:
0xdf000004
send-templ: sender: 223.0.0.4 lsp-id: 1
prev-hop: 0.0.0.0
in-if: <Local-API> out-if: <port-41>
explicit-route: 192.168.34.3=>192.168.23.2
detour: source-id: 223.0.0.4 avoid-node-id: 192.168.24.2


RSVP_PSB <rsvp_1>:
session-attr: name: R2-to-R4_R2-R4_detour flags: 0x0 setup-pri: 7
holding-pri: 7
session: end-point: 223.0.0.4 tunnel-id: 16385 ext-tunnel-id:
0xdf000002
send-templ: sender: 223.0.0.2 lsp-id: 1
prev-hop: 192.168.34.3
in-if: <port-41> out-if: <Local-API>
explicit-route:
detour: source-id: 223.0.0.2 avoid-node-id: 192.168.24.4


RSVP_PSB <rsvp_1>:
session-attr: name: R4-to-R3 flags: 0x0 setup-pri: 7 holding-pri: 0
session: end-point: 223.0.0.3 tunnel-id: 7 ext-tunnel-id: 0xdf000004
send-templ: sender: 223.0.0.4 lsp-id: 1
prev-hop: 0.0.0.0
in-if: <Local-API> out-if: <port-41>
explicit-route:


RSVP_PSB <rsvp_1>:
session-attr: name: R3-to-R4 flags: 0x0 setup-pri: 7 holding-pri: 0
session: end-point: 223.0.0.4 tunnel-id: 7 ext-tunnel-id: 0xdf000003
send-templ: sender: 223.0.0.3 lsp-id: 1
prev-hop: 192.168.34.3
```

```
in-if: <port-41> out-if: <Local-API>
explicit-route:
```

For brevity, the RSVP RSB listings are omitted in this case study.

Listing 8.26 illustrates the result of the traceroute test performed at Zephir-R2. The traceroute output validates that user traffic (source 223.0.0.2) is currently traversing network link R2–R4 to reach destination 223.0.0.4.

### Listing 8.26   Traceroute Test at Zephir-R2

```
Zephir-R2# traceroute 223.0.0.4 source 223.0.0.2

traceroute to 223.0.0.4 (223.0.0.4) from 223.0.0.2, 30 hops max, 40 byte
packets
1 223.0.0.4 (223.0.0.4) 0 ms 0 ms 0 ms
```

Listing 8.27 illustrates the result of the traceroute test performed at Eurus-R4. The traceroute output validates that return traffic (source 223.0.0.4) is currently traversing network link R2–R4 to reach destination 223.0.0.2.

### Listing 8.27   Traceroute Test at Eurus-R4

```
Eurus-R4# traceroute 223.0.0.2 source 223.0.0.4

traceroute to 223.0.0.2 (223.0.0.2) from 223.0.0.4, 30 hops max, 40 byte
packets
1 223.0.0.2 (223.0.0.2) 1 ms 0 ms 0 ms
```

### 8.9.4   Monitoring after Network Link R2–R4 Fails

As illustrated in Figure 8.14, when network link R2–R4 fails, the detour path taken by user traffic traversing the established LSP becomes R2–R3–R4.

Listing 8.28 illustrates the LSP information of Zephir-R2 (the PLR) after the link failure. The main LSP "R2–to–R4" is now in the detour state, indicating that Zephir-R2 is rerouting all user traffic heading for destination 223.0.0.4 and traversing failed link R2–R4 onto detour LSP "R2–to–R4_R2–R4_detour."

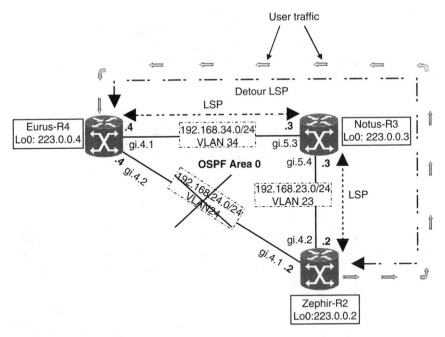

**Figure 8.14    Local Repairing with Detour LSP**

---

## Listing 8.28    Zephir-R2 LSP Information

```
Zephir-R2# mpls show label-switched-path brief


Ingress LSP:
                                                           Label
LSP Name                    To         From       State   Style  In    Out
R2-to-R4_R2-R4_detour       223.0.0.4  223.0.0.2  Up      FF     —     4097
R2-to-R3                    223.0.0.3  223.0.0.2  Up      FF     —     16
R2-to-R4                    223.0.0.4  223.0.0.2  Detour  FF     —     —


Transit LSP:
                                                           Label
LSP Name                    To         From       State   Style  In    Out


Egress LSP:
                                                           Label
LSP Name                    To         From       State   Style  In    Out
R4-to-R2_R4-R2_detour       223.0.0.2  223.0.0.4  Up      FF     16    —
R3-to-R2                    223.0.0.2  223.0.0.3  Up      FF     16    —
```

Listing 8.29 illustrates the LSP information of Eurus-R4 (the PLR) after the link failure. The main LSP "R4–to–R2" is now in the detour state, indicating that Eurus-R4 is rerouting all return traffic heading for destination 223.0.0.2 and traversing failed link R2–R4 onto detour LSP "R4–to–R2_R4–R2_detour."

---

### Listing 8.29   Eurus-R4 LSP Information

```
Eurus-R4# mpls show label-switched-path brief


Ingress LSP:

                                                           Label
LSP Name                 To         From       State  Style In     Out
R4-to-R2_R4-R2_detour    223.0.0.2  223.0.0.4  Up     FF    —      4097
R4-to-R3                 223.0.0.3  223.0.0.4  Up     FF    —      16
R4-to-R2                 223.0.0.2  223.0.0.4  Detour FF    —      —

Transit LSP:

                                                           Label
LSP Name                 To         From       State  Style In     Out

Egress LSP:

                                                           Label
LSP Name                 To         From       State  Style In     Out
R2-to-R4_R2-R4_detour    223.0.0.4  223.0.0.2  Up     FF    16     —
R3-to-R4                 223.0.0.4  223.0.0.3  Up     FF    16     —
```

---

Listing 8.30 illustrates the result of the traceroute test performed at Zephir-R2 after the link failure. The traceroute output validates that user traffic (source 223.0.0.2) is traversing detour path R2–R3–R4 to reach destination 223.0.0.4.

---

### Listing 8.30   Traceroute Test at Zephir-R2

```
Zephir-R2# traceroute 223.0.0.4 source 223.0.0.2

traceroute to 223.0.0.4 (223.0.0.4) from 223.0.0.2, 30 hops max, 40 byte
packets
1  192.168.23.3 (192.168.23.3) 3 ms 1 ms 0 ms
   MPLS Label1=4097 EXP1=0 TTL=1 S=1
2  223.0.0.4 (223.0.0.4) 2 ms 1 ms 0 ms
```

---

Listing 8.31 illustrates the result of the traceroute test performed at Eurus-R4 after the link failure. The traceroute output validates that return traffic (source 223.0.0.4) is traversing detour path R4–R3–R2 to reach destination 223.0.0.2.

---

### Listing 8.31   Traceroute Test at Eurus-R4

```
Eurus-R4# traceroute 223.0.0.2 source 223.0.0.4

traceroute to 223.0.0.2 (223.0.0.2) from 223.0.0.4, 30 hops max, 40 byte
packets
1 192.168.34.3 (192.168.34.3) 3 ms 1 ms 0 ms
  MPLS Label1=4097 EXP1=0 TTL=1 S=1
2 223.0.0.2 (223.0.0.2) 2 ms 0 ms 0 ms
```

---

## 8.10 SUMMARY

Network reliability is mandatory in high-speed networks such as Gigabit Ethernet MANs. Providing protections against link or node failures is an important requirement for the successful delivery of metro services. When a large quantity of mission-critical user traffic traverses through the SP's MAN, the effect of disruptions caused by network failure outages can become very significant and extremely costly.

Disruption can occur due to reasons such as congestion along a link or path, link failure, node failure, and making administrative changes to a path. This chapter discusses one of the most appealing features of MPLS-TE: the possibility of providing nondisruptive traffic across the LSP. In the event of outage, the upper-level applications will not be aware of any service disruption posed. With MPLS-TE, several cost-effective alternatives are available for path restoration:

- MPLS fast reroute—link protection (local repair technique)
- MPLS fast reroute—node protection (local repair technique)
- Path protection (global repair technique)

These options allow the metro SP to conceal occurring network faults from the metro subscribers and to satisfy their demanding service-level agreements. The two case studies in this chapter examine the path protection and link protection schemes.

In general, the duration of an outage can range from milliseconds, with minor service impact, to seconds, with possible call drops for IP telephony and session time-outs for connection-oriented transactions, to minutes and hours, with drastic social and business impact. Rerouting at the IP layer occurs after a period of routing convergence, which may require seconds to minutes to complete. Because MPLS is path oriented, it can potentially provide faster (typically 50 milliseconds or more) and more predictable protection and restoration capabilities than conventional hop-by-hop IP forwarding.

The TE and fast reroute features of MPLS and RSVP-TE make it possible to achieve recovery at the IP layer of a MAN prior to convergence. Therefore, the reliability aspect of MAN can be enhanced tremendously with MPLS-TE.

# Part 4
# Service Aspect of Metropolitan Area Networks

# Chapter 9
# Layer-3 and Layer-2 MPLS VPNs

## 9.1 INTRODUCTION

In the past, service providers (SPs) always had the dilemma of choosing between a switched or routed architectural design for their networks. In open system interconnection (OSI) terms, this means an impasse between operating functions at layer-2 (L2) or layer-3 (L3). This stalemate resurfaces once more when Multi-Protocol Label Switching (MPLS) is used to provide virtual private network (VPN) services over metropolitan area networks (MANs) for metro subscribers.

Before making a sensible choice between an MPLS VPN implemented at L2 or L3, it is important that we understand how these VPNs work as well as their benefits and limitations. We might think of the public network as a maze of streets and roads in a metropolis. A L3 MPLS VPN is like a separate subway system that must be boarded to traverse the city, whereas the L2 approach is similar to a sequence of car tunnels that go straight through the city without any transition to the subway.

Regardless of whether MPLS VPNs are L2 or L3 based, they play the critical role of segregating multiple users/customers with diverse service requirements into different logical groupings. In other words, the use of MPLS VPNs as user/customer classifiers set aside the foundation for service differentiation based on different users/customers, which in turn provides metro SPs the necessary impetus to incorporate more refined service differentiation mechanisms. Moreover, MPLS VPNs provide the fundamental constructs to converge disparate networks into a unified network core, thus forming the basis for any-to-any connectivity.

## 9.2 L3 MPLS VPN OVERVIEW

L3 MPLS VPN is based on [RFC2547] and the draft of [RFC2547bis]. Hence, they are commonly referred to as 2547 VPNs. How does a 2547 VPN work? From the architectural perspective, the 2547 VPN is divided into three main functional components (see Figure 9.1):

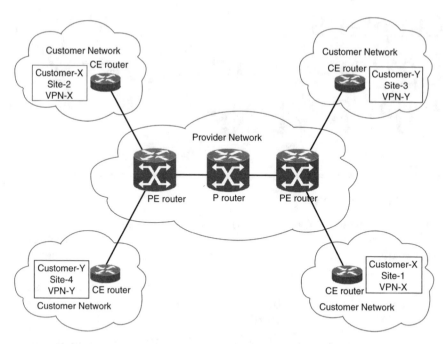

**Figure 9.1    Fundamental Building Blocks of a 2547 VPN**

- The customer edge (CE) router that forms part of the customer network (C-network) and interfaces to a PE router
- The provider edge (PE) router that serves as the edge router (or edge label switch router (LSR)) of the provider network (P-network) and interfaces to CE routers
- The provider (P) router, which is the core router (or core-LSR) in the P-network that provides transit transport across the provider or L3 MPLS VPN backbone but has no knowledge of VPNs

The relationship between the PE router and the CE router is the most unique feature of 2547 VPNs. Each 2547 VPN is a private IP network with each of the PE routers directly connected to the CE routers at the customer site. The CE router provides the PE router with route information of the private network and becomes a peer to the PE router (and not a peer to other remote CE routers). Put simply, the 2547 VPN is a true peer-to-peer model that coalesces the security and isolation among customers with simplified customer routing.

As illustrated in Figure 9.2, customer isolation is achieved through the concept of virtual routing and forwarding (VRF) instances in which the PE router is subdivided into virtual routers serving different VPNs or customer sites. Hence, a PE router must be capable of storing multiple VPN

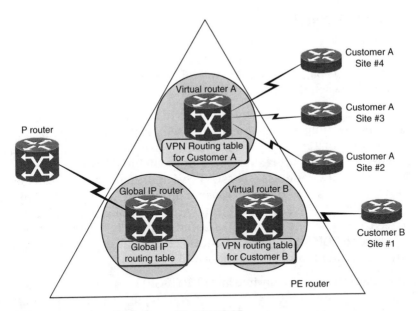

**Figure 9.2   Customer Isolation with VRF Tables**

routing tables, one for each customer connection, along with the regular Internet forwarding information exchanged between the P routers and the rest of the PE routers via a core interior gateway protocol (IGP) such as open shortest path first (OSPF) or integrated intermediate system to intermediate system (ISIS).

Because each customer is assigned an independent private routing table by the PE router, the deployment of overlapping address spaces in different customer sites is now possible. The route to each of these sites on the private network is distributed over the backbone (or the P-network to be exact) using Border Gateway Protocol (BGP), and MPLS is used to forward data packets over the backbone. In other words, MPLS handles the actual forwarding between the P and PE routers on a 2547 network. With MPLS responsible for the forwarding role, the P routers in the backbone no longer need to know about the VPN routes that are propagated from the customer sites.

## 9.3   ARCHITECTURAL COMPONENTS OF L3 MPLS VPN

The MPLS VPN architecture relies heavily on two novel components: route distinguisher (RD) and route target (RT). RDs enable overlapping customer address space, and RTs allow the implementation of complex VPN topologies that are difficult to implement with other VPN architectures.

### 9.3.1 Route Distinguisher

Even though BGP is chosen for its scaling capability, one important question still arises when BGP carries customer routes between the PE routers: How can BGP propagate several identical prefixes, belonging to different customers, between the PE routers?

The L3 MPLS VPN architecture supports overlapping customer address space by prepending the customer's IP prefix with a unique 64-bit prefix called the route distinguisher (RD). The RD is used to convert a nonunique 32-bit customer IPv4 address into a unique 96-bit VPNv4 address (also known as a VPN-IPv4 address). These VPNv4 addresses are only exchanged between PE routers, and they are transparent to the customer. This implies that the standard BGP [RFC1771] running between PE routers has to be extended to support the exchange of both traditional IPv4 and VPNv4 prefixes. Extended BGP that supports address families besides IPv4 addresses is known as multi-protocol BGP (MP-BGP [RFC2858]).

MP-BGP sessions are really internal BGP (IBGP) sessions except that MP-BGP is an extension to BGP that carries routing information about other protocols and address families (in this case, VPNv4 addresses). MP-BGP sessions can also be referred to as VPNv4 BGP sessions.

### 9.3.2 Route Target

Besides converting overlapping IPv4 addresses into globally unique VPNv4 addresses, the RD has no other special role in the MPLS VPN architecture. Nevertheless, simple nonoverlapping VPN topologies may require only one RD per customer. In such cases, the RD can also be considered a VPN identifier.

The problem becomes more obvious in the implementation of more complex VPN topologies when a customer site can belong to more than one VPN (typically variants of overlapping VPNs). The L3 MPLS VPN can be thought of as a community of interest or closed user group (CUG), which is simply a collection of sites sharing common routing information. Through this sharing of routing information, a site may belong to more than one VPN, resulting in differing routing requirements for sites that belong to different sets of VPNs. These routing requirements are supported with multiple VRF tables on the PE routers with the help of route targets (RTs), which are really BGP extended communities [DRAFT-BGP-EXT-COMM] attached to a VPNv4 route to indicate its VPN membership. RTs are carried in an MP-BGP update.

There are two types of RT: export and import. Export RTs are appended to a customer route when it is converted from an IPv4 route to a VPNv4 route. They are used to identify the set of remote sites to which a particular route should be exported and are configured separately for each VRF table in a PE router. Any number of export RTs can be attached to a single VPNv4 route. When VPNv4 routes are advertised to other PE routers, these routers need to select which routes to import into their VRF tables. This selection is based on import RTs. Each VRF table in a PE router can have various import RTs configured, indicating the set of VPNs from which a particular VRF table is accepting routes. In short, VRF tables use RTs to import desired routes and filter unwanted routes.

## 9.4 L3 MPLS VPN OPERATION

The entire L3 MPLS VPN operation is accomplished in two phases:

- *Phase 1*: Propagation of VPN routes and distribution of MPLS labels
- *Phase 2*: Packet forwarding

These two phases of operation are discussed in greater length in the following subsections.

### 9.4.1 Route Propagation and Label Distribution

Route propagation typically takes place in three different stages:

- *Stage 1*: PE routers receive IPv4 routing updates from CE routers and populate these routes into the appropriate VRF table. These routing updates are exchanged via PE-CE routing protocols running between the PE and CE routers. The PE-CE routing protocols include static routes, RIP version 2 (RIPv2), open shortest path first (OSPF), and external BGP (EBGP).
- *Stage 2*: PE routers export (or redistribute) VPN routes from VRF tables into MP-IBGP and propagate them as VPNv4 routes via MP-IBGP to other remote PE routers. A full mesh of MP-IBGP sessions is required between the PE routers, or BGP route reflectors [RFC2796] can be used to mitigate the IBGP full-mesh requirement. An MP-IBGP update is composed of the following:
  — VPNv4 address
  — BGP extended communities such as route targets and site of origin (optional)
  — MPLS VPN label used for VPNv4 packet forwarding
  — Any other BGP attributes, for instance, AS path, local preference, multi-exit discriminator (MED), or BGP standard community (see [RFC1771] and [RFC1997])
- *Stage 3*: The remote PE routers on receiving MP-IBGP updates will import the incoming VPNv4 routes into their respective VRF tables when the route targets (RTs) attached to these routes match with the import RTs configured in the VRF tables. The VPNv4 routes installed in VRF tables are then converted back to IPv4 routes and propagated to the CE routers. The optional site-of-origin (SOO)

attribute attached to the VPNv4 route controls the IPv4 route advertisement to the CE routers. A route installed into a VRF is not advertised to a CE router if the SOO attached to the route is identical to the SOO attribute associated with the CE router.

Two label distribution protocols and thus a two-level label stack are used for the L3 MPLS VPN operation. The egress PE routers allocate a VPN label to every VPN route received from their attached CE routers and to every summary route summarized within the PE router. These locally assigned VPN labels are then propagated together with the VPNv4 prefixes via MP-IBGP to all other remote PE routers (see stage 1 of route propagation). From the perspective of MPLS, each destination VPN is considered a forwarding equivalence class (FEC), and MP-IBGP is the label distribution protocol used for the allocation and binding of VPN labels to these FECs.

The VPN label must be allocated by the egress PE router, which also happens to be the BGP next-hop. The VPN label can only be identified by the egress PE router that originates it. Therefore, the BGP next-hop (the egress PE router) must be reachable via the common backbone IGP running between the P and PE routers. In other words, BGP next-hops are not advertised as BGP routes; instead, they are announced by the common backbone IGP running between the P and PE routers. In this case, the BGP next-hop is also considered an FEC, which must first be reached before proceeding to the designated VPN. Resource Reservation Protocol with TE extensions (RSVP-TE) (see chapter 6) or Label Distribution Protocol (LDP) [RFC3036] can be used as the label distribution protocol for the allocation and binding of labels to the respective BGP next-hops.

The VPN label and the BGP next-hop label together form the two-level label stack with the VPN label at the bottom and the BGP next-hop label on top. Figure 9.3 illustrates the route propagation and label distribution processes. In this example, a VPN label of value 58 is assigned to the network prefix 172.19.1.0/24 received from CE-1 by the egress PE router PE-1. The VPN label is propagated to the ingress PE router PE-2 together with prefix 172.19.1.0/24 in an MP-IBGP update, and a label stack is built in PE-2's VRF table. The PE and P routers learned the reachable BGP next-hop (PE-1's loopback address 192.168.1.9) through the backbone IGP, and labels are distributed through LDP or RSVP-TE corresponding to the BGP next-hop address 192.168.1.9.

### 9.4.2 Packet Forwarding

In packet forwarding, each VPN packet is marked with a label stack by the ingress PE router. The ingress PE router has two label stack entries (derived earlier in section 9.4.1) associated with each remote VPN route. The top label is bound to the BGP next-hop (belonging to the egress PE

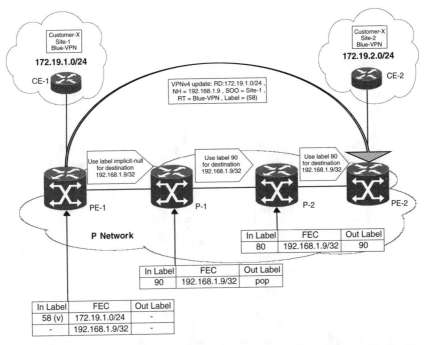

**Figure 9.3    L3 MPLS VPN Route Propagation and Label Distribution Example**

router) assigned by the next-hop P router. The bottom label is the VPN label. Because the egress PE router allocates a VPN label to every VPN route received from its attached CE routers, the labeled packet is forwarded in the direction of the egress PE router via a label switched path (LSP).

The egress PE router's upstream neighbor (penultimate hop) removes the top label on the packet (penultimate hop popping (PHP)) before forwarding it to the egress PE router. When the egress PE router receives the packet, it does a label lookup on the remaining bottom VPN label to determine an outgoing interface or VRF table, pops the label, and either forwards the packet toward the CE router through the outgoing interface or sends out the packet to the destination VPN site after performing another IP lookup in the VRF table to determine the correct L2 encapsulation. Figure 9.4 illustrates an example of the operation of the L3 MPLS VPN packet forwarding. This example is a continuation of the previous example shown in Figure 9.3.

In the example, the ingress PE router PE-2 receives a normal IP packet with a destination address of 172.19.1.1. PE-2 does an "IP prefix longest match" lookup from its "blue" VPN VRF table, finds the BGP next-hop PE-1 (with loopback address 192.168.1.9), and pushes the label stack <top BGP

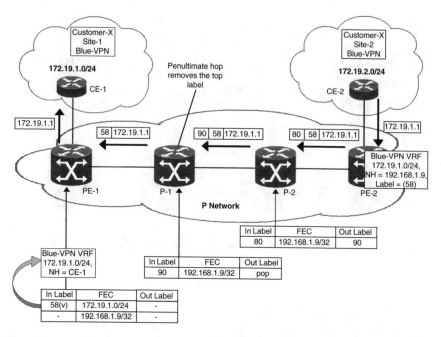

**Figure 9.4  L3 MPLS VPN Packet Forwarding Example**

next-hop label, bottom VPN label> = <80,58> derived in Figure 9.3. In this case, the BGP next-hop address 192.168.1.9 is reachable through the common backbone IGP running among the P and PE routers with an associated next-hop label 80.

Next, the packet is forwarded to P router P-2, which proceeds to switch the packet based solely on the top label. In this instance, label 80 is swapped with another next-hop label 90 and is then forwarded to P router P-1. P-1 is the penultimate hop, which does a PHP to pop the top label 90.

The egress PE router PE-1 then receives a labeled packet that contains only the bottom VPN label 58, which was previously assigned by itself to network prefix 172.19.1.0/24 received from CE-1. Based on this VPN label, PE-1 determines the outgoing interface to forward the packet to CE router CE-1, which is the next-hop to prefix 172.19.1.0/24. Finally, the VPN label is popped and the packet is forwarded toward CE-1 through the designated outgoing interface.

## 9.5  L2 MPLS VPN OVERVIEW

The L3 MPLS VPN implementation typically requires the use of a separate shared infrastructure for the core network, which will pose a migration issue if the current infrastructure only supports non-MPLS-based L2 VPNs.

L2 MPLS VPN is introduced to tackle this constraint. The L2 MPLS VPN model is used to extend, rather than replace, existing legacy L2 VPNs. Instead of building a separate IP network and running traffic across it, L2 MPLS VPNs take the existing L2 traffic and send it through point-to-point tunnels or LSPs on the MPLS core network.

Both L2 and L3 MPLS VPNs rely on MPLS transport through the core. The main difference lies in the association between the PE and CE routers. In a L2 MPLS VPN, the PE router is not a peer to the CE router, and it does not maintain separate routing tables. It simply maps incoming L2 traffic onto the appropriate LSPs. The L2 MPLS VPN still uses the overlay model (see [REF01]) commonly adopted by most traditional L2 VPN implementations. This is contrary to the L3 MPLS VPN, which adopts the peer-to-peer model (see [REF01]) instead.

## 9.6 MARTINI POINT-TO-POINT TUNNELING APPROACH

The method for establishing LSPs in an MPLS network that can handle a variety of L2 traffic is imperative to the L2 MPLS VPN model. Today, the industry is standardizing on the Martini drafts ([DRAFT-MARTINI-ENCAP] and [DRAFT-MARTINI-TRANSP]), which define point-to-point encapsulation mechanisms for Ethernet, Ethernet virtual local area network (VLAN), Frame Relay, Asynchronous Transfer Mode (ATM), point-to-point protocol (PPP), and high-level data link control (HDLC) traffic as well as the methods for transporting these L2 frames across the MPLS network. The Martini approach is just a fancy way of implementing bridging or L2 tunneling over traditional L2 media using MPLS; nevertheless, it also offers a simple way to consolidate disparate networks and provide any-to-any connectivity.

Together the two Martini drafts form the basis for point-to-point L2 services across an MPLS network, similar in concept to ATM. The point-to-point service is facilitated through a pair of LSPs in opposite directions, forming a single virtual pipe, which can be perceived as a trunk carrying multiple virtual channels belonging to different customers. This virtual pipe is analogous to a virtual path (VP) in ATM, which consists of multiple virtual channels (VCs) going to the same endpoint, each with its own unique identifier. A virtual path identifier (VPI) is used to identify the VP, and a virtual channel identifier (VCI) is used to represent each VC that is carried within the VP. In other words, each VPI denotes an ATM link connection, carrying multiple virtual channels, each of which is identified by a VCI.

### 9.6.1 VC Label

The Martini approach uses the ATM concept of VCs (in the context of Martini L2 VPN, VCs refer to virtual circuits instead of virtual channels) to link up identical L2 VPNs located at different sites. A pair of VC LSPs in reverse directions, each of which is represented by a VC label, is used to denote the VC. An L2 FEC is used to identify each L2 VPN. The L2 FEC can be based on port ID, VLAN ID, or both. These parameters are assigned to individual VC ID or group ID (optional) defined in the VC FEC element of type 128 (see [DRAFT-MARTINI-TRANSP]). Both the VC and group IDs are 32 bits in length. The group ID represents a collection of VC IDs. When the group ID is not available, the VC label is mapped to the VC ID. Otherwise, the VC label is mapped to the group ID.

The VC label bindings are distributed using LDP. The PE routers will establish an LDP session using the extended discovery mechanism [RFC3036] because they are most likely nonadjacent neighbors or remote peers. Once the LDP session has been established, VC labels are exchanged to identify the components of the individual L2 FEC. The signaling of the VC label is performed using LDP extensions to the Label Mapping Message [RFC3036]. Specifically, when a PE router is configured with a new L2 FEC, an LDP label (VC label) is locally generated, and using the LDP extensions to the Label Mapping Message, the information relating to the L2 FEC, such as VC ID and group ID, is forwarded to the remote LDP peer or PE router.

### 9.6.2 Tunnel Label

The tunnel LSP, represented by a tunnel label, is the point-to-point connection that is used to scale the core network by aggregating (or nesting) multiple VCs belonging to different customers into a single common tunnel. A model that implements a tunnel per customer or a tunnel per customer per service is bound to run into scalability issues early in the deployment. This is the rationale behind using a common tunnel. The signaling protocol used for the setup of the tunnel LSP is largely based on the service requirements. For instance, if TE (including the signaling of explicit paths) is required, RSVP-TE provides the solution. However, if it is acceptable to allow a traditional IGP to make hop-by-hop forwarding decisions and end-to-end path significance is not required, LDP can be used instead. As such, we have the option to deploy LDP over RSVP-TE if guaranteed services are mandatory, or LDP over LDP if best-effort delivery suffices.

### 9.6.3 L2 MPLS VPN Label Stack Operation

A two-level label stack is used to create hierarchies separating the common tunnel LSP and the VC LSPs that are nested within. Tunnel LSPs are associated with tunnel labels and VC LSPs are associated with VC labels. On the two-level label stack, the tunnel label sits on top while the VC label rests at the bottom.

The tunnel label determines the path through the core network. It serves as a trunk aggregating traffic from multiple metro subscribers and is signaled with LDP for best-effort delivery only or RSVP-TE for guaranteed

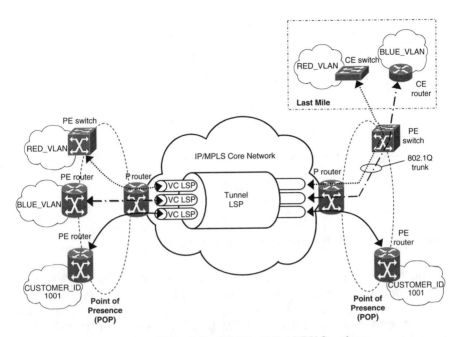

**Figure 9.5    Deployment of Martini's Point-to-Point VPN Services**

services with TE capability. The VC label identifies the L2 VPN (VLAN ID) or the user connection (port ID) at the endpoint. It acts like a per-customer per-site circuit. As such, all services look like a VC to the core network and are provisioned by associating each endpoint with a common VC ID. The VC label is signaled with LDP.

Figure 9.5 illustrates the deployment of Martini's point-to-point VPN services between two points of presence (POPs). Of the three VC LSPs, two are mapped to two respective 802.1Q VLAN IDs (RED_VLAN and BLUE_VLAN), while the other is mapped to a port ID (CUSTOMER_ID 1001).

RSVP-TE can be used within the metro SP's IP/MPLS core network in Figure 9.5 to establish the tunnel LSP with TE capability, while remote LDP is used to establish the VC LSPs between the PE devices (routers or switches). These LDP (VC) LSPs are tunneled (or nested) within the RSVP-TE LSPs between the P routers (LDP over RSVP-TE).

### 9.6.4    Encapsulation for Ethernet

Due to its relatively low cost and simplicity, Ethernet has become the most commonly deployed L2 transport medium in recent MAN implementations. In Martini's defined encapsulation for Ethernet/Ethernet 802.1Q VLAN, the original Ethernet frame is stripped of the preamble and frame check sequence (FCS) at the ingress point. Next, the tunnel and VC labels

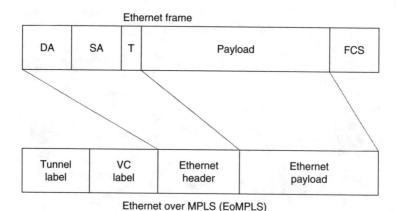

Figure 9.6   Martini's Ethernet Encapsulation Mechanism

are pushed onto the remaining frame. The final frame is further encapsulated with a new Ethernet header and then sent on its way. Runts or frames with errors are dropped on input. Figure 9.6 illustrates the encapsulation mechanism for Ethernet/Ethernet 802.1Q VLAN defined in [DRAFT-MARTINI-ENCAP], while Figure 9.7 illustrates the two-level label stack operation of such a frame. This text only examines Martini's defined encapsulation for Ethernet/Ethernet 802.1Q VLAN. The other encapsulation mechanisms listed in [DRAFT-MARTINI-ENCAP], such as Frame Relay, ATM, HDLC, and PPP, are beyond the scope of this text.

## 9.7   CASE STUDY 9.1: L3 MPLS VPNs WITH STATIC ROUTES AND OSPF

Case study 9.1 examines the setup of L3 MPLS VPNs using static routes and OSPF for the PE-CE routing.

### 9.7.1   Case Overview and Network Topology

Case study 9.1 is built on top of case study 8.2 (see section 8.9), but with some adjustments made to the original setting and new requirements:

- The Notus district has a new commercial estate 3.
- Likewise, the Eurus district has a new commercial estate 4.
- Company ABC has a main office in commercial estate 3 and a branch office in commercial estate 2 (situated at the Zephir district). Both offices must be able to communicate with each other via the MAN as if they are in a single private network.
- Company XYZ has its main branch in commercial estate 3 and a subbranch in commercial estate 4. Both sites must be able to communicate with each other via the MAN as if they are in a single private network.

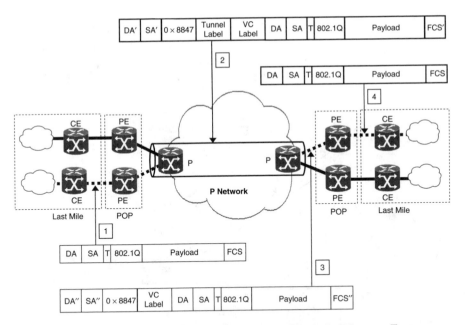

**Figure 9.7    Two-Level Label Stack Operation of a Martini's Ethernet Frame**

Figure 9.8 illustrates the network diagram for this case study. Two L3 VPNs are implemented in this case: "blue" VPN for Company ABC and "green" VPN for Company XYZ. The blue VPN spans across the Zephir and Notus districts, while the green VPN spans across the Notus and Eurus districts.

Zephir-R2, Notus-R3, and Eurus-R4 are the PE routers, whereas Polydeukes-R12, Aether-R13, Aeolus-R33, and Iris-R14 are the CE routers. RS8600 and RS38000 metro routers are used for the PE routers as before (see chapter 7), while RS3000 metro routers are used for the CE routers. The VPN connectivity requirement for the case study is shown in Figure 9.9:

- Aeolus-R33 should be able to see prefixes 192.168.14.0/24 and 223.0.0.24/32 from Iris-R14.
- Iris-R14 should be able to see prefixes 192.168.33.0/24 and 223.0.0.33/32 from Aeolus-R33.
- Polydeukes-R12 should be able to see prefixes 192.168.13.0/24 and 223.0.0.13/32 from Aether-R13.
- Aether-R13 should be able to see prefixes 192.168.12.0/24 and 223.0.0.24/32 from Polydeukes-R12.

### 9.7.2    Network Configurations

Listing 9.1 illustrates the network configuration (with in-line headers) for CE router Polydeukes-R12 situated at commercial estate 2. The loopback

251

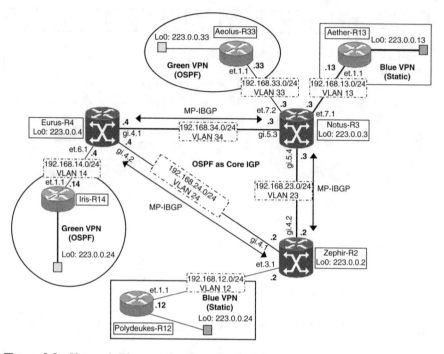

**Figure 9.8   Network Diagram for Case Study 9.1**

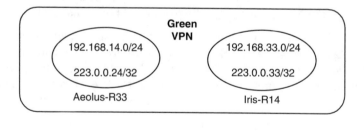

**Figure 9.9   VPN Connectivity Requirement for Case Study 9.1**

address 223.0.0.24 is configured to be the same as that of Iris-R14 to demonstrate that overlapping address space is possible with L3 MPLS VPNs.

With static routing implemented at the PE-CE, a default route is configured on Polydeukes-R12 and a specific static route is configured on its corresponding PE router Zephir-R2 (see Listing 9.5). Static routes are recommended in sites where the metro SP is required to enforce control over customer routing for some specific central services. However, if new or changing routes occur frequently within the site, static routing is not recommended.

---

### Listing 9.1   Polydeukes-R12 Configuration

```
system set name Polydeukes-R12


VLAN configuration
vlan create vlan12 id 12 ip
vlan add ports et.1.1 to vlan12


IP address configuration
interface create ip port-11 vlan vlan12 address-netmask
192.168.12.12/24


Create loopback address
interface add ip lo0 address-netmask 223.0.0.24/32


Set loopback address as router identifier
ip-router global set router-id 223.0.0.24


Default route configuration

ip add route default gateway 192.168.12.2
```

---

Listing 9.2 illustrates the network configuration (with in-line headers) for one of the two CE routers, Aether-R13, situated at commercial estate 3. With static routing implemented at the PE-CE, a default route is configured on Aether-R13 and a specific static route is configured on its corresponding PE router, Notus-R3 (see Listing 9.6).

---

### Listing 9.2   Aether-R13 Configuration

```
system set name Aether-R13
```

**VLAN configuration**
```
vlan create vlan13 id 13 ip
vlan add ports et.1.1 to vlan13
```

**IP address configuration**
```
interface create ip port-11 vlan vlan13 address-netmask
192.168.13.13/24
```

**Create loopback address**
```
interface add ip lo0 address-netmask 223.0.0.13/32
```

**Set loopback address as router identifier**
```
ip-router global set router-id 223.0.0.13
```

**Default route configuration**

```
ip add route default gateway 192.168.13.3
```

---

Listing 9.3 illustrates the network configuration (with in-line headers) for CE router Aeolus-R33, also situated at commercial estate 3. OSPF is implemented as the PE-CE routing protocol for this router.

OSPF was designed to support hierarchical networks that have a central backbone, and the network running OSPF is divided into areas, which could correspond to individual sites, or to multiple sites, such as a region, thus making OSPF a very scalable solution for the PE-CE routing. OSPF is usually deployed in a very large customer network. OSPF and its associations with L3 MPLS VPN are further discussed in section 9.9.

---

### Listing 9.3    Aeolus-R33 Configuration

```
system set name Aeolus-R33
```

**VLAN configuration**
```
vlan create vlan33 id 33 ip
vlan add ports et.1.1 to vlan33
```

**IP address configuration**
```
interface create ip port-11 vlan vlan33 address-netmask
192.168.33.33/24
```

**Create loopback address**
```
interface add ip lo0 address-netmask 223.0.0.33/32
```

```
Set loopback address as router identifier
ip-router global set router-id 223.0.0.33
```

```
OSPF configuration
```

```
ospf create area backbone
ospf add interface port-11 to-area backbone
ospf add stub-host 223.0.0.33 to-area backbone cost 10
ospf start
```

Listing 9.4 illustrates the network configuration (with in-line headers) for CE router Iris-R14 situated at commercial estate 4. The loopback address 223.0.0.24 is configured to be the same as that of Polydeukes-R12 to demonstrate that overlapping address space is possible with L3 MPLS VPNs. OSPF is implemented as the PE-CE routing protocol for this router.

### Listing 9.4   Iris-R14 Configuration

```
system set name Iris-R14
```

```
VLAN configuration
vlan create vlan14 id 14 ip
vlan add ports et.1.1 to vlan14
```

```
IP address configuration
interface create ip port-11 vlan vlan14 address-netmask
192.168.14.14/24
```

```
Create loopback address
interface add ip lo0 address-netmask 223.0.0.24
```

```
Set loopback address as router identifier
ip-router global set router-id 223.0.0.24
```

```
OSPF configuration
```

```
ospf create area backbone
ospf add stub-host 223.0.0.24 to-area backbone cost 10
ospf add interface port-11 to-area backbone
ospf start
```

Listing 9.5 illustrates the network configuration (with in-line headers and comments) for PE router Zephir-R2, which interconnects CE router Polydeukes-R12. Self-explanatory comments (in italics) are embedded in between the configuration lines to provide clarification to the configuration commands and procedures.

OSPF is the backbone or core IGP running between the three PE routers. The MPLS-TE/RSVP-TE configuration is adapted directly from Listing 8.14 and remains intact as before. Besides establishing and maintaining LSPs across the core network, the LSPs' setup by RSVP-TE is used by BGP for resolving next-hops. In this case, RSVP-TE is used as the label distribution protocol for the allocation and binding of labels to the respective BGP next-hops. This BGP next-hop label is stacked on top of the VPN label associated with a customer site. RSVP-TE is also used to provide the desired L3 TE functionalities such as link protection in the core network (see case study 8.2).

In addition:

- MP-IBGP is run between the PE routers to advertise VPNv4 prefixes and their corresponding VPN labels from the customer sites.
- A blue VRF instance is created for the blue VPN and is configured with RD = 65000:102 and RT = 65000:102 for both import and export routing policies.
- The interface connecting Zephir-R2 (PE) to Polydeukes-R12 (CE) is added to the blue VRF instance.
- Zephir-R2 is configured to route statically with Polydeukes-R12.

---

## Listing 9.5   Zephir-R2 Configuration

```
system set name Zephir-R2
```

**VLAN configuration**
```
vlan create vlan23 id 23 port-based
vlan create vlan24 id 24 port-based
vlan create Polydeukes id 200 ip
vlan add ports gi.4.2 to vlan23
vlan add ports gi.4.1 to vlan24
vlan add ports et.3.1 to Polydeukes
```

**IP address configuration**
```
interface create ip port-42 vlan vlan23 address-netmask 192.168.23.2/24
interface create ip port-41 vlan vlan24 address-netmask 192.168.24.2/24
interface create ip Polydeukes address-netmask 192.168.12.2/24 vlan
Polydeukes
```

**Create loopback address**

```
interface add ip lo0 address-netmask 223.0.0.2/32
```

**Set loopback address as router identifier**

```
ip-router global set router-id 223.0.0.2
```

**Backbone IGP configuration**

```
ospf create area backbone
ospf add stub-host 223.0.0.2 to-area backbone cost 10
ospf add interface port-41 to-area backbone
ospf add interface port-42 to-area backbone
ospf set traffic-engineering on
ospf start
```

**MPLS-TE configuration**

```
mpls set global local-repair-enable node-prefer-protection
mpls add interface port-41
mpls add interface port-42
mpls set interface port-41 no-php
mpls set interface port-42 no-php
mpls create path R2-R4 num-hops 1
mpls set path R2-R4 hop 1 ip-addr 192.168.24.4 type strict
mpls create label-switched-path R2-to-R4 to 223.0.0.4
mpls create label-switched-path R2-to-R3 to 223.0.0.3 no-cspf
mpls set label-switched-path R2-to-R4 fast-reroute
mpls set label-switched-path R2-to-R4 primary R2-R4
mpls start
```

**RSVP configuration**

```
rsvp add interface port-41
rsvp add interface port-42
rsvp start
```

**BGP configuration**

```
! -- Set the autonomous system number to 65000
ip-router global set autonomous-system 65000

! -- Enables BGP to use LSPs set up by RSVP-TE for resolving next-hops
ip-router global set install-lsp-routes bgp

! -- Create BGP peer group provider
bgp create peer-group provider autonomous-system 65000
bgp set peer-group provider local-address 223.0.0.2
```

```
! -- Add BGP peers to peer group provider
bgp add peer-host 223.0.0.3 group provider
bgp add peer-host 223.0.0.4 group provider

! -- Enables MP-BGP to support both conventional IPv4 and VPN-IPv4
prefixes
bgp set peer-group provider vpnv4-unicast ipv4-unicast

bgp start

! -- Define import RT value for blue vrf and its match condition
community-list blue-import permit 10 target:65000:102
route-map blue-import permit 10 match-community-list blue-import

! -- Define export RT value for blue vrf and its set condition
ip-router policy create community-list blue-export target:65000:102
route-map blue-export permit 10 set-community-list blue-export

L3 MPLS VPN configuration

! -- Create blue vrf for blue VPN and configure its RD value
routing-instance blue vrf set route-distinguisher "65000:102"

! -- Add the interface connecting Polydeukes-R12 to blue vrf
routing-instance blue vrf add interface Polydeukes

! -- Configure import routing policy for blue vrf
routing-instance blue vrf set vrf-import blue-import in-sequence 1

! -- Configure export routing policy for blue vrf
routing-instance blue vrf set vrf-export blue-export out-sequence 1

! -- Configure Zephir-R2 to route statically with Polydeukes-R12
routing-instance blue ip add route 223.0.0.24/32 gateway 192.168.12.12
```

Listing 9.6 illustrates the network configuration (with in-line headers and comments) for PE router Notus-R3, which interconnects CE routers Aether-R13 and Aeolus-R33. Self-explanatory comments (in italics) are embedded in between the configuration lines to provide clarification to the configuration commands and procedures.

The MPLS-TE/RSVP-TE configuration is adapted directly from Listing 8.15 and remains intact as before. In addition:

- A blue VRF instance is created for the blue VPN and is configured with RD = 65000:102 and RT = 65000:102 for both import and export routing policies.
- The interface connecting Notus-R3 (PE) to Aether-R13 (CE) is added to the blue VRF instance.
- Notus-R3 is configured to route statically with Aether-R13.
- A green VRF instance is created for the green VPN and is configured with RD = 65000:101 and RT = 65000:101 for both import and export routing policies.
- The interface connecting Notus-R3 (PE) to Aeolus-R33 (CE) is added to the green VRF instance.
- Notus-R3 is configured to route OSPF with Aeolus-R33.

## Listing 9.6   Notus-R3 Configuration

```
system set name Notus-R3

VLAN configuration
vlan create vlan34 id 34 port-based
vlan create vlan23 id 23 port-based
vlan create Aeolus id 330 ip
vlan create Aether id 130 ip

vlan add ports gi.5.3 to vlan34
vlan add ports gi.5.4 to vlan23
vlan add ports et.7.1 to Aether
vlan add ports et.7.2 to Aeolus

IP address configuration
interface create ip port-53 vlan vlan34 address-netmask 192.168.34.3/24
interface create ip port-54 vlan vlan23 address-netmask 192.168.23.3/24
interface create ip Aeolus address-netmask 192.168.33.3/24 vlan Aeolus
interface create ip Aether address-netmask 192.168.13.3/24 vlan Aether

Create loopback address
interface add ip lo0 address-netmask 223.0.0.3/32

Set loopback address as router identifier
ip-router global set router-id 223.0.0.3

Backbone IGP configuration
ospf create area backbone
ospf add stub-host 223.0.0.3 to-area backbone cost 10
ospf add interface port-53 to-area backbone
```

```
ospf add interface port-54 to-area backbone
ospf set traffic-engineering on
ospf start
```

**MPLS-TE configuration**
```
mpls set global local-repair-enable node-prefer-protection
mpls add interface port-53
mpls add interface port-54
mpls set interface port-53 no-php
mpls set interface port-54 no-php
mpls create label-switched-path R3-to-R2 to 223.0.0.2 no-cspf
mpls create label-switched-path R3-to-R4 to 223.0.0.4 no-cspf
mpls start
```

**RSVP configuration**
```
rsvp add interface port-53
rsvp add interface port-54
rsvp start
```

**BGP configuration**

```
! -- Set the autonomous system number to 65000
ip-router global set autonomous-system 65000

! -- Enables BGP to use LSPs set up by RSVP-TE for resolving next-hops
ip-router global set install-lsp-routes bgp

! -- Create BGP peer group provider
bgp create peer-group provider autonomous-system 65000
bgp set peer-group provider local-address 223.0.0.3

! -- Add BGP peers to peer group provider
bgp add peer-host 223.0.0.2 group provider
bgp add peer-host 223.0.0.4 group provider

! -- Enables MP-BGP to support both conventional IPv4 and VPN-IPv4
prefixes
bgp set peer-group provider vpnv4-unicast ipv4-unicast

bgp start

! -- Define import RT value for blue vrf and its match condition
community-list blue-import permit 10 target:65000:102
route-map blue-import permit 10 match-community-list blue-import
```

```
! -- Define export RT value for blue vrf and its set condition
ip-router policy create community-list blue-export target:65000:102
route-map blue-export permit 10 set-community-list blue-export
```

```
! -- Define import RT value for green vrf and its match condition
community-list green-import permit 10 target:65000:101
route-map green-import permit 10 match-community-list green-import
```

```
! -- Define export RT value for green vrf and its set condition
ip-router policy create community-list green-export target:65000:101
route-map green-export permit 10 set-community-list green-export
```

```
! -- Create route map "bgproutes" that matches all BGP routes
route-map bgproutes permit 10 match-route-type bgp
```

**L3 MPLS VPN configuration**

```
! -- Create blue vrf for blue VPN and configure its RD value
routing-instance blue vrf set route-distinguisher "65000:102"
```

```
! -- Add the interface connecting Aether-R13 to blue vrf
routing-instance blue vrf add interface Aether
```

```
! -- Configure import routing policy for blue vrf
routing-instance blue vrf set vrf-import blue-import in-sequence 1
```

```
! -- Configure export routing policy for blue vrf
routing-instance blue vrf set vrf-export blue-export out-sequence 1
```

```
! -- Configure Notus-R3 to route statically with Aether-R13
routing-instance blue ip add route 223.0.0.13/32 gateway 192.168.13.13
```

```
! -- Create green vrf for green VPN and configure its RD value
routing-instance green vrf set route-distinguisher "65000:101"
```

```
! -- Add the interface connecting Aeolus-R33 to green vrf
routing-instance green vrf add interface Aeolus
```

```
! -- Configure import routing policy for green vrf
routing-instance green vrf set vrf-import green-import in-sequence 1
```

```
! -- Configure export routing policy for green vrf
routing-instance green vrf set vrf-export green-export out-sequence 1
```

```
! -- Configure Notus-R3 to route OSPF with Aeolus-R33
```

261

```
routing-instance green vrf add interface lo0
routing-instance green vrf set router-id 223.0.0.3
routing-instance green ospf create area backbone
routing-instance green ospf add stub-host 223.0.0.3 to-area backbone
cost 10
routing-instance green ospf add interface Aeolus to-area backbone

! -- Set OSPF domain identifier
routing-instance green ospf set domain-id 0.0.0.0

! -- Set VPN route tag (OSPF tag field)
routing-instance green ospf set vpn-route-tag 65000

! -- Specify OSPF should also learn BGP routes for green vrf using route
map "bgproutes"
routing-instance green ospf set route-map-vpn bgproutes

routing-instance green ospf start
```

Listing 9.7 illustrates the network configuration (with in-line headers and comments) for PE router Eurus-R4, which interconnects CE router Iris-R14. Self-explanatory comments (in italics) are embedded in between the configuration lines to provide clarification to the configuration commands and procedures.

The MPLS-TE/RSVP-TE configuration is adapted directly from Listing 8.16 and remains intact as before. In addition:

- A green VRF instance is created for the green VPN and is configured with RD = 65000:101 and RT = 65000:101 for both import and export routing policies.
- The interface connecting Eurus-R4 (PE) to Iris-R14 (CE) is added to the green VRF instance.
- Eurus-R4 is configured to route OSPF with Iris-R14.

### Listing 9.7    Eurus-R4 Configuration

```
system set name Eurus-R4

VLAN configuration
vlan create vlan34 id 34 port-based
vlan create vlan24 id 24 port-based
vlan create Iris id 400 ip
vlan add ports gi.4.1 to vlan34
```

```
vlan add ports gi.4.2 to vlan24
vlan add ports et.6.1 to Iris
```

**IP address configuration**
```
interface create ip port-42 vlan vlan24 address-netmask 192.168.24.4/24
interface create ip port-41 vlan vlan34 address-netmask 192.168.34.4/24
interface create ip Iris address-netmask 192.168.14.4/24 vlan Iris
```

**Create loopback address**
```
interface add ip lo0 address-netmask 223.0.0.4/32
```

**Set loopback address as router identifier**
```
ip-router global set router-id 223.0.0.4
```

**Backbone IGP configuration**
```
ospf create area backbone
ospf add stub-host 223.0.0.4 to-area backbone cost 10
ospf add interface port-42 to-area backbone
ospf add interface port-41 to-area backbone
ospf set traffic-engineering on
ospf start
```

**MPLS-TE configuration**
```
mpls set global local-repair-enable node-prefer-protection
mpls add interface port-41
mpls add interface port-42
mpls set interface port-41 no-php
mpls set interface port-42 no-php
mpls create path R4-R2 num-hops 1
mpls set path R4-R2 hop 1 ip-addr 192.168.24.2 type strict
mpls create label-switched-path R4-to-R2 to 223.0.0.2
mpls create label-switched-path R4-to-R3 to 223.0.0.3 no-cspf
mpls set label-switched-path R4-to-R2 fast-reroute
mpls set label-switched-path R4-to-R2 primary R4-R2
mpls start
```

**RSVP configuration**
```
rsvp add interface port-41
rsvp add interface port-42
rsvp start
```

**BGP configuration**

```
! -- Set the autonomous system number to 65000
ip-router global set autonomous-system 65000
```

```
! -- Enables BGP to use LSPs set up by RSVP-TE for resolving next-hops
ip-router global set install-lsp-routes bgp
```

```
! -- Create BGP peer group provider
bgp create peer-group provider autonomous-system 65000
bgp set peer-group provider local-address 223.0.0.4
```

```
! -- Add BGP peers to peer group provider
bgp add peer-host 223.0.0.2 group provider
bgp add peer-host 223.0.0.3 group provider
```

```
! -- Enables MP-BGP to support both conventional IPv4 and VPN-IPv4
prefixes
bgp set peer-group provider vpnv4-unicast ipv4-unicast
```

```
bgp start
```

```
! -- Define import RT value for green vrf and its match condition
community-list green-import permit 10 target:65000:101
route-map green-import permit 10 match-community-list green-import
```

```
! -- Define export RT value for green vrf and its set condition
route-map green-export permit 10 set-community-list green-export
ip-router policy create community-list green-export target:65000:101
```

```
! -- Create route map "bgproutes" that matches all BGP routes
route-map bgproutes permit 10 match-route-type bgp
```

```
L3 MPLS VPN configuration
```

```
! -- Create green vrf for green VPN and configure its RD value
routing-instance green vrf set route-distinguisher "65000:101"
```

```
! -- Add the interface connecting Iris-R14 to green vrf
routing-instance green vrf add interface Iris
```

```
! -- Configure import routing policy for green vrf
routing-instance green vrf set vrf-import green-import in-sequence 1
```

```
! -- Configure export routing policy for green vrf
routing-instance green vrf set vrf-export green-export out-sequence 1
```

```
! -- Configure Eurus-R4 to route OSPF with Iris-14
routing-instance green vrf add interface lo0
```

```
routing-instance green vrf set router-id 223.0.0.4
routing-instance green ospf create area backbone
routing-instance green ospf add stub-host 223.0.0.4 to-area backbone
cost 10
routing-instance green ospf add interface Iris to-area backbone

! -- Set OSPF domain identifier
routing-instance green ospf set domain-id 0.0.0.0

! -- Set VPN route tag (OSPF tag field)
routing-instance green ospf set vpn-route-tag 65000

! -- Specify OSPF should also learn BGP routes for green vrf using route
map "bgproutes"
routing-instance green ospf set route-map-vpn bgproutes

routing-instance green ospf start
```

### 9.7.3 VPN Connectivity Verification

Listing 9.8 illustrates the blue VPN routes that are advertised from Notus-R3 to Zephir-R2 via MP-IBGP. Notus-R3 is the egress PE router for prefixes 192.168.13.0/24 and 223.0.0.13/32. VPN label 18 is assigned to prefix 192.168.13.0/24, and VPN label 19 is assigned to prefix 223.0.0.13/32.

#### Listing 9.8   Blue VPN Routes Advertised from Notus-R3

```
Notus-R3# bgp show peer-host 223.0.0.2 instance blue advertised-routes
Local router ID is 223.0.0.3
Status codes: > - best, * - valid, i - internal, t - stale
s - suppressed, d - damped
Origin codes: i - IGP, e - EGP, ? - incomplete
```

| Network | Next Hop | Metric | LocPrf | Label | Path |
|---|---|---|---|---|---|
| *> i192.168.13.0/24 | 192.168.13.3 | 1 | 100 | 18 | i |
| *> i223.0.0.13/32 | 223.0.0.3 | | 100 | 19 | ? |

Listing 9.9 illustrates the blue VPN routes from Notus-R3 that are received at Zephir-R2. The listing validates that prefixes 192.168.13.0/24 and 223.0.0.13/32 have indeed been advertised from Notus-R3 and are received correctly at Zephir-R2.

---

### Listing 9.9    Blue VPN Routes Received at Zephir-R2

```
Zephir-R2# bgp show peer-host 223.0.0.3 instance blue received-routes
Local router ID is 223.0.0.2
Status codes: > - best, * - valid, i - internal, t - stale
s - suppressed, d - damped
Origin codes: i - IGP, e - EGP, ? - incomplete
```

| Network | Next Hop | Metric | LocPrf | Label | Path |
|---|---|---|---|---|---|
| *> i192.168.13.0/24 | 223.0.0.3 | 0 | 100 | 18 | i |
| *> i223.0.0.13/32 | 223.0.0.3 | 0 | 100 | 19 | ? |

---

Listing 9.10 illustrates a default route in the IPv4 routing table of CE router Poludeukes-R12 pointing to next-hop 192.168.12.2 (Zephir-R2).

---

### Listing 9.10    Polydeukes-R12 IPv4 Routing Table

```
Polydeukes-R12# ip show route
```

| Destination | Gateway | Owner | Netif |
|---|---|---|---|
| Default | 192.168.12.2 | Static | Port-11 |
| 192.168.12.0/24 | Directly connected | — | Port-11 |
| 223.0.0.24 | 223.0.0.24 | — | lo0 |

---

Listing 9.11 illustrates the successful outcome of the ping test performed from Polydeukes-R12 to IP address 223.0.0.13 (Aether-R13) in the blue VPN.

---

### Listing 9.11    Ping to 223.0.0.13 from Polydeukes-R12

```
Polydeukes-R12# ping 223.0.0.13
PING 223.0.0.13: 36 bytes of data
5 second timeout, 1 repetition
36 bytes from 223.0.0.13: icmp_seq=0 ttl=253 time=3.010 ms

-- 223.0.0.13 ping statistics --
1 packets transmitted, 1 packets received, 0.00% packet loss
round-trip min/avg/max/dev = 3.010/3.010/3.010/0.000 ms
```

---

Listing 9.12 illustrates the unsuccessful outcome of the ping test performed from Polydeukes-R12 to IP address 223.0.0.33 (Aeolus-R33) in the green VPN.

### Listing 9.12   Ping to 223.0.0.33 from Polydeukes-R12

```
Polydeukes-R12# ping 223.0.0.33
PING 223.0.0.33: 36 bytes of data
5 second timeout, 1 repetition

--- 223.0.0.33 ping statistics ---
1 packets transmitted, 0 packets received, 100.00% packet loss
round-trip min/avg/max/dev = 0.000/0.000/0.000/0.000 ms
```

Listing 9.13 illustrates the blue VPN routes that are advertised from Zephir-R2 to Notus-R3 via MP-IBGP. Zephir-R2 is the egress PE router for prefixes 192.168.12.0/24 and 223.0.0.24/32 (Polydeukes-R12). VPN label 17 is assigned to prefix 192.168.12.0/24, and VPN label 18 is assigned to prefix 223.0.0.24/32 (Polydeukes-R12).

### Listing 9.13   Blue VPN Routes Advertised from Zephir-R2

```
Zephir-R2# bgp show peer-host 223.0.0.3 instance blue advertised-routes

Local router ID is 223.0.0.2
Status codes: > - best, * - valid, i - internal, t - stale
s - suppressed, d - damped
Origin codes: i - IGP, e - EGP, ? - incomplete
```

| Network | Next Hop | Metric | LocPrf | Label | Path |
|---|---|---|---|---|---|
| *> i192.168.12.0/24 | 192.168.12.2 | 1 | 100 | 17 | i |
| *> i223.0.0.24/32 | 223.0.0.2 | | 100 | 18 | ? |

Listing 9.14 illustrates the blue VPN routes from Zephir-R2 that are received at Notus-R3. The listing validates that prefixes 192.168.12.0/24 and 223.0.0.24/32 (Polydeukes-R12) have indeed been advertised from Zephir-R2 and are received correctly at Notus-R3.

---

### Listing 9.14    Blue VPN Routes Received at Notus-R3

```
Notus-R3# bgp show peer-host 223.0.0.2 instance blue received-routes
Local router ID is 223.0.0.3
Status codes: > - best, * - valid, i - internal, t - stale
s - suppressed, d - damped
Origin codes: i - IGP, e - EGP, ? - incomplete
```

| Network | Next Hop | Metric | LocPrf | Label | Path |
|---|---|---|---|---|---|
| *> i192.168.12.0/24 | 223.0.0.2 | 0 | 100 | 17 | i |
| *> i223.0.0.24/32 | 223.0.0.2 | 0 | 100 | 18 | ? |

---

Listing 9.15 illustrates a default route in the IPv4 routing table of CE router Aether-R13 pointing to next-hop 192.168.13.3 (Notus-R3).

---

### Listing 9.15    Aether-R13 IPv4 Routing Table

```
Aether-R13# ip show route
```

| Destination | Gateway | Owner | Netif |
|---|---|---|---|
| Default | 192.168.13.3 | Static | Port-11 |
| 192.168.13.0/24 | Directly connected | — | Port-11 |
| 223.0.0.13 | 223.0.0.13 | — | lo0 |

---

Listing 9.16 illustrates the successful outcome of the ping test performed from Aether-R13 to IP address 223.0.0.24 (Polydeukes-R12) in the blue VPN.

---

### Listing 9.16    Ping to 223.0.0.24 from Aether-R13

```
Aether-R13# ping 223.0.0.24
PING 223.0.0.24: 36 bytes of data
5 second timeout, 1 repetition
36 bytes from 223.0.0.24: icmp_seq=0 ttl=253 time=3.066 ms

--- 223.0.0.24 ping statistics ---
1 packets transmitted, 1 packets received, 0.00% packet loss
round-trip min/avg/max/dev = 3.066/3.066/3.066/0.000 ms
```

---

Listing 9.17 illustrates the unsuccessful outcome of the ping test performed from Aether-R13 to IP address 223.0.0.33 (Aeolus-R33) in the green VPN.

---

### Listing 9.17    Ping to 223.0.0.33 from Aether-R13

```
Aether-R13# ping 223.0.0.33
PING 223.0.0.33: 36 bytes of data
5 second timeout, 1 repetition

--- 223.0.0.33 ping statistics ---
1 packets transmitted, 0 packets received, 100.00% packet loss
round-trip min/avg/max/dev = 0.000/0.000/0.000/0.000 ms
```

---

Listing 9.18 further verifies that the previous ping response (Internet control message protocol (ICMP) echo-reply), as illustrated in Listing 9.16, is from host address 223.0.0.24 in Polydeukes-R12. The path that is taken to reach this address is R3–R4–R2 after link R2–R3 fails. While traversing this path, the previously assigned VPN label 18 for this address is at the bottom of the MPLS stack (S = 1) and the BGP next-hop (or RSVP-TE) label 4097 is at the top of the MPLS stack (S = 0). They are popped accordingly at the egress hop (Zephir-R2).

---

### Listing 9.18    Traceroute to 223.0.0.24 from Aether-R13 after Link R2–R3 Fails

```
Aether-R13# traceroute 223.0.0.24

traceroute to 223.0.0.24 (223.0.0.24), 30 hops max, 40 byte packets
1  192.168.13.3 (192.168.13.3) 0 ms 0 ms 0 ms
2  192.168.34.4 (192.168.34.4) 3 ms 1 ms 0 ms
   MPLS Label1=4097 EXP1=0 TTL=1 S=0
   MPLS Label2=18 EXP2=0 TTL=1 S=1
3  192.168.24.2 (192.168.24.2) 1 ms 0 ms 0 ms
4  223.0.0.24 (223.0.0.24) 2 ms 1 ms 1 ms
```

---

Together, Listings 9.8 to 9.18 verify that the blue VPN connectivity or CUG between the Zephir district and the Notus district is up and working. They also prove that overlapping addresses can be implemented for different VPNs.

Listing 9.19 illustrates the green VPN routes that are advertised from Eurus-R4 to Notus-R3 via MP-IBGP. Eurus-R4 is the egress PE router for prefixes 192.168.14.0/24 and 223.0.0.24/32 (Iris-R14). VPN label 17 is assigned to prefix 192.168.14.0/24, and VPN label 19 is assigned to prefix 223.0.0.24/32 (Iris-14).

---

**Listing 9.19   Green VPN Routes Advertised from Eurus-R4**

```
Eurus-R4# bgp show peer-host 223.0.0.3 instance green advertised-routes

Local router ID is 223.0.0.4
Status codes: > - best, * - valid, i - internal, t - stale
s - suppressed, d - damped
Origin codes: i - IGP, e - EGP, ? - incomplete
```

| Network | Next Hop | Metric | LocPrf | Label | Path |
|---------|----------|--------|--------|-------|------|
| *> i192.168.14.0/24 | 192.168.14.4 | 1 | 100 | 17 | i |
| *> i223.0.0.24/32 | 223.0.0.4 | 31 | 100 | 19 | i |

---

Listing 9.20 illustrates the green VPN routes from Eurus-R4 that are received at Notus-R3. The listing validates that prefixes 192.168.14.0/24 and 223.0.0.24/32 (Iris-R14) have indeed been advertised from Eurus-R4 and are received correctly at Notus-R3.

---

**Listing 9.20   Green VPN Routes Received at Notus-R3**

```
Notus-R3# bgp show peer-host 223.0.0.4 instance green received-routes
Local router ID is 223.0.0.3
Status codes: > - best, * - valid, i - internal, t - stale
s - suppressed, d - damped
Origin codes: i - IGP, e - EGP, ? - incomplete
```

| Network | Next Hop | Metric | LocPrf | Label | Path |
|---------|----------|--------|--------|-------|------|
| *> i192.168.14.0/24 | 223.0.0.4 | 0 | 100 | 17 | i |
| *> i223.0.0.24/32 | 223.0.0.4 | 31 | 100 | 19 | i |

---

The IPv4 routing table of CE router Aeolus-R33 in Listing 9.21 registers prefixes 192.168.14.0/24 and 223.0.0.24/32, indicating that it has successfully received these routes advertised from PE router Notus-R3 via OSPF. Note that prefix 223.0.0.24/32 is reflected as an OSPF inter-area (IA) route (with type 3 link-state advertisement (LSA)), whereas prefix 192.168.14.0/24 is reflected as an OSPF AS external (ASE) route (with type 5 LSA). Although OSPF has been configured for prefix 192.168.14.0/24 at Eurus-R4, the prefix belongs to a directly connected interface; thus, it is considered an AS external route when received by Notus-R3.

**Listing 9.21   Aeolus-R33 IPv4 Routing Table**

```
Aeolus-R33# ip show route
```

| Destination | Gateway | Owner | Netif |
|---|---|---|---|
| **192.168.14.0/24** | **192.168.33.3** | **OSPF_ASE** | **Port-11** |
| 192.168.33.0/24 | Directly connected | – | Port-11 |
| 223.0.0.3 | 192.168.33.3 | OSPF | Port-11 |
| **223.0.0.24** | **192.168.33.3** | **OSPF_IA** | **Port-11** |
| 223.0.0.33 | 223.0.0.33 | – | lo0 |

Listing 9.22 illustrates the green VPN routes that are advertised from Notus-R3 to Eurus-R4 via MP-IBGP. Notus-R3 is the egress PE router for prefixes 192.168.33.0/24 and 223.0.0.33/32. VPN label 17 is assigned to prefix 192.168.33.0/24, and VPN label 21 is assigned to prefix 223.0.0.33/32.

**Listing 9.22   Green VPN Routes Advertised from Notus-R3**

```
Notus-R3# bgp show peer-host 223.0.0.4 instance green advertised-routes
Local router ID is 223.0.0.3
Status codes: > - best, * - valid, i - internal, t - stale
s - suppressed, d - damped
Origin codes: i - IGP, e - EGP, ? - incomplete
```

| Network | Next Hop | Metric | LocPrf | Label | Path |
|---|---|---|---|---|---|
| *> i192.168.33.0/24 | 192.168.33.3 | 1 | 100 | 17 | i |
| *> i223.0.0.33/32 | 223.0.0.3 | 31 | 100 | 21 | i |

Listing 9.23 illustrates the green VPN routes from Notus-R3 that are received at Eurus-R4. The listing validates that prefixes 192.168.33.0/24 and

223.0.0.33/32 have indeed been advertised from Notus-R3 and are received correctly at Eurus-R4.

---

### Listing 9.23    Green VPN Routes Received at Eurus-R4

```
Eurus-R4# bgp show peer-host 223.0.0.3 instance green received-routes
Local router ID is 223.0.0.4
Status codes: > - best, * - valid, i - internal, t - stale
s - suppressed, d - damped
Origin codes: i - IGP, e - EGP, ? - incomplete
```

| Network | Next Hop | Metric | LocPrf | Label | Path |
|---------|----------|--------|--------|-------|------|
| *> i192.168.33.0/24 | 223.0.0.3 | 0 | 100 | 17 | i |
| *> i223.0.0.33/32 | 223.0.0.3 | 31 | 100 | 21 | i |

---

The IPv4 routing table of CE router Iris-R14 in Listing 9.24 registers prefixes 192.168.33.0/24 and 223.0.0.33/32, indicating that it has successfully received these routes advertised from PE router Eurus-R4 via OSPF. Note that prefix 223.0.0.33/32 is reflected as an OSPF IA route (with type 3 LSA), whereas prefix 192.168.33.0/24 is reflected as an OSPF ASE route (with type 5 LSA). Although OSPF has been configured for prefix 192.168.33.0/24 at Notus-R3, the prefix belongs to a directly connected interface; thus, it is considered an AS external route when received by Eurus-R4.

---

### Listing 9.24    Iris-R14 IPv4 Routing Table

```
Iris-R14# ip show route
```

| Destination | Gateway | Owner | Netif |
|-------------|---------|-------|-------|
| 192.168.14.0/24 | Directly connected | — | Port-11 |
| 192.168.33.0/24 | 192.168.14.4 | OSPF_ASE | Port-11 |
| 223.0.0.4 | 192.168.14.4 | OSPF | Port-11 |
| 223.0.0.24 | 223.0.0.24 | — | lo0 |
| 223.0.0.33 | 192.168.14.4 | OSPF_IA | Port-11 |

---

Listing 9.25 further verifies the traceroute performed at Aeolus-R33 for host address 223.0.0.24, which leads to Iris-R14. The path that is taken to reach this address is R3–R2–R4 after link R3–R4 fails. While traversing this path, the previously assigned VPN label 19 for this address is at the bottom of the MPLS stack (S = 1) and the BGP next-hop (or RSVP-TE) label 4097 is at the top of the MPLS stack (S = 0). They are popped accordingly at the egress hop (Eurus-R4).

---

**Listing 9.25   Traceroute to 223.0.0.24 from Aeolus-R33 after Link R3–R4 Fails**

```
Aeolus-R33# traceroute 223.0.0.24

traceroute to 223.0.0.24 (223.0.0.24), 30 hops max, 40 byte packets
1 192.168.33.3 (192.168.33.3) 0 ms 0 ms 0 ms
2 192.168.23.2 (192.168.23.2) 3 ms 1 ms 0 ms
   MPLS Label1=4097 EXP1=0 TTL=1 S=0
   MPLS Label2=19 EXP2=0 TTL=1 S=1
3 192.168.24.4 (192.168.24.4) 1 ms 0 ms 0 ms
4 223.0.0.24 (223.0.0.24) 2 ms 1 ms 1 ms
```

---

Together, Listings 9.19 to 9.25 verify that the green VPN connectivity or CUG between the Notus district and the Eurus district is up and working. They also prove that overlapping addresses can be implemented for different VPNs.

## 9.8   CASE STUDY 9.2: L3 MPLS VPNs WITH RIPv2 AND EBGP

Case study 9.2 examines the setup of L3 MPLS VPNs using RIPv2 and EBGP for the PE-CE routing.

### 9.8.1   Case Overview and Network Topology

Case study 9.2 uses the same setting as case study 9.1 except for the following:

- RIPv2 is used as the PE-CE routing protocol between Notus-R3 and Aeolus-R33 for green VPN.
- RIPv2 is used as the PE-CE routing protocol between Eurus-R4 and Iris-R14 for green VPN.
- EBGP is used as the PE-CE routing protocol between Notus-R3 and Aether-R13 (AS 65033).
- EBGP is used as the PE-CE routing protocol between Zephir-R2 and Polydeukes-R12 (AS 65022).

Figure 9.10 illustrates the network diagram for this case study.

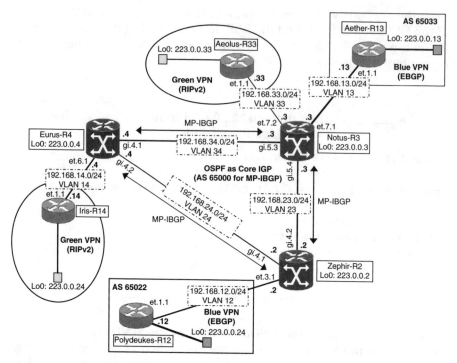

**Figure 9.10   Network Diagram for Case Study 9.2**

## 9.8.2   Network Configurations

Listing 9.26 is based on the previous configuration from Listing 9.1. The PE-CE routing protocol is changed to EBGP. For brevity, only this portion of change is shown. Polydeukes-R12 is configured to be in AS 65022 and is peered to its EBGP neighbor Zephir-R2 in AS 65000. Polydeukes-R12 also advertises prefix 223.0.0.24/32 to Zephir-R2.

EBGP is usually used for multi-homed sites where optimal routing is mandatory. Deploying EBGP as the PE-CE routing protocol allows the continuity of BGP policies between customer sites because the BGP attributes set by one customer site are transparently advertised to other customer sites.

---

### Listing 9.26   Polydeukes-R12 Configuration

```
system set name Polydeukes-R12

! -- Set the autonomous system number to 65022
ip-router global set autonomous-system 65022
```

```
! -- Create route map "allroutes" that permits all routes
route-map allroutes permit 10

! -- Create peer group "to-provider"
bgp create peer-group to-provider autonomous-system 65000

! -- Add EBGP peer Zephir-R2 to peer group "to-provider"
bgp add peer-host 192.168.12.2 group to-provider

! -- Allow the advertisement of all BGP routes to EBGP peer Zephir-R2
bgp set peer-host 192.168.12.2 route-map-out allroutes out-sequence 1

! -- Specifically advertise prefix 223.0.0.24/32
bgp advertise network 223.0.0.24/32

bgp start
```

Listing 9.27 is based on the previous configuration from Listing 9.2. The PE-CE routing protocol is changed to EBGP. For brevity, only this portion of change is shown. Aether-R13 is configured to be in AS 65033 and is peered to its EBGP neighbor Notus-R3 in AS 65000. Aether-R13 also advertises prefix 223.0.0.13/32 to Notus-R3.

### Listing 9.27   Aether-R13 Configuration

```
system set name Aether-R13

! -- Set the autonomous system number to 65033
ip-router global set autonomous-system 65033

! -- Create route map "allroutes" that permits all routes
route-map allroutes permit 10

! -- Create peer group "to-provider"
bgp create peer-group to-provider autonomous-system 65000

! -- Add EBGP peer Notus-R3 to peer group "to-provider"
bgp add peer-host 192.168.13.3 group to-provider

! -- Allow the advertisement of all BGP routes to EBGP peer Notus-R3
bgp set peer-host 192.168.13.3 route-map-out allroutes out-sequence 1
```

```
! -- Specifically advertise prefix 223.0.0.13/32
bgp advertise network 223.0.0.13/32

bgp start
```

Listing 9.28 is based on the previous configuration from Listing 9.3. The PE-CE routing protocol is changed to RIPv2. For brevity, only this portion of change is shown. Aeolus-R33 is configured to route RIPv2 with Notus-R3.

RIPv2 allows for dynamic routing between the PE and CE routers. This option is usually used in stub sites where the SP does not manage the CE routers, where every CE router needs to know all routes from other CE routers with the same VPN membership, or where there is more than one IP prefix per customer site.

### Listing 9.28  Aeolus-R33 Configuration

```
system set name Aeolus-R33

rip add interface all
rip set interface all version 2
rip start
```

Listing 9.29 is based on the previous configuration from Listing 9.4. The PE-CE routing protocol is changed to RIPv2. For brevity, only this portion of change is shown. Iris-R14 is configured to route RIPv2 with Eurus-R4.

### Listing 9.29  Iris-R14 Configuration

```
system set name Iris-R14

rip add interface all
rip set interface all version 2
rip start
```

Listing 9.30 is based on the previous configuration from Listing 9.5. The PE-CE routing protocol is changed to EBGP. For brevity, only this portion of change is shown. Zephir-R2 is in AS 65000 and is peered to its EBGP neighbor Polydeukes-R12 in AS 65022.

---

### Listing 9.30    Zephir-R2 Configuration

```
system set name Zephir-R2

! -- Create route map "allroutes" that permits all routes
route-map allroutes permit 10

! -- Create peer group "to-customer" and set the autonomous system
number to 65022
routing-instance blue bgp create peer-group to-customer autonomous-
system 65022

! -- Add EBGP peer Polydeukes-R12 to peer group "to-customer"
routing-instance blue bgp add peer-host 192.168.12.12 group to-customer

! -- Allow the advertisement of all BGP routes to EBGP peer Polydeukes-
R12
routing-instance blue bgp set peer-host 192.168.12.12 route-map-out
allroutes out-sequence 1
routing-instance blue bgp start
```

---

Listing 9.31 is based on the previous configuration from Listing 9.6:

- The PE-CE routing protocol between Notus-R3 and Aether-R13 is changed to EBGP. Notus-R3 is in AS 65000 and is peered to its EBGP neighbor Aether-R13 in AS 65033.
- The PE-CE routing protocol between Notus-R3 and Aeolus-R33 is changed to RIPv2.

For brevity, only the above-stated changes are shown.

---

### Listing 9.31    Notus-R3 Configuration

```
system set name Notus-R3

! -- Create route map "allroutes" that permits all routes
route-map allroutes permit 10

! -- Create peer group "to-customer" and set the autonomous system
number to 65033
routing-instance blue bgp create peer-group to-customer autonomous-
system 65033

! -- Add EBGP peer Aether-R13 to peer group "to-customer"
routing-instance blue bgp add peer-host 192.168.13.13 group to-customer
```

```
! -- Allow the advertisement of all BGP routes to EBGP peer Aether-R13
routing-instance blue bgp set peer-host 192.168.13.13 route-map-out
allroutes out-sequence 1
routing-instance blue bgp start


! -- Configure Notus-R3 to route RIPv2 with Aeolus-R33
routing-instance green rip add interface Aeolus
routing-instance green rip set interface Aeolus version 2


! -- Specify RIPv2 should also learn BGP routes for green vrf using route
map "bgproutes"
routing-instance green rip set route-map-out bgproutes
routing-instance green rip start
```

Listing 9.32 is based on the previous configuration from Listing 9.7. The PE-CE routing protocol is changed to RIPv2. For brevity, only this portion of change is shown. Eurus-R4 is configured to route RIPv2 with Iris-R14.

### Listing 9.32    Eurus-R4 Configuration

```
system set name Eurus-R4

! -- Configure Notus-R3 to route RIPv2 with Iris-R14
routing-instance green rip add interface Iris
routing-instance green rip set interface Iris version 2


! -- Specify RIPv2 should also learn BGP routes for green vrf using route
map "bgproutes"
routing-instance green rip set route-map-out bgproutes
routing-instance green rip start
```

### 9.8.3    VPN Connectivity Verification

The IPv4 routing table of CE router Polydeukes-R12 in Listing 9.33 registers prefixes 192.168.13.0/24 and 223.0.0.13/32, indicating that it has successfully received these routes advertised from PE router Zephir-R2 via EBGP.

### Listing 9.33    Polydeukes-R12 IPv4 Routing Table

```
Polydeukes-R12# ip show routes
```

| Destination | Gateway | Owner | Netif |
|---|---|---|---|
| 192.168.12.0/24 | Directly connected | – | Port–11 |
| **192.168.13.0/24** | **192.168.12.2** | **BGP** | **Port–11** |
| **223.0.0.13** | **192.168.12.2** | **BGP** | **Port–11** |
| 223.0.0.24 | 223.0.0.24 | – | lo0 |

The BGPv4 routing table of Polydeukes-R12 in Listing 9.34 further examines the BGP routes illustrated in Listing 9.33. Prefix 223.0.0.13/32 originates from AS 65033 of Aether-R13, while prefix 192.168.13.0/24 originates from AS 65000 because its associated interface is directly connected to Notus-R3.

### Listing 9.34 Polydeukes-R12 BGPv4 Routing Table

```
Polydeukes-R12# bgp show routes all

Local router ID is 223.0.0.24
Status codes: > - best, * - valid, i - internal, t - stale
s - suppressed, d - damped
Origin codes: i - IGP, e - EGP, ? - incomplete
```

| | Network | Next Hop | Metric | LocPrf | Label | Path | | |
|---|---|---|---|---|---|---|---|---|
| * | 192.168.12.0/24 | 192.168.12.2 | | 100 | | 65000 | i | |
| *> | 192.168.13.0/24 | 192.168.12.2 | | 100 | | 65000 | i | |
| *> | 223.0.0.13/32 | 192.168.12.2 | | 100 | | 65000 | 65033 | i |

The IPv4 routing table of CE router Aether-R13 in Listing 9.35 registers prefixes 192.168.12.0/24 and 223.0.0.24/32, indicating that it has successfully received these routes advertised from PE router Notus-R3 via EBGP.

### Listing 9.35 Aether-R13 IPv4 Routing Table

```
Aether-R13# ip show routes
```

| Destination | Gateway | Owner | Netif |
|---|---|---|---|
| **192.168.12.0/24** | **192.168.13.3** | **BGP** | **Port–11** |
| 192.168.13.0/24 | Directly connected | – | Port–11 |
| 223.0.0.13 | 223.0.0.13 | – | lo0 |
| **223.0.0.24** | **192.168.13.3** | **BGP** | **Port–11** |

The BGPv4 routing table of Aether-R13 in Listing 9.36 further examines the BGP routes illustrated in Listing 9.35. Prefix 223.0.0.24/32 originates from AS 65022 of Polydeukes-R12, while prefix 192.168.12.0/24 originates from AS 65000 because its associated interface is directly connected to Zephir-R2.

### Listing 9.36   Aether-R13 BGPv4 Routing Table

```
Aether-R13# bgp show routes all

Local router ID is 223.0.0.13
Status codes: > - best, * - valid, i - internal, t - stale
s - suppressed, d - damped
Origin codes: i - IGP, e - EGP, ? - incomplete
```

| Network | Next Hop | Metric | LocPrf | Label | Path |
|---------|----------|--------|--------|-------|------|
| *> 192.168.12.0/24 | 192.168.13.3 | | 100 | | 65000 i |
| *  192.168.13.0/24 | 192.168.13.3 | | 100 | | 65000 i |
| *> 223.0.0.24/32 | 192.168.13.3 | | 100 | | 65000 65022 i |

The IPv4 routing table of CE router Aeolus-R33 in Listing 9.37 registers prefixes 192.168.14.0/24 and 223.0.0.24/32, indicating that it has successfully received these routes advertised from PE router Notus-R3 via RIPv2.

### Listing 9.37   Aeolus-R33 IPv4 Routing Table

```
Aeolus-R33# ip show routes
```

| Destination | Gateway | Owner | Netif |
|-------------|---------|-------|-------|
| 192.168.14.0/24 | 192.168.33.3 | RIP | Port-11 |
| 192.168.33.0/24 | Directly connected | — | Port-11 |
| 223.0.0.24 | 192.168.33.3 | RIP | Port-11 |
| 223.0.0.33 | 223.0.0.33 | — | lo0 |

The IPv4 routing table of CE router Iris-R14 in Listing 9.38 registers prefixes 192.168.33.0/24 and 223.0.0.33/32, indicating that it has successfully received these routes advertised from PE router Eurus-R4 via RIPv2.

## Listing 9.38   Iris-R14 IPv4 Routing Table

```
Iris-R14# ip show routes
```

| Destination | Gateway | Owner | Netif |
|---|---|---|---|
| 192.168.14.0/24 | Directly connected | – | Port-11 |
| **192.168.33.0/24** | **192.168.14.4** | **RIP** | **Port-11** |
| 223.0.0.24 | 223.0.0.24 | – | lo0 |
| **223.0.0.33** | **192.168.14.4** | **RIP** | **Port-11** |

Together Listings 9.33 to 9.36 verify that the blue VPN connectivity or CUG between the Zephir district and the Notus district is up and working, while Listings 9.37 and 9.38 verify that the green VPN connectivity or CUG between the Notus district and the Eurus district is up and working.

### 9.9   OSPF AND L3 MPLS VPN

#### 9.9.1   Overview

Traditionally, an elaborate OSPF network consists of a backbone area (Area 0) and a number of areas connected to this backbone area via an area border router (ABR). With OSPF as the L3 MPLS VPN PE-CE routing protocol, a third level in the hierarchy of the OSPF model is introduced. This third level is called the MPLS VPN super backbone. In simple cases, the MPLS VPN super backbone is regarded as the traditional Area 0, implying that the MPLS VPN super backbone assumes the role of Area 0 when Area 0 is not configured on the customer network. On the other hand, the super backbone also allows customers to use Area 0 on their respective sites. Each site can have a separate Area 0 as long as it is connected to the super backbone. The result is the same as a partitioned Area 0.

#### 9.9.2   OSPF versus L3 MPLS VPN Routing Model

OSPF partitions an IP network into areas and interconnects all these areas via Area 0. From the MPLS VPN perspective, areas could correspond to individual customer sites. From the customer viewpoint, the MPLS VPN backbone is BGP based with IGPs running at the edge of the customer sites. Thus, redistribution between IGP and BGP is performed so that MP-IBGP can propagate customer routes across the MPLS VPN backbone.

Assuming that OSPF is the IGP in this case, a local subnet belonging to an individual customer site is announced to the local PE router as either an OSPF type 1 or type 2 LSA. This OSPF route is redistributed into BGP and becomes a MP-IBGP route, which in turn is propagated to other remote PE routers. This MP-IBGP route is then redistributed back as an AS external (type 5 LSA) route into the other remote customer sites running OSPF.

The main issue here pertains to the OSPF LSA type not being preserved when the OSPF route is redistributed into MP-IBGP. All OSPF routes originating from a local site are inserted as AS external (type 5 LSA) routes into other remote sites. Consequently, OSPF route summarization and stub areas will be difficult to implement. MPLS VPN must therefore extend the classic OSPF-BGP routing model to alleviate this constraint. Because OSPF Area 0 might extend into individual sites, the MPLS VPN backbone must appear as a super backbone or pseudo-Area 0 to the OSPF customers rather than a BGP backbone.

### 9.9.3 BGP Extended Communities and MPLS OSPF

BGP extended community attributes [DRAFT-BGP-EXT-COMM] are used to propagate the OSPF domain ID (mandatory), the OSPF route type (mandatory), and the OSPF router ID (optional) across the MP-IBGP backbone. This section defines these attributes [DRAFT-OSPF-BGP-MPLS] in detail:

- *OSPF domain ID extended community attribute*: This attribute is mandatory unless the VRF instance has a domain ID of value zero (default). This attribute is encoded with a 2-byte type field, which can be $0 \times 0005$ ($0 \times 8005$), $0 \times 0105$, or $0 \times 0205$. The OSPF process ID is carried in the value field.
- *OSPF route type extended community attribute*: This attribute is mandatory. It is encoded with a 2-byte type field, which has a value of $0 \times 0306$ ($0 \times 8000$). The remaining 6 bytes of the attribute are encoded as follows:
  - *OSPF area number*: This subattribute is 4 bytes in length and is composed of a 32-bit area number. The area number is used to identify an area in the same OSPF domain where routes are being propagated and is relative to that particular OSPF domain. For Area 0 or AS external routes, the value is zero. A nonzero value identifies the route as being internal to the OSPF domain and within the identified area.
  - *OSPF route type*: This subattribute is 1 byte in length and indicates the OSPF route type (LSA type). A value of 1 or 2 indicates intraarea routes; a value of 3 indicates summary routes (type 4 LSAs are ignored by PE routers); a value of 5 indicates AS external routes (in this case, the area number must be 0); a value of 7 indicates not-so-stubby area (NSSA) routes; and a value of 129 indicates sham-link endpoint addresses. In the MPLS VPN environment, several VPN customer sites can be interconnected via the same OSPF area. If these sites are connected over an intraarea backdoor link in addition to the MPLS VPN backbone, all

traffic will pass over the backdoor link instead of over the MPLS VPN backbone. This is because OSPF always prefers intra-area routes to inter-area (or external) routes. A sham-link is used to correct this default OSPF behavior in an MPLS VPN. It represents an intra-area (unnumbered point-to-point) connection between PEs. All other routers in the area will use the sham-link to calculate intra-area routes to the remote site.

— *Options*: This subattribute is 1 byte in length. Currently, this field is used only if the OSPF route type value is 5 or 7. Setting the least significant bit in the field indicates that the route carries a type 2 external metric.

- *OSPF router ID extended community attribute*: This attribute is optional. It is encoded with a 2-byte type field and has a value of 0 × 0107 (0 × 8001). The router ID specifies the OSPF router ID of the particular VRF instance from which the route is exported and is carried in the first 4 bytes of the value field.

### 9.9.4  Extending OSPF in L3 MPLS VPN

The following goals need to be taken into consideration when extending the applicability of OSPF in L3 MPLS VPN:

- OSPF between sites should not use normal OSPF-BGP redistribution.
- OSPF continuity must be provided across the MPLS VPN backbone.
- Internal OSPF routes must remain as internal OSPF routes.
- External routes must remain as external routes.
- OSPF metrics must be preserved.

The solution is to regard the MPLS VPN backbone as a third-level hierarchy. The MPLS VPN backbone is acting as a pseudo-Area 0 (or an OSPF super backbone) above the regular OSPF Area 0. MP-IBGP is still implemented between the PE routers. This OSPF super backbone spans between the PE routers and behaves exactly like Area 0 in traditional OSPF. The PE routers are advertised as an area border router (ABR), and in the case of nonstub areas, the PE routers also act as an autonomous system border router (ASBR).

The local PE router, configured for MP-IBGP, uses extended BGP communities (see section 9.9.3) to carry the OSPF domain ID and OSPF route type (LSA type) across the super backbone to the other remote PE routers. At the same time, the OSPF cost is copied into the BGP multi-exit discriminator (MED) attribute and is advertised to the remote PE routers as well.

Routes redistributed from MP-IBGP into OSPF will appear as inter-area (type 3 LSA) routes in the other OSPF sites if the original routes were intra-area (type 1 or type 2 LSA) or inter-area (type 3 LSA) routes, and as AS external (type 5 LSA) routes if the original routes were AS external (type 5

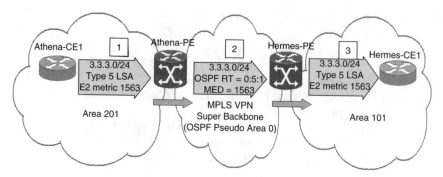

**Figure 9.11  Propagation of OSPF AS External Route**

LSA) routes. Route summarization can now be performed at the redistribution point (or area boundary in this case) by the PE router.

### 9.9.5  Propagation of OSPF AS External Routes

OSPF AS external routes are propagated in the same manner as internal OSPF routes across the MPLS VPN super backbone. The external metric and route type are preserved as well. Figure 9.11 illustrates the propagation of OSPF AS external routes across the MPLS VPN super backbone between Athena-PE and Hermes-PE:

- In this scenario, Athena-CE1 in Area 201 advertises subnet 3.3.3.0/24 as an OSPF AS external (type 5 LSA) route to Athena-PE.
- On redistribution into MP-IBGP, the OSPF route type (RT) extended community attribute is denoted as 0:5:1. The value 0 represents AS external routes (see section 9.9.3), the value 5 indicates OSPF type 5 LSA, and the value 1 indicates the route carries an external type 2 (E2) metric. Note that the notations of OSPF route type are documented in [DRAFT-OSPF-BGP-MPLS], and this example is based on those notations defined in this Internet Engineering Task Force (IETF) draft.
- The OSPF external metric 1563 is also copied to BGP MED.
- Hermes-PE then inserts this route into Area 101 as an OSPF AS external (type 5 LSA) route with a type 2 (E2) external metric of 1563 and propagates it to Hermes-CE1.

Routes from the MPLS VPN super backbone that do not originate from OSPF are still subject to standard redistribution behavior when they are inserted into OSPF.

### 9.9.6  OSPF Down Bit

There is more than one PE router (Hermes-PE1 and Hermes-PE2) at the egress of the MPLS VPN super backbone in this example (see Figure 9.12):

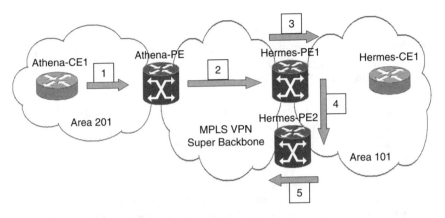

**Figure 9.12   OSPF-BGP Routing Loop Illustration**

- *Step 1*: Athena-CE1 in Area 201 announces the local subnet as an OSPF intra-area route to Athena-PE.
- *Step 2*: The OSPF route is received by Athena-PE, redistributed into MP-IBGP, and propagated across the MPLS VPN super backbone to Hermes-PE1 (first) and Hermes-PE2.
- *Step 3*: Hermes-PE1 inserts the route as an OSPF inter-area route into Area 101.
- *Step 4*: The OSPF route is propagated across Area 101 to Hermes-PE2.
- *Step 5*: Hermes-PE2 redistributes the route back to MP-IBGP and an OSPF-BGP routing loop situation arises.

The OSPF down bit, an additional bit that has been included in the options field of the OSPF LSA header, can be used to prevent the above routing loop scenario from happening. As illustrated in Figure 9.13:

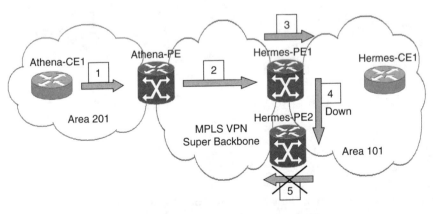

**Figure 9.13   Routing Loop Prevention with OSPF Down Bit**

- *Step 1*: Athena-CE1 in Area 201 announces the local subnet to Athena-PE without setting the down bit.
- *Step 2*: The OSPF route is received by Athena-PE, redistributed into MP-IBGP, and propagated across the MPLS VPN backbone.
- *Step 3*: Hermes-PE1 inserts the route as an OSPF inter-area route into Area 101.
- *Step 4*: The OSPF route is propagated across Area 101 to Hermes-PE2 with the down bit set.
- *Step 5*: The OSPF route is not redistributed back to MP-IBGP by Hermes-PE2.

PE routers set the down bit when redistributing routes from MP-IBGP into OSPF and will never redistribute OSPF routes with the down bit set back into MP-IBGP. In this example, Hermes-PE1 set the OSPF down bit when redistributing the route from MP-IBGP into Area 101. Hermes-PE2 will never redistribute this OSPF route with the down bit set back to MP-IBGP. Hence, the OSPF-BGP routing loop problem is resolved with the OSPF down bit.

### 9.9.7  OSPF Tag Field

Even with the OSPF down bit set, cross-domain routing loops (or double redistribution loops) are still possible, as shown in Figure 9.14:

- *Step 1*: A non-OSPF route is redistributed as an OSPF AS external route into OSPF domain 401 by Athena-PE.
- *Step 2*: This OSPF route is propagated to Morpheus-CE with the down bit set.

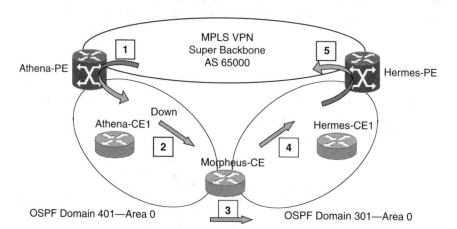

**Figure 9.14  OSPF Cross-Domain Routing Loop Illustration**

- *Step 3*: Morpheus-CE redistributes the AS external route into OSPF domain 301 with the down bit cleared.
- *Step 4*: The route is propagated to Hermes-PE without the down bit.
- *Step 5*: The route is redistributed back to MP-IBGP by Hermes-PE.

When the CE router, in this case Morpheus-CE, does the redistribution between OSPF domains 401 and 301, the down bit is lost. For this purpose, the tag field in OSPF AS external routes is used to prevent this cross-domain routing loop. As illustrated in Figure 9.15:

- *Step 1*: A non-OSPF route is redistributed as an OSPF AS external route into OSPF domain 401 by Athena-PE.
- *Step 2*: This OSPF route is propagated to Morpheus-CE with the down bit set and the tag field set to 65000, the AS number of the MPLS VPN super backbone.
- *Step 3*: Morpheus-CE redistributes the AS external route into OSPF domain 301. The down bit is cleared, but the value of the tag field is retained.
- *Step 4*: The route is propagated to Hermes-PE with the tag value of 65000.
- *Step 5*: Because the tag value matches the AS number belonging to Hermes-PE, the route is not redistributed back to MP-IBGP.

PE routers set the tag field automatically to the BGP AS number to which they belong when redistributing non-OSPF routes from MP-IBGP into OSPF. The tag field remains intact when OSPF AS external routes are redistributed between OSPF domains. PE routers will never redistribute OSPF routes with the tag field equal to their BGP AS number back to MP-IBGP.

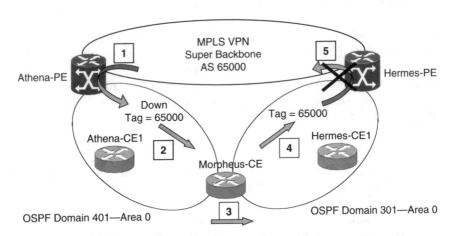

**Figure 9.15    Cross-Domain Routing Loop Prevention with OSPF Tag Field**

In this example, when Hermes-PE receives the OSPF AS external route, the tag field of the route matches its BGP AS number, so the route is not redistributed back to MP-IBGP. Therefore, the cross-domain routing loop issue is resolved using the OSPF tag field.

## 9.10 CASE STUDY 9.3: DEPLOYING OSPF ON CUSTOMER SIDE WITHOUT AREA 0 (USING CISCO ROUTERS)

Case study 9.3 examines the L3 MPLS VPN scenario where OSPF without Area 0 is deployed at the customer edge. The entire L3 MPLS VPN setup is implemented using Cisco 2621 (CE) and 3640 (PE) routers running IOS Software Release 12.2(11T). The scenario presented in this case study is based on a hypothetical Frame Relay environment. All the names used are fictitious, and all IP addresses as well as configurations are provided in the case study for illustrative purposes only. Nevertheless, all the configurations were tested and verified in a lab environment and can be deployed in the field.

### 9.10.1 Case Overview and Network Topology

The setting of this case study is based on the implementation of L3 MPLS VPN over Frame Relay where OSPF is configured as the PE-CE routing protocol on the customer side without Area 0. The L3 MPLS VPN deployment uses Cisco routers spanning across four different customer sites: Site-11, Site-12, Site-21, and Site-22. The network diagram for this case study is shown in Figure 9.16, and the overall OSPF area hierarchy is illustrated in Figure 9.17.

The L3 MPLS VPN implementation allows the four different sites to transparently interconnect through a metro SP's Frame Relay network. Athena-PE and Hermes-PE are the PE routers, whereas Athena-CE1 (Site-11), Athena-CE2 (Site-21), Hermes-CE1 (Site-12), and Hermes-CE2 (Site-22) are the CE routers. The main objective is to support two different IP VPNs: VPN 101 and VPN 102. Each of these VPNs should appear to its users as a private network, separate from all other networks. Within a VPN, each site can send IP packets to any other site belonging to the same VPN. The VPN connectivity requirement for the case study is shown in Figure 9.18:

- Hermes-CE1 should see prefixes 130.130.130.0/24, 8.8.8.8/32, and 3.3.3.0/24 from Athena-CE1.
- Athena-CE1 should see prefixes 200.200.200.0/24, 6.6.6.6/32, and 7.7.7.0/24 from Hermes-CE1.
- Hermes-CE2 should see prefixes 100.100.100.0/24 and 9.9.9.9/32 from Athena-CE2.
- Athena-CE2 should see prefixes 202.202.202.0/24 and 5.5.5.5/32 from Hermes-CE2.

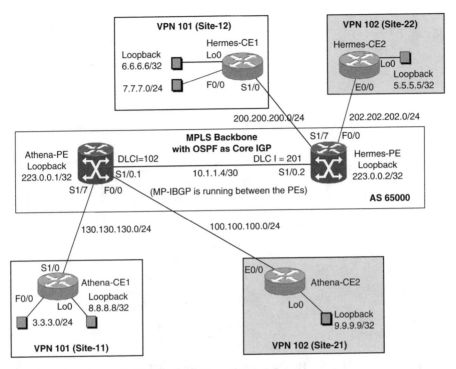

**Figure 9.16   Network Diagram for Case Study 9.3**

## 9.10.2   Network Configurations

Listing 9.39 illustrates the network configuration for the Athena-CE1 router. Self-explanatory comments (in italics) are embedded in between the configuration lines to provide clarification to the configuration commands and procedures.

Area 0 is not defined in Athena-CE1. The FastEthernet0/0, Loopback0, and Serial1/0 interfaces are all part of Area 201. This implies that Athena-CE1 is an OSPF internal router.

---

### Listing 9.39   Athena-CE1 Configuration

```
hostname Athena-CE1

interface Loopback0
ip address 8.8.8.8 255.255.255.255

interface FastEthernet0/0
ip address 3.3.3.3 255.255.255.0
```

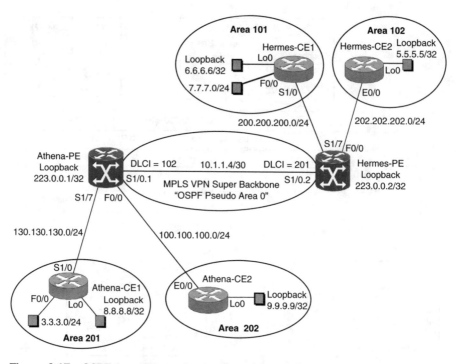

**Figure 9.17    OSPF Area Hierarchy for Case Study 9.3**

```
interface Serial1/0
bandwidth 64
ip address 130.130.130.2 255.255.255.0

! -- Standard OSPF Configuration
router ospf 1

! -- FastEthernet0/0 is part of Area 201
network 3.3.3.0 0.0.0.255 area 201

! -- Loopback0 is part of Area 201
network 8.8.8.8 0.0.0.0 area 201

! -- Serial1/0 is part of Area 201
network 130.130.130.0 0.0.0.255 area 201
```

Listing 9.40 illustrates the network configuration for the Athena-CE2 router. Self-explanatory comments (in italics) are embedded in between

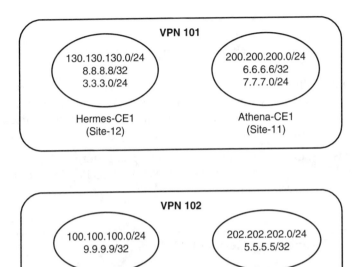

**Figure 9.18   VPN Connectivity Requirement for Case Study 9.3**

the configuration lines to provide clarification to the configuration commands and procedures.

Area 0 is not defined in Athena-CE2. The Loopback0 and Ethernet0/0 interfaces are all part of Area 202. This implies that Athena-CE2 is an OSPF internal router.

---

**Listing 9.40   Athena-CE2 Configuration**

```
hostname Athena-CE2

interface Loopback0
ip address 9.9.9.9 255.255.255.255

interface Ethernet0/0
ip address 100.100.100.2 255.255.255.0

! -- Standard OSPF Configuration
router ospf 1

! -- Loopback0 is part of Area 202
network 9.9.9.9 0.0.0.0 area 202
```

```
! -- Ethernet0/0 is part of Area 202
network 100.100.100.0 0.0.0.255 area 202
```

Listing 9.41 illustrates the network configuration for the Hermes-CE1 router. Self-explanatory comments (in italics) are embedded in between the configuration lines to provide clarification to the configuration commands and procedures.

Area 0 is not defined in Hermes-CE1. The FastEthernet0/0, Loopback0, and Serial1/0 interfaces are all part of Area 101. This implies that Hermes-CE1 is an OSPF internal router.

### Listing 9.41    Hermes-CE1 Configuration

```
hostname Hermes-CE1

interface Loopback0
ip address 6.6.6.6 255.255.255.255

interface FastEthernet0/0
ip address 7.7.7.7 255.255.255.0

interface Serial1/0
bandwidth 64
ip address 200.200.200.2 255.255.255.0

! -- Standard OSPF Configuration
router ospf 1

! -- Loopback0 is part of Area 101
network 6.6.6.6 0.0.0.0 area 101

! -- FastEthernet0/0 is part of Area 101
network 7.7.7.0 0.0.0.255 area 101

! -- Serial1/0 is part of Area 101
network 200.200.200.0 0.0.0.255 area 101
```

Listing 9.42 illustrates the network configuration for the Hermes-CE2 router. Self-explanatory comments (in italics) are embedded in between the configuration lines to provide clarification to the configuration commands and procedures.

Area 0 is not defined in Hermes-CE2. The Loopback0 and Ethernet0/0 interfaces are all part of Area 102. This implies that Hermes-CE2 is an OSPF internal router.

---

**Listing 9.42   Hermes-CE2 Configuration**

```
hostname Hermes-CE2
!
interface Loopback0
ip address 5.5.5.5 255.255.255.255
!
interface Ethernet0/0
ip address 202.202.202.2 255.255.255.0

! -- Standard OSPF Configuration
router ospf 1

! -- Loopback0 is part of Area 102
network 5.5.5.5 0.0.0.0 area 102

! -- Ethernet0/0 is part of Area 102
network 202.202.202.0 0.0.0.255 area 102
```

---

Listing 9.43 illustrates the network configuration for the Athena-PE router. Self-explanatory comments (in italics) are embedded in between the configuration lines to provide clarification to the configuration commands and procedures. The configuration commands and procedures for Athena-PE are summarized as follows:

- OSPF (single-area implementation) is the backbone IGP running between the two PE routers over point-to-point Frame Relay. The VPN prefix information from Athena-CE1 and Athena-CE2 is advertised via MP-IBGP from Athena-PE to Hermes-PE and is reinjected into the OSPF areas at the customer edge of Hermes-PE as OSPF summary network links (type 3 LSAs). The PE routers are area border routers (ABRs).

- For MPLS to be operating properly, Cisco Express Forwarding (CEF) has to be enabled first using the "ip cef " command.
- MPLS is then enabled on the point-to-point Frame-Relay subinterface Serial1/0.1 using the "mpls ip" or "tag-switching ip" command.
- Once MPLS is set up, create VRF instance for VPN 101 connected on Serial1/7, and create VRF instance for VPN 102 connected on

FastEthernet0/0, using the "ip vrf <VPN routing/forwarding instance name>" command.

- When doing this, specify the correct route distinguisher used for VPN 101 and VPN 102 using the "rd <VPN route distinguisher>" command.
- Set up the import and export route targets using the "route-target [export | import | both] <target VPN extended community>" command.
- Configure the forwarding details for the Serial1/7 and FastEthernet0/0 interfaces using the "ip vrf forwarding <table name>" command. Remember to set up the IP address after doing this.
- Configure the OSPF as the PE-CE routing protocol for VPN 101 and VPN 102 using the "router ospf <process ID> vrf <VPN routing/forwarding instance name>" command. This is only applicable to the VRF forwarding interfaces Serial1/7 and FastEthernet0/0. Once an OSPF process ID is assigned to a VRF instance, the same process ID is always used for that particular VRF instance.
- Serial1/7 is configured as part of Area 201 that encompasses VPN 101, and FastEthernet0/0 is configured as part of Area 202 that encompasses VPN 102.
- Note that the current Cisco IOS software implementation limits the overall number of routing protocols in a router to 32. Two routing methods are predefined (static and connected), and two routing protocols are required for proper MPLS VPN backbone operation (MP-IBGP and backbone IGP). Therefore, the number of PE-CE routing processes is restricted to 28.
- Redistribute MP-IBGP into OSPF. Remember to configure OSPF with the same process ID for the same VPN residing on the two PE routers. This is because MP-IBGP supports multiple independent OSPF VPNs and it uses OSPF domain ID, which is derived from the process ID to differentiate one VPN from another (see section 9.9.3). If the OSPF domain ID is different, the redistributed MP-IBGP routes will be regarded as AS external routes instead of inter-area routes.
- Configure MP-IBGP between the two PE routers. Declare Hermes-PE as the IBGP peer.
- The "no auto-summary" command sends subprefix routing information across classful network boundaries. By default, BGP does not accept subnets redistributed from IGP. To advertise and carry subnet routes in BGP, this command is required, which is enabled by default.
- Enter the IPv4 address family mode for each VPN present at this PE router using the "address-family ipv4 vrf <VPN routing/forwarding instance name>" command and redistribute the OSPF routing information.

- Enter the VPNv4 address family mode for this PE router using the "address-family vpnv4" command, and activate the MP-IBGP peers (Hermes-PE in this case) for the advertisement of VPNv4 network layer reachability information (NLRI). Specify that an extended community must be used. This is mandatory.

---

### Listing 9.43    Athena-PE Configuration

```
hostname Athena-PE

! -- Create VRF instance for VPN 101
ip vrf vpn101

! -- Specify the route distinguisher 65000:101 for VRF vpn101
rd 65000:101

! -- Set up the import and export route targets for VRF vpn101
route-target export 65000:101
route-target import 65000:101

! -- Create VRF instance for VPN 102
ip vrf vpn102

! -- Specify the route distinguisher 65000:102 for VRF vpn102
rd 65000:102

! -- Set up the import and export route targets for VRF vpn102
route-target export 65000:102
route-target import 65000:102

! -- Enable Cisco Express Forwarding (CEF)
ip cef

interface Loopback0
ip address 223.0.0.1 255.255.255.255

interface FastEthernet0/0

! -- Assign FastEthernet0/0 to VRF vpn102
ip vrf forwarding vpn102
ip address 100.100.100.1 255.255.255.0

interface Serial1/0
encapsulation frame-relay
```

```
frame-relay lmi-type ansi

interface Serial1/0.1 point-to-point
bandwidth 64
ip address 10.1.1.5 255.255.255.252


! -- Enable MPLS on Frame Relay subinterface Serial1/0.1
tag-switching ip
frame-relay interface-dlci 102


interface Serial1/7
bandwidth 64


! -- Assign Serial1/7 to VRF vpn101
ip vrf forwarding vpn101
ip address 130.130.130.1 255.255.255.0


! -- Configure standard OSPF as the backbone IGP between Athena-PE and
Hermes-PE
router ospf 1
network 10.1.1.4 0.0.0.3 area 0
network 223.0.0.1 0.0.0.0 area 0


! -- Configure OSPF with process ID 101 as the PE-CE routing protocol
for VRF vpn101
router ospf 101 vrf vpn101


! -- Redistribute MP-IBGP into OSPF 101
redistribute bgp 65000 subnets


! -- Serial1/7 is part of Area 201
network 130.130.130.0 0.0.0.255 area 201


! -- Configure OSPF with process ID 102 as the PE-CE routing protocol
for VRF vpn102
router ospf 102 vrf vpn102


! -- Redistribute MP-IBGP into OSPF 102
redistribute bgp 65000 subnets


! -- FastEthernet0/0 is part of Area 202
network 100.100.100.0 0.0.0.255 area 202


! -- Configure MP-IBGP for AS 65000
router bgp 65000
no synchronization
```

```
! -- Declare Hermes-PE as IBGP peer
neighbor 223.0.0.2 remote-as 65000
neighbor 223.0.0.2 update-source Loopback0
no auto-summary

! -- The IPv4 address family mode for VRF vpn102
address-family ipv4 vrf vpn102

! -- Redistribute OSPF 102 into MP-IBGP
redistribute ospf 102
no auto-summary
no synchronization
exit-address-family

! -- The IPv4 address family mode for VRF vpn101
address-family ipv4 vrf vpn101

! -- Redistribute OSPF 101 into MP-IBGP
redistribute ospf 101
no auto-summary
no synchronization
exit-address-family

! -- The VPNv4 address family mode for Athena-PE
address-family vpnv4

! -- Activate Hermes-PE for the advertisement of VPNv4 NLRI
neighbor 223.0.0.2 activate

! -- Specify that BGP extended community must be used
neighbor 223.0.0.2 send-community extended
no auto-summary
exit-address-family
```

Listing 9.44 illustrates the network configuration for the Hermes-PE router. Self-explanatory comments (in italics) are embedded in between the configuration lines to provide clarification to the configuration commands and procedures.

### Listing 9.44  Hermes-PE Configuration

```
hostname Hermes-PE
```

```
! -- Create VRF instance for VPN 101
ip vrf vpn101

! -- Specify the route distinguisher 65000:101 for VRF vpn101
rd 65000:101

! -- Set up the import and export route targets for VRF vpn101
route-target export 65000:101
route-target import 65000:101

! -- Create VRF instance for VPN 102
ip vrf vpn102

! -- Specify the route distinguisher 65000:102 for VRF vpn102
rd 65000:102

! -- Set up the import and export route targets for VRF vpn102
route-target export 65000:102
route-target import 65000:102

! -- Enable Cisco Express Forwarding (CEF)
ip cef

interface Loopback0
ip address 223.0.0.2 255.255.255.255

interface FastEthernet0/0

! -- Assign FastEthernet0/0 to VRF vpn102
ip vrf forwarding vpn102
ip address 202.202.202.1 255.255.255.0

interface Serial1/0
encapsulation frame-relay
frame-relay lmi-type ansi

interface Serial1/0.2 point-to-point
ip address 10.1.1.6 255.255.255.252

! -- Enable MPLS on Frame Relay subinterface Serial1/0.2
tag-switching ip
frame-relay interface-dlci 201

interface Serial1/7
bandwidth 64
```

```
! -- Assign Serial1/7 to VRF vpn101
ip vrf forwarding vpn101
ip address 200.200.200.1 255.255.255.0
clockrate 64000


! -- Configure standard OSPF as the backbone IGP between Hermes-PE and
Athena-PE
router ospf 1
network 10.1.1.4 0.0.0.3 area 0
network 223.0.0.2 0.0.0.0 area 0


! -- Configure OSPF with process ID 101 as the PE-CE routing protocol
for VRF vpn101
router ospf 101 vrf vpn101


! -- Redistribute MP-IBGP into OSPF 101
redistribute bgp 65000 subnets


! -- Serial1/7 is part of Area 101
network 200.200.200.0 0.0.0.255 area 101


! -- Configure OSPF with process ID 102 as the PE-CE routing protocol
for VRF vpn102
router ospf 102 vrf vpn102


! -- Redistribute MP-IBGP into OSPF 102
redistribute bgp 65000 subnets


! -- FastEthernet0/0 is part of Area 102
network 202.202.202.0 0.0.0.255 area 102


! -- Configure MP-IBGP for AS 65000
router bgp 65000
no synchronization


! -- Declare Athena-PE as IBGP peer
neighbor 223.0.0.1 remote-as 65000
neighbor 223.0.0.1 update-source Loopback0
no auto-summary


! -- The IPv4 address family mode for VRF vpn102
address-family ipv4 vrf vpn102


! -- Redistribute OSPF 102 into MP-IBGP
redistribute ospf 102
no auto-summary
```

```
no synchronization
exit-address-family


! -- The IPv4 address family mode for VRF vpn101
address-family ipv4 vrf vpn101


! -- Redistribute OSPF 101 into MP-IBGP
redistribute ospf 101
no auto-summary
no synchronization
exit-address-family


! -- The VPNv4 address family mode for Hermes-PE
address-family vpnv4


! -- Activate Athena-PE for the advertisement of VPNv4 NLRI
neighbor 223.0.0.1 activate


! -- Specify that BGP extended community must be used
neighbor 223.0.0.1 send-community extended
no auto-summary
exit-address-family
```

### 9.10.3 Monitoring OSPF in L3 MPLS VPN Environments

In this case study, the MPLS VPN super backbone assumes the role of a pseudo-Area 0 because the traditional Area 0 is not defined for the customer's OSPF network. To have a better understanding of the MPLS VPN super backbone behavior, a step-by-step analysis is performed for the following:

- Route propagation from Athena-PE to Hermes-PE
- Route propagation from Hermes-PE to Athena-PE

**9.10.3.1 Route Propagation from Athena-PE to Hermes-PE.** Figure 9.19 illustrates the route propagation from Athena-PE to Hermes-PE. The customer routes are originated from Site-11 (advertised by Athena-CE1) and Site-21 (advertised by Athena-CE2).

OSPF Area 201 is implemented for VPN 101 at Athena-CE1. Network prefixes 3.3.3.0/24 and 8.8.8.8/32 are announced by Athena-CE1 to Athena-PE, as illustrated in Listing 9.45. Athena-PE then propagates these routes to Hermes-PE. Network prefix 130.130.130.0/24 is advertised to Hermes-PE as well because Area 201 also encompasses it. Note that prefix 130.130.130.0/24 is reflected as directly connected in Athena-PE's VPNv4 forwarding table for VPN 101 (VRF vpn101).

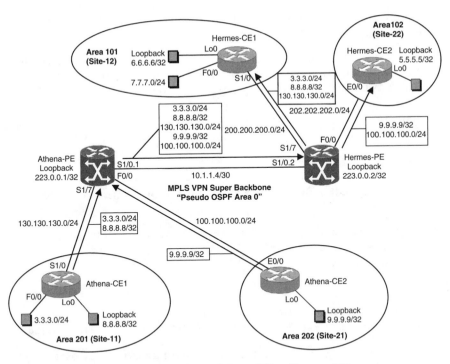

**Figure 9.19    Route Propagation from Athena-PE to Hermes-PE**

---

### Listing 9.45    Athena-PE VPNv4 Forwarding Table for VPN 101

```
Athena-PE#show ip route vrf vpn101

B  200.200.200.0/24 [200/0] via 223.0.0.2, 00:13:25
     3.0.0.0/24 is subnetted, 1 subnets
O  3.3.3.0 [110/1563] via 130.130.130.2, 00:20:09, Serial1/7
     6.0.0.0/32 is subnetted, 1 subnets
B  6.6.6.6 [200/1563] via 223.0.0.2, 00:12:10
     7.0.0.0/24 is subnetted, 1 subnets
B  7.7.7.0 [200/1563] via 223.0.0.2, 00:12:10
     8.0.0.0/32 is subnetted, 1 subnets
O  8.8.8.8 [110/1563] via 130.130.130.2, 00:20:09, Serial1/7
     130.130.0.0/24 is subnetted, 1 subnets
C  130.130.130.0 is directly connected, Serial1/7
```

---

The OSPF routes from Site-11 are advertised by Athena-CE1 to Athena-PE as type 2 LSAs. Listing 9.46 illustrates the BGP VPNv4 information for prefix 3.3.3.0/24. The OSPF route type (RT) extended community attribute is denoted as 201:2:0. The value 201 represents area number 201, and the value 2 indicates OSPF type 2 LSA.

The OSPF domain ID corresponds to 0.0.0.101. This value is derived from OSPF process ID 101 and is used to differentiate VPN 101 (OSPF domain 101) from VPN 102 (OSPF domain 102). In addition, the OSPF cost, which is {1562 (Athena-PE's 64-kbps serial link to Athena-CE1) + 1} = 1563, is copied into the BGP MED attribute. By default, MED is set to the value of the OSPF distance associated with the route, plus 1 (see [DRAFT-OSPF-BGP-MPLS]).

---

**Listing 9.46    BGP VPNv4 Information for Prefix 3.3.3.0/24**

```
Athena-PE#show ip bgp vpnv4 vrf vpn101 3.3.3.3
BGP routing table entry for 65000:101:3.3.3.0/24, version 13
Paths: (1 available, best #1, table vpn101)
Advertised to non peer-group peers:
223.0.0.2
Local
130.130.130.2 from 0.0.0.0 (223.0.0.1)
Origin incomplete, metric 1563, localpref 100, weight 32768, valid,
sourced, best
Extended Community: RT:65000:101 OSPF DOMAIN ID:0.0.0.101 OSPF
RT:201:2:0
```

---

Listing 9.47 illustrates the BGP VPNv4 information for prefix 8.8.8.8/32. The OSPF RT, OSPF domain ID, and MED are identical to those listed in Listing 9.46.

---

**Listing 9.47    BGP VPNv4 Information for Prefix 8.8.8.8/32**

```
Athena-PE#show ip bgp vpnv4 vrf vpn101 8.8.8.8
BGP routing table entry for 65000:101:8.8.8.8/32, version 14
Paths: (1 available, best #1, table vpn101)
Advertised to non peer-group peers:
223.0.0.2
Local
130.130.130.2 from 0.0.0.0 (223.0.0.1)
Origin incomplete, metric 1563, localpref 100, weight 32768, valid,
sourced, best
Extended Community: RT:65000:101 OSPF DOMAIN ID:0.0.0.101 OSPF
RT:201:2:0
```

---

Similarly at Athena-CE2, OSPF Area 202 is implemented for VPN 102. Network prefix 9.9.9.9/32 is announced by Athena-CE2 to Athena-PE, as illustrated in Listing 9.48. Athena-PE then propagates this route to Hermes-PE. Network prefix 100.100.100.0/24 is advertised to Hermes-PE as well because Area 202 also encompasses it. Note that prefix 100.100.100.0/24 is reflected as directly connected in Athena-PE's VPNv4 forwarding table for VPN 102 (VRF vpn102).

---

**Listing 9.48   Athena-PE VPNv4 Forwarding Table for VPN 102**

```
Athena-PE#show ip route vrf vpn102

B  202.202.202.0/24 [200/0] via 223.0.0.2, 00:13:35
     100.0.0.0/24 is subnetted, 1 subnets
C  100.100.100.0 is directly connected, FastEthernet0/0
     5.0.0.0/32 is subnetted, 1 subnets
B  5.5.5.5 [200/11] via 223.0.0.2, 00:10:34
     9.0.0.0/32 is subnetted, 1 subnets
O  9.9.9.9 [110/11] via 100.100.100.2, 00:17:20, FastEthernet0/0
```

---

The OSPF routes from Site-21 are advertised by Athena-CE2 to Athena-PE as type 2 LSAs. Listing 9.49 illustrates the BGP VPNv4 information for prefix 9.9.9.9/32. The OSPF route type (RT) extended community attribute is denoted as 202:2:0. The value 202 represents area number 202, and the value 2 indicates OSPF type 2 LSA.

The OSPF domain ID corresponds to 0.0.0.102. This value is derived from OSPF process ID 102 and is used to differentiate VPN 102 (OSPF domain 102) from VPN 101 (OSPF domain 101). In addition, the OSPF cost, which is {10 (Athena-PE's Ethernet link to Athena-CE2) + 1} = 11, is copied into the BGP MED attribute.

---

**Listing 9.49   BGP VPNv4 Information for Prefix 9.9.9.9/32**

```
Athena-PE#show ip bgp vpnv4 vrf vpn102 9.9.9.9
BGP routing table entry for 65000:102:9.9.9.9/32, version 18
Paths: (1 available, best #1, table vpn102)
Advertised to non peer-group peers:
223.0.0.2
Local
100.100.100.2 from 0.0.0.0 (223.0.0.1)
Origin incomplete, metric 11, localpref 100, weight 32768, valid,
sourced, best
  Extended Community: RT:65000:102 OSPF DOMAIN ID:0.0.0.102 OSPF
RT:202:2:0
```

---

Listing 9.50 illustrates the BGP VPNv4 table of Hermes-PE, which registers prefixes 3.3.3.0/24, 8.8.8.8/32, and 130.130.130.0/24 as belonging to VPN 101 (VRF vpn101), and prefixes 9.9.9.9/32 and 100.100.100.0/24 as belonging to VPN 102 (VRF vpn102). All these prefixes are carried across the super backbone from Athena-PE via IBGP, or rather by MP-IBGP. Note that the BGP MED value is 0 for both prefixes 130.130.130.0/24 and 100.100.100.0/24 because these two subnets are directly connected (with a metric value 0) to Athena-PE.

---

### Listing 9.50   Hermes-PE BGP VPNv4 Forwarding Table

```
Hermes-PE#show ip bgp vpnv4 all
BGP table version is 36, local router ID is 223.0.0.2
Status codes: s suppressed, d damped, h history, * valid, > best, i -
internal
Origin codes: i - IGP, e - EGP, ? - incomplete
```

| Network | Next Hop | Metric | LocPrf | Weight | Path |
|---|---|---|---|---|---|
| Route Distinguisher: 65000:101 (default for vrf vpn101) | | | | | |
| *> i3.3.3.0/24 | 223.0.0.1 | 1563 | 100 | 0 | ? |
| *> 6.6.6.6/32 | 200.200.200.2 | 1563 | | 32768 | ? |
| *> 7.7.7.0/24 | 200.200.200.2 | 1563 | | 32768 | ? |
| *> i8.8.8.8/32 | 223.0.0.1 | 1563 | 100 | 0 | ? |
| *> i130.130.130.0/24 | 223.0.0.1 | 0 | 100 | 0 | ? |
| *> 200.200.200.0 | 0.0.0.0 | 0 | | 32768 | ? |
| Route Distinguisher: 65000:102 (default for vrf vpn102) | | | | | |
| *> 5.5.5.5/32 | 202.202.202.2 | 11 | | 32768 | ? |
| *> i9.9.9.9/32 | 223.0.0.1 | 11 | 100 | 0 | ? |
| *> i100.100.100.0/24 | 223.0.0.1 | 0 | 100 | 0 | ? |
| *> 202.202.202.0 | 0.0.0.0 | 0 | | 32768 | ? |

---

As illustrated in Listing 9.51, the IPv4 routing table of Hermes-CE1 indicates that subnets 3.3.3.0/24, 8.8.8.8/32, and 130.130.130.0/24 are inserted into Area 101 as inter-area (IA) routes (type 3 LSAs). This implies that the original type 2 LSAs were converted to type 3 LSAs by Hermes-PE before they were advertised to Area 101. The cost metrics for the IA routes are aggregated from the ones carried across the super backbone via BGP MED:

- The cost for subnet 3.3.3.0/24 is {1562 (64-kbps serial link) + 1563 (from the BGP MED in Listing 9.46)} = 3125.
- The cost for subnet 8.8.8.8/32 is {1562 (64-kbps serial link) + 1563 (from the BGP MED in Listing 9.47)} = 3125.
- The cost for subnet 130.130.130.0/24 is {1562 (64-kbps serial link) + 1} = 1563.

### Listing 9.51   Hermes-CE1 IPv4 Routing Table

```
Hermes-CE1#show ip route

C 200.200.200.0/24 is directly connected, Serial1/0
  3.0.0.0/24 is subnetted, 1 subnets
O IA 3.3.3.0 [110/3125] via 200.200.200.1, 00:35:39, Serial1/0
  6.0.0.0/32 is subnetted, 1 subnets
C 6.6.6.6 is directly connected, Loopback0
  7.0.0.0/24 is subnetted, 1 subnets
C 7.7.7.0 is directly connected, FastEthernet0/0
  8.0.0.0/32 is subnetted, 1 subnets
O IA 8.8.8.8 [110/3125] via 200.200.200.1, 00:35:39, Serial1/0
  130.130.0.0/24 is subnetted, 1 subnets
O IA 130.130.130.0 [110/1563] via 200.200.200.1, 00:35:39, Serial1/0
```

As illustrated in Listing 9.52, the IPv4 routing table of Hermes-CE2 indicates that subnets 9.9.9.9/32 and 100.100.100.0/24 are inserted into Area 102 as IA routes (type 3 LSAs). This implies that the original type 2 LSAs were converted to type 3 LSAs by Hermes-PE before they were advertised to Area 102. The cost metrics for the IA routes are aggregated from the ones carried across the super backbone via BGP MED:

- The cost for subnet 9.9.9.9/32 is {10 (Ethernet link) + 11 (from the BGP MED in Listing 9.49)} = 21.
- The cost for subnet 100.100.100.0/24 is {10 (Ethernet Link) + 1} = 11.

### Listing 9.52   Hermes-CE2 IPv4 Routing Table

```
Hermes-CE2#show ip route

C 202.202.202.0/24 is directly connected, Ethernet0/0
  100.0.0.0/24 is subnetted, 1 subnets
O IA 100.100.100.0 [110/11] via 202.202.202.1, 00:36:41, Ethernet0/0
  5.0.0.0/16 is subnetted, 1 subnets
C 5.5.0.0 is directly connected, Loopback0
  9.0.0.0/32 is subnetted, 1 subnets
O IA 9.9.9.9 [110/21] via 202.202.202.1, 00:36:41, Ethernet0/0
```

**9.10.3.2   Route Propagation from Hermes-PE to Athena-PE.** Figure 9.20 illustrates the route propagation from Hermes-PE to Athena-PE. The customer routes are originated from Site-12 (advertised by Hermes-CE1) and Site-22 (advertised by Hermes-CE2).

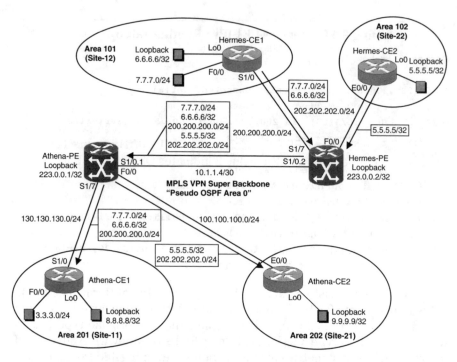

**Figure 9.20   Route Propagation from Hermes-PE to Athena-PE**

OSPF Area 101 is implemented for VPN 101 at Hermes-CE1. Network prefixes 6.6.6.6/32 and 7.7.7.0/24 are announced by Hermes-CE1 to Hermes-PE, as illustrated in Listing 9.53. Hermes-PE then propagates these routes to Athena-PE. Network prefix 200.200.200.0/24 is advertised as well because Area 101 also encompasses it. Note that prefix 200.200.200.0/24 is reflected as directly connected in Hermes-PE's VPNv4 forwarding table for VPN 101 (VRF vpn101).

---

**Listing 9.53   Hermes-PE VPNv4 Forwarding Table for VPN 101**

```
Hermes-PE#show ip route vrf vpn101

C 200.200.200.0/24 is directly connected, Serial1/7
  3.0.0.0/24 is subnetted, 1 subnets
B 3.3.3.0 [200/1563] via 223.0.0.1, 00:04:06
  6.0.0.0/32 is subnetted, 1 subnets
O 6.6.6.6 [110/1563] via 200.200.200.2, 00:26:41, Serial1/7
  7.0.0.0/24 is subnetted, 1 subnets
O 7.7.7.0 [110/1563] via 200.200.200.2, 00:26:41, Serial1/7
```

```
   8.0.0.0/32 is subnetted, 1 subnets
B  8.8.8.8 [200/1563] via 223.0.0.1, 00:04:06
   130.130.0.0/24 is subnetted, 1 subnets
B  130.130.130.0 [200/0] via 223.0.0.1, 00:04:06
```

The OSPF routes from Site-12 are advertised by Hermes-CE1 to Hermes-PE as type 2 LSAs. Listing 9.54 illustrates the BGP VPNv4 information for prefix 6.6.6.6/32. The OSPF RT is denoted as 101:2:0. The value 101 represents area number 101, and the value 2 indicates OSPF type 2 LSA.

The OSPF domain ID corresponds to 0.0.0.101. This value is derived from OSPF process ID 101 and is used to differentiate VPN 101 (OSPF domain 101) from VPN 102 (OSPF domain 102). In addition, OSPF cost, which is {1562 (Hermes-PE's 64-kbps serial link to Hermes-CE1) + 1} = 1563, is copied into the BGP MED attribute.

### Listing 9.54   BGP VPNv4 Information for Prefix 6.6.6.6/32

```
Hermes-PE#show ip bgp vpnv4 vrf vpn101 6.6.6.6
BGP routing table entry for 65000:101:6.6.6.6/32, version 18
Paths: (1 available, best #1, table vpn101)
Advertised to non peer-group peers:
223.0.0.1
Local
200.200.200.2 from 0.0.0.0 (223.0.0.2)
Origin incomplete, metric 1563, localpref 100, weight 32768, valid,
sourced, best
Extended Community: RT:65000:101 OSPF DOMAIN ID:0.0.0.101 OSPF
RT:101:2:0
```

Listing 9.55 illustrates the BGP VPNv4 information for prefix 7.7.7.0/24. The OSPF RT, OSPF domain ID, and MED are identical to those listed in Listing 9.54.

### Listing 9.55   BGP VPNv4 Information for Prefix 7.7.7.0/24

```
Hermes-PE#show ip bgp vpnv4 vrf vpn101 7.7.7.7
BGP routing table entry for 65000:101:7.7.7.0/24, version 19
Paths: (1 available, best #1, table vpn101)
Advertised to non peer-group peers:
223.0.0.1
Local
```

```
200.200.200.2 from 0.0.0.0 (223.0.0.2)
```

Origin incomplete, **metric 1563,** localpref 100, weight 32768, valid, sourced, best

Extended Community: RT:65000:101 **OSPF DOMAIN ID:0.0.0.101 OSPF RT:101:2:0**

---

Similarly at Hermes-CE2, OSPF Area 102 is implemented for VPN 102. Network prefix 5.5.5.5/32 is announced by Hermes-CE2 to Hermes-PE, as illustrated in Listing 9.56. Hermes-PE then propagates this route to Athena-PE. Network prefix 202.202.202.0/24 is advertised to Athena-PE as well because Area 102 also encompasses it. Note that prefix 202.202.202.0/24 is reflected as directly connected in Hermes-PE's VPNv4 forwarding table for VPN 102 (VRF vpn102).

---

### Listing 9.56   Hermes-PE VPNv4 Forwarding Table for VPN 102

```
Hermes-PE#show ip route vrf vpn102

C  202.202.202.0/24 is directly connected, FastEthernet0/0
   100.0.0.0/24 is subnetted, 1 subnets
B  100.100.100.0 [200/0] via 223.0.0.1, 00:04:15
   5.0.0.0/32 is subnetted, 1 subnets
O  5.5.5.5 [110/11] via 202.202.202.2, 00:25:04, FastEthernet0/0
   9.0.0.0/32 is subnetted, 1 subnets
B  9.9.9.9 [200/11] via 223.0.0.1, 00:04:15
```

---

The OSPF routes from Site-22 are advertised by Hermes-CE2 to Hermes-PE as type 2 LSAs. Listing 9.57 illustrates the BGP VPNv4 information for prefix 5.5.5.5/32. The OSPF RT is denoted as 102:2:0. The value 102 represents Area 102, and the value 2 indicates OSPF type 2 LSA.

The OSPF domain ID corresponds to 0.0.0.102. This value is derived from OSPF process ID 102 and is used to differentiate VPN 102 (OSPF domain 102) from VPN 101(OSPF domain 101). In addition, the OSPF cost, which is {10 (Hermes-PE's Ethernet link to Hermes-CE2) + 1} = 11, is copied into the BGP MED attribute.

---

### Listing 9.57   BGP VPNv4 Information for Prefix 5.5.5.5/32

```
Hermes-PE#show ip bgp vpnv4 vrf vpn102 5.5.5.5
BGP routing table entry for 65000:102:5.5.5.5/32, version 21
Paths: (1 available, best #1, table vpn102)
```

```
Advertised to non peer-group peers:
223.0.0.1
Local
202.202.202.2 from 0.0.0.0 (223.0.0.2)
Origin incomplete, metric 11, localpref 100, weight 32768, valid,
sourced, best
Extended Community: RT:65000:102 OSPF DOMAIN ID:0.0.0.102 OSPF
RT:102:2:0
```

Listing 9.58 illustrates the BGP VPNv4 table of Athena-PE, which registers prefixes 6.6.6.6/32, 7.7.7.0/24, and 200.200.200.0/24 as belonging to VPN 101 (VRF vpn101), and prefixes 5.5.5.5/32 and 202.202.202.0/24 as belonging to VPN 102 (VRF vpn102). All these prefixes are carried across the super backbone from Hermes-PE via MP-IBGP. Note that the BGP MED value is 0 for both prefixes 200.200.200.0/24 and 202.202.202.0/24 because these two subnets are directly connected (with a metric value of 0) to Hermes-PE.

### Listing 9.58   Athena-PE BGP VPNv4 Forwarding Table

```
Athena-PE#show ip bgp vpnv4 all
BGP table version is 30, local router ID is 223.0.0.1
Status codes: s suppressed, d damped, h history, * valid, > best, i -
internal
Origin codes: i - IGP, e - EGP, ? - incomplete
```

| Network | Next Hop | Metric | LocPrf | Weight | Path |
|---|---|---|---|---|---|
| Route Distinguisher: 65000:101 (default for vrf vpn101) | | | | | |
| *> 3.3.3.0/24 | 130.130.130.2 | 1563 | | 32768 | ? |
| *> i6.6.6.6/32 | 223.0.0.2 | 1563 | 100 | 0 | ? |
| *> i7.7.7.0/24 | 223.0.0.2 | 1563 | 100 | 0 | ? |
| *> 8.8.8.8/32 | 130.130.130.2 | 1563 | | 32768 | ? |
| *> 130.130.130.0/24 | 0.0.0.0 | 0 | | 32768 | ? |
| *> i200.200.200.0 | 223.0.0.2 | 0 | 100 | 0 | ? |
| Route Distinguisher: 65000:102 (default for vrf vpn102) | | | | | |
| *> i5.5.5.5/32 | 223.0.0.2 | 11 | 100 | 0 | ? |
| *> 9.9.9.9/32 | 100.100.100.2 | 11 | | 32768 | ? |
| *> 100.100.100.0/24 | 0.0.0.0 | 0 | | 32768 | ? |
| *> i202.202.202.0 | 223.0.0.2 | 0 | 100 | 0 | ? |

As illustrated in Listing 9.59, the IPv4 routing table of Athena-CE1 indicates that subnets 6.6.6.6/32, 7.7.7.0/24, and 200.200.200.0/24 are inserted into Area 201 as IA routes (type 3 LSAs). This implies that the original type

2 LSAs were converted to type 3 LSAs by Athena-PE before they were advertised to Area 201. The cost metrics for the IA routes are aggregated from the ones that are carried across the super backbone via the BGP MED:

- The cost for subnet 6.6.6.6/32 is {1562 (64-kbps serial link) + 1563 (from the BGP MED in Listing 9.54)} = 3125.
- The cost for subnet 7.7.7.0/24 is {1562 (64-kbps serial link) + 1563 (from the BGP MED in Listing 9.55)} = 3125.
- The cost for subnet 200.200.200.0/24 is {1562 (64-kbps serial link) + 1} = 1563.

---

### Listing 9.59    Athena-CE1 IPv4 Routing Table

```
Athena-CE1#show ip route

O IA 200.200.200.0/24 [110/1563] via 130.130.130.1, 00:09:32, Serial1/0
   3.0.0.0/24 is subnetted, 1 subnets
C 3.3.3.0 is directly connected, FastEthernet0/0
   6.0.0.0/32 is subnetted, 1 subnets
O IA 6.6.6.6 [110/3125] via 130.130.130.1, 00:08:17, Serial1/0
   7.0.0.0/24 is subnetted, 1 subnets
O IA 7.7.7.0 [110/3125] via 130.130.130.1, 00:08:17, Serial1/0
   8.0.0.0/24 is subnetted, 1 subnets
C 8.8.8.0 is directly connected, Loopback0
   130.130.0.0/24 is subnetted, 1 subnets
C 130.130.130.0 is directly connected, Serial1/0
```

---

As illustrated in Listing 9.60, the IPv4 routing table of Athena-CE2 indicates that subnets 5.5.5.5/32 and 202.202.202.0/24 are inserted into Area 202 as IA routes. This implies that the original type 2 LSAs were converted to type 3 LSAs by Athena-PE before they were advertised to Area 202. The cost metrics for these routes are aggregated from the ones that are carried across the super backbone via the BGP MED:

- The cost for subnet 5.5.5.5/32 is {10 (Ethernet link) + 11 (from the BGP MED in Listing 9.57)} = 21.
- The cost for subnet 202.202.202.0/24 is {10 (Ethernet link) + 1} = 11.

---

### Listing 9.60    Athena-CE2 IPv4 Routing Table

```
Athena-CE2#show ip route

O IA 202.202.202.0/24 [110/11] via 100.100.100.1, 00:06:48, Ethernet0/0
   100.0.0.0/24 is subnetted, 1 subnets
```

```
C 100.100.100.0 is directly connected, Ethernet0/0
   5.0.0.0/32 is subnetted, 1 subnets
O IA 5.5.5.5 [110/21] via 100.100.100.1, 00:03:47, Ethernet0/0
C 9.0.0.0/8 is directly connected, Loopback0
```

### 9.10.4  VPN Connectivity Verification

The ping and traceroute commands are used to verify the VPN connectivity (see Figure 9.18) in this case study.

Listing 9.61 illustrates the successful outcome of the ping and traceroute tests performed at Athena-CE1 for host addresses 6.6.6.6 and 7.7.7.7 found in OPSF Area 101 of VPN 101.

### Listing 9.61  Ping and Traceroute Tests at Athena-CE1

```
Athena-CE1#ping 6.6.6.6

Type escape sequence to abort.
Sending 5, 100-byte ICMP Echos to 6.6.6.6, timeout is 2 seconds:
!!!!!
Success rate is 100 percent (5/5), round-trip min/avg/max = 120/126/144
ms

Athena-CE1#traceroute 6.6.6.6

Type escape sequence to abort.
Tracing the route to 6.6.6.6

1 130.130.130.1 40 msec 28 msec 24 msec
2 200.200.200.1 [MPLS: Label 21 Exp 0] 136 msec 144 msec 140 msec
3 200.200.200.2 64 msec * 64 msec

Athena-CE1#ping 7.7.7.7

Type escape sequence to abort.
Sending 5, 100-byte ICMP Echos to 7.7.7.7, timeout is 2 seconds:
!!!!!
Success rate is 100 percent (5/5), round-trip min/avg/max = 120/126/144
ms

Athena-CE1#traceroute 7.7.7.7

Type escape sequence to abort.
Tracing the route to 7.7.7.7
```

```
1 130.130.130.1 44 msec 28 msec 28 msec
2 200.200.200.1 [MPLS: Label 22 Exp 0] 136 msec 144 msec 136 msec
3 200.200.200.2 64 msec * 64 msec
```

Listing 9.62 illustrates the successful outcome of the ping and traceroute tests performed at Hermes-CE1 for host addresses 8.8.8.8 and 3.3.3.3 found in OPSF Area 201 of VPN 101.

### Listing 9.62   Ping and Traceroute Tests at Hermes-CE1

**Hermes-CE1#ping 8.8.8.8**

```
Type escape sequence to abort.
Sending 5, 100-byte ICMP Echos to 8.8.8.8, timeout is 2 seconds:
!!!!!
Success rate is 100 percent (5/5), round-trip min/avg/max = 120/130/160
ms
```

**Hermes-CE1#traceroute 8.8.8.8**

```
Type escape sequence to abort.
Tracing the route to 8.8.8.8
```

```
1 200.200.200.1 28 msec 24 msec 24 msec
2 130.130.130.1 [MPLS: Label 23 Exp 0] 88 msec 116 msec 104 msec
3 130.130.130.2 76 msec * 72 msec
```

**Hermes-CE1#ping 3.3.3.3**

```
Type escape sequence to abort.
Sending 5, 100-byte ICMP Echos to 3.3.3.3, timeout is 2 seconds:
!!!!!
Success rate is 100 percent (5/5), round-trip min/avg/max = 120/130/160
ms
```

**Hermes-CE1#traceroute 3.3.3.3**

```
Type escape sequence to abort.
Tracing the route to 3.3.3.3
```

```
1 200.200.200.1 32 msec 24 msec 24 msec
2 130.130.130.1 [MPLS: Label 22 Exp 0] 88 msec 116 msec 100 msec
3 130.130.130.2 80 msec * 72 msec
```

Listing 9.62 illustrates the successful outcome of the ping and traceroute tests performed at Athena-CE2 for host address 5.5.5.5 found in OPSF Area 102 of VPN 102.

---

### Listing 9.62   Ping and Traceroute Tests at Athena-CE2

```
Athena-CE2#ping 5.5.5.5

Type escape sequence to abort.
Sending 5, 100-byte ICMP Echos to 5.5.5.5, timeout is 2 seconds:
!!!!!
Success rate is 100 percent (5/5), round-trip min/avg/max = 60/61/64 ms

Athena-CE2#traceroute 5.5.5.5

Type escape sequence to abort.
Tracing the route to 5.5.5.5

1 100.100.100.1 4 msec  4 msec  0 msec
2 202.202.202.1 64 msec 60 msec 64 msec
3 202.202.202.2 32 msec *  32 msec
```

---

Listing 9.63 illustrates the successful outcome of the ping and traceroute tests performed at Hermes-CE2 for host address 9.9.9.9 found in OPSF Area 202 of VPN 102.

---

### Listing 9.63   Ping and Traceroute Tests at Hermes-CE2

```
Hermes-CE2#ping 9.9.9.9

Type escape sequence to abort.
Sending 5, 100-byte ICMP Echos to 9.9.9.9, timeout is 2 seconds:
!!!!!
Success rate is 100 percent (5/5), round-trip min/avg/max = 60/61/64 ms

Hermes-CE2#traceroute 9.9.9.9

Type escape sequence to abort.
Tracing the route to 9.9.9.9

1 202.202.202.1 0 msec  0 msec  4 msec
2 100.100.100.1 60 msec 60 msec 60 msec
3 100.100.100.2 32 msec *  32 msec
```

---

## 9.11  SUMMARY

The L3 MPLS VPN model allows metro SPs to provide VPN services to customers using an IP backbone. By using MPLS for forwarding VPN traffic and
MP-IBGP for distributing VPN routes, this approach increases scalability
and flexibility for the metro SP while removing the need for customers to
build and run their own IP backbone using globally unique IPv4 addresses.
The three case studies in this chapter examine the setup of L3 MPLS VPNs
using various PE-CE routing protocols such as static routes, RIPv2, EBGP,
and OSPF. When deploying OSPF at the customer edge, the MPLS VPN backbone becomes a pseudo-Area 0 (or an OSPF super backbone) above the
regular OSPF Area 0.

The L2 MPLS VPN implementation supports the encapsulation and
transport of L2 protocol data units (PDUs) across an MPLS network
through the use of Martini's point-to-point tunnels. MPLS labels are used
instead of network layer encapsulation to tunnel L2 frames across an MPLS
backbone network. The direct interoperability with existing L2 VPN
deployments is the main strength of the Martini approach. Thus, L2 MPLS
VPN is well posed to extend and scale legacy L2 VPN setups. The case studies for L2 MPLS VPNs are presented in the next chapter.

To sum up, with MPLS, metro SPs can offer tiered L3 or L2 VPN services
coupled with different quality-of-service (QoS) and reliability guarantees
for each individual metro subscriber. These simple and yet robust VPN services not only fulfill the metro subscribers' common VPN needs, but also
become powerful service-delivery constructs for the metro SPs.

Each metro subscriber usually has its own unique set of business
requirements and will pay accordingly as its traffic is prioritized differently
in the MAN. If only a uniform type of service is offered, the metro SPs will
only be able to charge a flat rate for bandwidth. To realize new revenue
opportunities, metro SPs must offer different classes of service to the subscribers.

The application of MPLS VPNs as user/customer classifiers lays down
the foundation for service differentiation based on different users/custom-
ers, which in turn provides metro SPs the flexibility to incorporate more
refined service differentiation mechanisms (see chapter 11) to these
users/customers.

# Chapter 10
# Virtual Private LAN Services

## 10.1 INTRODUCTION

Ethernet has become the most universal and extensively deployed local area network (LAN) technology due to its relatively low cost and simple setup. As the throughput of Ethernet increased from 10 Mbps to 10 Gbps (IEEE 802.3ae), it gained popularity as a metropolitan area network (MAN) service, commonly referred to as metro Ethernet service.

With Martini's approach, metro Ethernet service offerings are somewhat limited to point-to-point connections between multiple sites within the same metro area. Martini's point-to-point tunneling implementation alone is insufficient to ensure the scalability required during network expansion because Ethernet by nature is not point to point. The main challenge is to move beyond point-to-point connectivity for sites within a single metro area to multipoint connectivity for intra- or inter-metro sites.

Virtual private LAN services (VPLS) comes into the picture because it allows metro service providers (SPs) with an existing Multi-Protocol Label Switching (MPLS) infrastructure to offer geographically dispersed Ethernet multipoint services (EMS), also known as Ethernet private LAN services, defined by the Metropolitan Ethernet Forum (MEF). VPLS literally extends the physical reach of Ethernet to function as a wide area network (WAN) solution (see section 10.8.1).

VPLS is a layer-2 (L2) VPN architecture that enables EMS over a packet-switched network infrastructure such as an IP/MPLS core network. The VPN subscribers or end users get an emulated LAN segment with a single L2 broadcast domain and perceive the service as a virtual private Ethernet switch that forwards frames to their respective destinations within the VPN.

As EMS is based on Ethernet transparent bridging, which is non-IP based, it is often referred to as transparent LAN service. In other words, all sites belonging to an enterprise user who has subscribed to EMS will appear as if they are connected to the same Ethernet LAN, regardless of whether the sites are in a specific metro area or spanning across multiple metro areas.

## 10.2 MULTIPOINT L2 VPN SERVICES

One of the key features of EMS is that each customer edge (CE) node communicates directly with all other remote CE nodes associated with the EMS (same concept as a virtual LAN (VLAN) or L2 VPN). This any-to-any or multipoint L2 VPN connectivity service is very useful to enterprise applications such as IP telephony that need to operate on a peer-to-peer basis.

VPLS is also a multipoint L2 VPN technology that allows an end user to directly access multiple destination sites associated with the same VPN membership through a single physical or logical connection. It also requires the network to make a forwarding decision derived from the destination address of the packet. Within the context of VPLS, the forwarding decision is based on the destination MAC address of the Ethernet frame.

A multipoint service is desirable because full connectivity between multiple sites is achieved with fewer connections and optimal forwarding. For the same level of connectivity, a point-to-point technology would require more connections and probably the use of suboptimal packet forwarding (for instance, in a hub-and-spoke topology), thus introducing additional transmission delay.

## 10.3 SERVICE OFFERINGS

VPLS creates an important emerging market opportunity for metro SPs to offer multipoint L2 VPN services that interconnect both intra- and inter-metro sites. The service offerings supported by VPLS include:

- *Distributed network access point (NAP)*: Metro SPs can offer a geographically dispersed virtual NAP across a high-speed MPLS backbone with the virtual L2 switch created by VPLS. This allows metro SPs to offer transparent private peering between multiple Internet service providers (ISPs).
- *Provision of L2 connectivity between multiple corporate sites*: This enables enterprise applications with nonroutable L2 mechanisms, such as server cluster heartbeats, to be geographically distributed, thus enhancing business continuity and availability.
- *Interconnected multiple sites running non-IP applications*: Multiple customer sites running legacy applications that are still using nonroutable protocols, such as NetBEUI, for communication purposes can be interconnected.
- *Provision of any-to-any L2 VPN service between sites*: This is particularly useful for enterprise customers who would prefer any-to-any L2 VPN connectivity between sites instead of layer-3 (L3) VPN.

## 10.4    FUNCTIONAL COMPONENTS OF VPLS

In its simplest form, a VPLS consists of several sites connected to provider edge (PE) devices that implement the emulated LAN service. These PE devices make forwarding decisions between sites and encapsulate the Ethernet frames across an IP/MPLS network using a "pseudowire," which is really a virtual circuit (VC), or an Ethernet emulated circuit to be exact. From the perspective of MPLS, a pseudowire refers to a VC label switched path (LSP).

VPLS relies on the same encapsulation defined by Martini for point-to-point Ethernet over MPLS (see section 9.6.4). The frame preamble and frame check sequence (FCS) are stripped, and the remaining payload is encapsulated with a VC label and a tunnel label.

A virtual switching instance (VSI) is used at each PE device to implement the forwarding decisions of each VPLS domain or emulated LAN. A VSI can be considered a virtual transparent bridge. Each VSI is interconnected with a full mesh of pseudowires to forward the Ethernet frames between PE devices. Figure 10.1 illustrates the components of a VPLS interconnecting three points of presence (POPs).

Because a full mesh of pseudowires (or LSPs) needs to be built between all the VSIs on each of the PE devices in a particular VPLS domain, the amount of administrative effort required to establish this mesh of pseudowires can be a concern whenever a new PE device or VSI is added. This scalability constraint is addressed in a later section.

## 10.5    FRAME FORWARDING

VPLS enables the PE device to act as a self-learning transparent bridge with one media access control (MAC) address table per VSI. In other words, the VSI on the PE device has a VSI forwarding table that is constructed through learning the source MAC addresses of Ethernet frames as they enter specific physical/logical ports or pseudowires, functioning in the same manner as a legacy L2 Ethernet switch.

When an Ethernet frame enters via a user-facing ingress port on the PE device, the destination MAC address is looked up in the VSI forwarding table, and the frame is forwarded onto the pseudowire that will deliver it to the correct PE device attaching the remote site. If the destination is a broadcast, multicast, or unknown unicast MAC address, the Ethernet frame will be replicated and flooded out of all ports associated with that VSI, except the ingress port where it just entered.

Once the PE device receives a response back from the host on a specific port that owns that MAC address, the VSI forwarding table is updated accordingly. MAC addresses that have not been used for a specific period

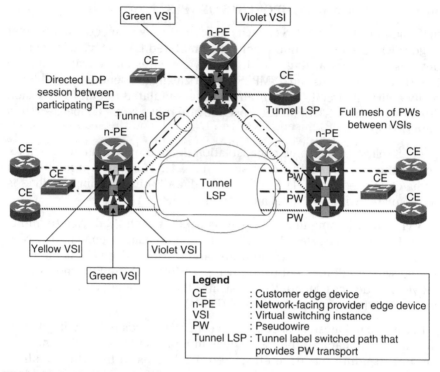

**Figure 10.1   VPLS Components**

of time are aged out to control the VSI forwarding table size. PE devices use split-horizon forwarding on the pseudowires to ensure a loop-free topology. This way the full mesh of pseudowires provides direct connectivity between the PE devices in a VPLS domain, and the Spanning Tree Protocol (SPT) is no longer required to provide a loop-free topology.

## 10.6   VPLS VERSUS MARTINI POINT-TO-POINT L2 VPN SERVICE

Martini's approach enables point-to-point L2 VPN services, and VPLS complements Martini's model with a multipoint service offering based on Ethernet. These two kinds of services impose different requirements for the PE devices. A point-to-point service relies on a VC (or pseudowire) that PE devices set up to transport L2 frames between two attachment circuits (or L2 forwarding equivalence classes (FECs)). The mapping between attachment circuits and VCs is static and one to one. A multipoint service requires the PE device to perform a lookup on the frame contents (typically, MAC addresses) to determine the VC to be used to forward the frame to the destination. This lookup creates the multipoint nature of a VPLS domain. The VC signaling and encapsulation characteristics performed by the PE devices are identical. The operation of PE devices is transparent from the type of service implemented by the devices.

## 10.7  VPLS IMPLEMENTATION

There are two leading VPLS drafts currently being discussed in the Internet Engineering Task Force (IETF) that the majority of vendors are considering from an implementation perspective. The first approach is Label Distribution Protocol (LDP) based [DRAFT-VPLS-LDP], and the second is Border Gateway Protocol (BGP) based [DRAFT-VPLS-BGP]. These models can be further described by two functional attributes:

- *Autodiscovery*: This attribute determines how multiple PE devices participating in a VPLS domain can automatically locate each other.
- *Signaling*: This attribute determines what protocol is used to distribute MPLS labels between the PE devices and setup pseudowires (VC LSPs).

The [DRAFT-VPLS-LDP] does not describe any particular discovery mechanism, whereas [DRAFT-VPLS-BGP] uses BGP for the autodiscovery of PE devices. The BGP-based approach is beyond the scope of this text, but the LDP-based approach is discussed in the following subsection.

### 10.7.1  LDP-Based Approach

Because no discovery protocol has been defined for the LDP-based VPLS implementation, the metro SP must know explicitly which PE devices are associated with which VSI. In other words, for every VSI present on a PE device, the metro SP will have to statically configure that PE device with the addresses of all other PEs that are part of that VSI. Put simply, static configuration of PE associations within a VPLS domain is required in the LDP-based approach.

The MPLS label distribution and pseudowire or VC setup uses the same LDP signaling mechanism defined for Martini's point-to-point services. Using a directed LDP session, each PE device advertises a VC label mapping that is used as part of the label stack imposed on the Ethernet frames by the ingress PE device during packet forwarding. The L2 VPN information is carried in a Label Mapping Message [RFC3036] sent in the downstream-unsolicited mode, which contains the VC FEC type/length/value (TLV) (see [DRAFT-MARTINI-TRANSP]). To establish a full mesh of pseudowires, all PE devices in a VPLS domain must have a full mesh of LDP sessions. Once an LDP session has been established between two PE devices, all pseudowires are signaled over this session.

This VPLS implementation model does not require the exchange of reachability (MAC addresses) information via LDP. Such information is learned from the forwarding plane [REF01] using standard address learning, aging, and filtering mechanisms defined for Ethernet bridging (see section 10.5). On the other hand, LDP, which is used for setting up and tearing down the VCs, can also be used to indicate to a remote PE device that some or all MAC addresses learned over a VC need to be withdrawn from the VSI when there is a topology change (for instance, failure of the primary link for a dual-homed bridging-capable multi-tenant unit or customer site). This MAC address withdrawal mechanism [DRAFT-VPLS-LDP] provides a convergence optimization over the normal address aging that would eventually flush the invalid addresses at a longer time interval.

## 10.8  SCALING LDP-BASED VPLS

Packet replication and the amount of address information are the two main scalability concerns for PE devices implementing LDP-based VPLS. The ingress PE device needs to perform packet replication and flooding for packets that have broadcast, multicast, or destination-unknown unicast addresses. As the number of PE devices in a VPLS domain grows, the number of replicated packets that need to be generated also increases. Depending on the hardware architecture, packet replication can have a profound impact on processing and memory resources. Moreover, the number of MAC addresses that may be learned from the forwarding plane may grow very quickly if many hosts connect to the VPLS domain all at once.

The VPLS architecture that we have discussed hitherto is basically flat or nonhierarchical. The above-mentioned scaling issues can be mitigated through the deployment of a hierarchical VPLS model.

### 10.8.1  Hierarchical VPLS

Hierarchical VPLS (H-VPLS) further partitions the flat VPLS network into Ethernet edge domains (see Figure 10.2). Two types of PE devices are defined for this model: user-facing provider edge (u-PE) and network provider edge (n-PE). The IP/MPLS core interconnects the n-PE devices, which in turn are connected to the u-PE devices. The CE devices are directly connected to the u-PE devices, which aggregate VPLS traffic before it reaches the n-PE devices, where the VPLS forwarding takes place based on the VSI.

As illustrated in Figure 10.2, the architecture of H-VPLS uses standard Ethernet switches at the edge and IP/MPLS within the core. The following standards need to be supported at the respective edge domains:

- IEEE 802.1Q at the CE and u-PE
- IEEE 802.1AD provider bridges draft at the u-PE and n-PE
- LDP-based VPLS at the IP/MPLS core between the n-PEs

Based on the above-stated standards, the CE and u-PE devices will have to function as standard Ethernet switches that support IEEE 802.1Q. The

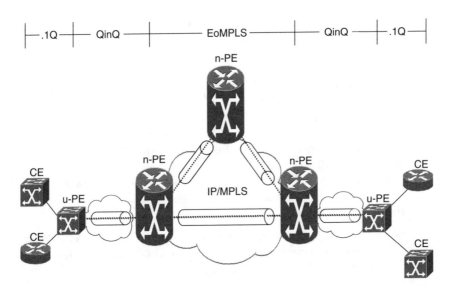

**Figure 10.2  Hierarchical VPLS**

u-PE devices will also have to support VLAN tag stacking, also known as a double 802.1Q or Q-in-Q encapsulation, described within the IEEE 802.1AD provider bridges draft. The Q-in-Q encapsulation is used to aggregate traffic between the u-PE and n-PE devices. The Q-in-Q trunk becomes an access port to a VSI on an n-PE device. Correspondingly, the n-PE devices will have to support Q-in-Q encapsulation and VPLS.

The H-VPLS model allows metro SPs to interconnect dispersed metro Ethernet domains, thus extending the geographical coverage of metro Ethernet services. Because the IEEE 802.1Q specifications define a 12-bit field used for the VLAN ID, each of these networks can support at most 4,096 VLANs. As one unique VLAN ID is required per subscriber, the total number of metro subscribers that can be provisioned on a flat VPLS network is limited. H-VPLS helps to scale metro Ethernet services beyond the 4,096-subscriber limit imposed by the VLAN address space using VLAN tag stacking (Q-in-Q).

On the other hand, using Ethernet as the edge technology contributes to the scalability of VPLS by reducing the signaling overhead and packet replication requirements because the full mesh of pseudowires is confined to only between the n-PE devices. Having an Ethernet access network also simplifies the operation of the edge domain and considerably brings down the cost of the edge devices.

## 10.9  COMPARISON BETWEEN VPLS AND L3 MPLS VPN

VPLS and L3 MPLS VPN enable two very different services:

- VPLS offers a multipoint Ethernet service that can support multiple higher-level protocols. Although L3 MPLS VPN also offers a multipoint service, it is limited to the transport of IP traffic and all traffic that can be carried over IP.
- Both VPLS and L3 MPLS VPN support multiple link technologies such as Ethernet, Frame Relay, Asynchronous Transfer Mode (ATM), point-to-point protocol (PPP), and high-level data control (HDLC) for the CE to PE connections. VPLS, however, imposes additional requirements such as bridged encapsulation on the CE devices to support non-Ethernet links.
- With L3 MPLS VPN, the IP routing design and operation required from the VPN user are minimal. VPLS leaves full control of IP routing to the VPN user.

VPLS and L3 MPLS VPN are two alternatives to implement a VPN. Whichever is the more appropriate VPN technology will depend on the specific service requirements of the VPN customer.

## 10.10   CASE STUDY 10.1: VIRTUAL LEASED LINE SERVICE

Case study 10.1 describes the implementation of virtual leased line (VLL) service connectivity using Martini's point-to-point tunneling approach.

### 10.10.1  Case Overview and Network Topology

The following setting for case study 10.1 is acquired directly from case study 9.1 (see section 9.7):

- Commercial estate 3 in the Notus district
- Commercial estate 4 in the Eurus district
- Company XYZ has a main branch in commercial estate 3 and a subbranch in commercial estate 4. Both sites must be able to communicate with each other via the MAN as if they are interconnected via a virtual leased line.

In addition, Company RST has a main office in commercial estate 2 and a branch office in commercial estate 4. Both offices must be able to communicate with each other via the MAN as if they are interconnected via a virtual leased line.

These companies have existing LAN services in their premises, and they are only interested in scaling these services. Figure 10.3 illustrates the network diagram for this case study. Two L2 VPNs or VLANs are implemented in this case: "blue" VLAN for Company RST and "green" VLAN for Company XYZ. The blue VLAN extends across the Zephir and Eurus districts, while the green VPN spans across the Notus and Eurus districts.

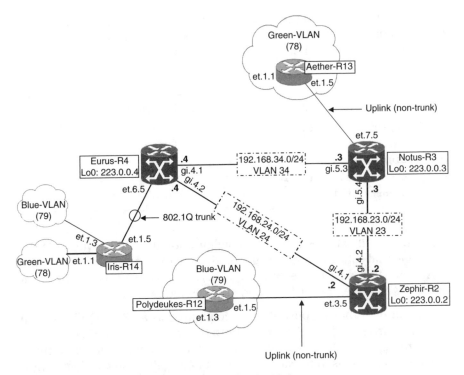

**Figure 10.3   Network Diagram for Case Study 10.1**

Zephir-R2, Notus-R3, and Eurus-R4 (Hub) are L3 PE switches, whereas Polydeukes-R12, Aether-R13, and Iris-R14 are L2 CE switches. RSVP-TE is used within the IP/MPLS MAN to establish the tunnel LSP with TE capability between the PE switches, while LDP is used to establish the VC LSPs. These LDP LSPs are tunneled within the RSVP LSPs between the PE switches (LDP over RSVP-TE).

## 10.10.2 Network Configurations

Listing 10.1 illustrates the VLAN configuration for L2 CE switch Polydeukes-R12 in commercial estate 2. The blue VLAN with VLAN ID 79 is created, and Ethernet ports et.1.3 and et.1.5 are added to this VLAN.

---

**Listing 10.1   Polydeukes-R12 VLAN Configuration**

```
system set name Polydeukes-R12

vlan create blue-vlan id 79 port-based
vlan add ports et.1.3 to blue-vlan
vlan add ports et.1.5 to blue-vlan
```

---

Listing 10.2 illustrates the VLAN configuration for L2 CE switch Aether-R13 in commercial estate 3. The green VLAN with VLAN ID 78 is created, and Ethernet ports et.1.1 and et.1.5 are added to this VLAN.

---

### Listing 10.2   Aether-R13 VLAN Configuration

```
system set name Aether-R13

vlan create green-vlan id 78 port-based
vlan add ports et.1.1 to green-vlan
vlan add ports et.1.5 to green-vlan
```

---

Listing 10.3 illustrates the VLAN configuration for L2 CE switch Iris-R14 in commercial estate 4. Both the green VLAN with VLAN ID 78 and the blue VLAN with VLAN ID 79 are created. Ethernet port et.1.1 is added to the green VLAN, while port et.1.3 is added to the blue VLAN.

In addition, Ethernet port et.1.5 is configured as an 802.1Q VLAN trunk port that carries traffic belonging to both the green and blue VLANs. In this case, traffic from the default VLAN (VLAN ID 1) is excluded. Note that a trunk port can be used to carry multiple VLAN traffic using a single physical port. This is different from an access port, which can only carry traffic from a specific VLAN.

---

### Listing 10.3   Iris-R14 VLAN Configuration

```
system set name Iris-R14

vlan create green-vlan id 78 port-based
vlan create blue-vlan id 79 port-based
vlan add ports et.1.1 to green-vlan
vlan add ports et.1.3 to blue-vlan

vlan make trunk-port et.1.5 exclude-default-vlan
vlan add ports et.1.5 to green-vlan
vlan add ports et.1.5 to blue-vlan
```

---

Listing 10.4 illustrates the network configuration (with in-line headers and comments) for L3 PE switch Zephir-R2, which interconnects CE switch Poly-deukes-R12. Self-explanatory comments (in italics) are embedded in between the configuration lines to provide clarification to the configuration commands and procedures. Because the OSPF, MPLS-TE, and RSVP-TE configurations are the same as those in case study 9.1 (see Listing 9.5), they are omitted for brevity. The main focus of this listing is on the VLAN and LDP configurations:

- The L2 FEC of concern to Zephir-R2 is a combination of customer VLAN ID 79 (blue VLAN) and incoming access port et.3.5. The VLAN is created with the "vlan create" and "vlan add ports" commands, while the "ldp map ports" command is used to map incoming access port et.3.5 to a logical customer ID number of value 77.
- Because only directly connected peers are automatically discovered when LDP is enabled on an LSR, remote LDP has to be used instead to establish the VC LSPs between LSRs (PE switches) that are not directly connected. Remote LDP peering typically takes place between the reachable loopback addresses of LDP peers and is achieved by enabling LDP on the loopback interface lo0 of the local LDP peer (Zephir-R2) and specifying the remote LDP peer (Eurus-R4) with the "ldp add remote-peer" command. The router ID of the remote LDP peer must also be a loopback address.
- The FEC-to-label mapping is advertised via LDP in the downstream-unsolicited mode to the remote peers. The "ldp add l2-fec" command is used to specify the FEC (VLAN ID 79 and customer ID 77) to be mapped to the VC label and sent to remote LDP peer Eurus-R4. Note that the VLAN ID 79 is assigned to the VC ID, and the customer ID 77 is assigned to the group ID. The VC label is actually mapped to the group ID (customer ID 77) in this case (see section 9.6.1).
- Set up tunnel LSP with RSVP-TE (see Listing 9.5 for RSVP-TE configuration).

---

### Listing 10.4   Zephir-R2 Configuration

```
system set name Zephir-R2

! -- Configure router loopback
interface add ip lo0 address-netmask 223.0.0.2/32
ip-router global set router-id 223.0.0.2

VLAN configuration
vlan create vlan23 id 23 port-based
vlan create vlan24 id 24 port-based
vlan add ports gi.4.2 to vlan23
vlan add ports gi.4.1 to vlan24

interface create ip port-42 vlan vlan23 address-netmask 192.168.23.2/24
```

```
interface create ip port-41 vlan vlan24 address-netmask 192.168.24.2/24

! -- Configure blue VLAN with a VLAN ID of 79
vlan create blue-vlan id 79 port-based
vlan add ports et.3.5 to blue-vlan

LDP configuration

! -- Enable LDP on loopback interface lo0
ldp add interface lo0

! -- Map incoming port et.3.5 to customer ID 77
ldp map ports et.3.5 customer-id 77

! -- Add Eurus-R4 as remote LDP peer
ldp add remote-peer 223.0.0.4

! -- Send label mapping for customer ID 77 and VLAN ID 79 to Eurus-R4
ldp add l2-fec customer-id 77 vlan 79 to-peer 223.0.0.4

ldp start
```

---

Listing 10.5 illustrates the network configuration (with in-line headers and comments) for L3 PE switch Notus-R3, which interconnects CE switch Aether-R13. Self-explanatory comments (in italics) are embedded in between the configuration lines to provide clarification to the configuration commands and procedures. Because the OSPF, MPLS-TE, and RSVP-TE configurations are the same as those in case study 9.1 (see Listing 9.6), they are omitted for brevity. The main focus of this listing is on the VLAN and LDP configurations:

- The L2 FEC of concern to Notus-R3 is a combination of customer VLAN ID 78 (green VLAN) and incoming access port et.7.5. The VLAN is created with the "vlan create" and "vlan add ports" commands, while the "ldp map ports" command is used to map incoming access port et.7.5 to a logical customer ID number of value 77.
- Remote LDP peering is achieved through enabling LDP on the loopback interface lo0 of the local LDP peer (Notus-R3) and specifying the remote LDP peer (Eurus-R4) with the "ldp add remote-peer" command.
- The "ldp add l2-fec" command is used to specify the FEC (VLAN ID 78 and customer ID 77) to be mapped to the VC label and sent to remote LDP peer Eurus-R4.

- Set up tunnel LSP with RSVP-TE (see Listing 9.6 for RSVP-TE configuration).

---

## Listing 10.5   Notus-R3 Configuration

```
system set name Notus-R3


! -- Configure router loopback
interface add ip lo0 address-netmask 223.0.0.3/32
ip-router global set router-id 223.0.0.3


VLAN configuration
vlan create vlan34 id 34 port-based
vlan create vlan23 id 23 port-based
vlan add ports gi.5.3 to vlan34
vlan add ports gi.5.4 to vlan23


interface create ip port-53 vlan vlan34 address-netmask 192.168.34.3/24
interface create ip port-54 vlan vlan23 address-netmask 192.168.23.3/24


! -- Configure green VLAN with a VLAN ID of 78
vlan create green-vlan id 78 port-based
vlan add ports et.7.5 to green-vlan


LDP configuration


! -- Enable LDP on loopback interface lo0
ldp add interface lo0


! -- Map incoming port et.7.5 to customer ID 77
ldp map ports et.7.5 customer-id 77


! -- Add Eurus-R4 as remote LDP peer
ldp add remote-peer 223.0.0.4


! -- Send label mapping for customer ID 77 and VLAN ID 78 to Eurus-R4
ldp add 12-fec customer-id 77 vlan 78 to-peer 223.0.0.4


ldp start
```

---

L3 PE switch Eurus-R4 assumes the role of a hub in the case study. Listing 10.6 illustrates the network configuration (with in-line headers and comments) for Eurus-R4, which interconnects CE switch Iris-R14.

Self-explanatory comments (in italics) are embedded in between the configuration lines to provide clarification to the configuration commands and procedures. Because the OSPF, MPLS-TE, and RSVP-TE configurations are the same as those in case study 9.1 (see Listing 9.7), they are omitted for brevity. The main focus of this listing is on the VLAN and LDP configurations:

- The two L2 FECs of concern to Eurus-R4 are:
  - A combination of customer VLAN ID 78 (green VLAN) and incoming trunk port et.6.5. The VLAN is created with the "vlan create" and "vlan add ports" commands.
  - A combination of customer VLAN ID 79 (blue VLAN) and incoming trunk port et.6.5. The VLAN is created with the "vlan create" and "vlan add ports" commands.
  - The "ldp map ports" command is used to map incoming trunk port et.6.5 to a logical customer ID number of value 77.
- Remote LDP peering is achieved through enabling LDP on the loopback interface lo0 of the local LDP peer (Eurus-R4) and specifying the remote LDP peers (Zephir-R2 and Notus-R3) with the "ldp add remote-peer" command.
- The "ldp add l2-fec" command is used to specify the FEC-to-label mapping for:
  - VLAN ID 78 and customer ID 77 to be advertised to remote LDP peer Notus-R3
  - VLAN ID 79 and customer ID 77 to be advertised to remote LDP peer Zephir-R2
- Set up tunnel LSP with RSVP-TE (see Listing 9.7 for RSVP-TE configuration).

---

### Listing 10.6   Eurus-R4 Configuration

```
system set name Eurus-R4


! -- Configure router loopback
interface add ip lo0 address-netmask 223.0.0.4/32
ip-router global set router-id 223.0.0.4


VLAN configuration
vlan create vlan34 id 34 port-based
vlan create vlan24 id 24 port-based
vlan add ports gi.4.1 to vlan34
vlan add ports gi.4.2 to vlan24


interface create ip port-42 vlan vlan24 address-netmask 192.168.24.4/24
interface create ip port-41 vlan vlan34 address-netmask 192.168.34.4/24
```

```
! -- Make et.6.5 a trunk port that excludes traffic from default VLAN
vlan make trunk-port et.6.5 exclude-default-vlan


! -- Configure green VLAN with a VLAN ID of 78
vlan create green-vlan id 78 port-based
vlan add ports et.6.5 to green-vlan


! -- Configure blue VLAN with a VLAN ID of 79
vlan create blue-vlan id 79 port-based
vlan add ports et.6.5 to blue-vlan


LDP configuration

! -- Enable LDP on loopback interface lo0
ldp add interface lo0


! -- Map incoming port et.6.5 to customer ID 77
ldp map ports et.6.5 customer-id 77


! -- Add Zephir-R2 as remote LDP peer
ldp add remote-peer 223.0.0.2


! -- Add Notus-R3 as remote LDP peer
ldp add remote-peer 223.0.0.3


! -- Send label mapping for customer ID 77 and VLAN ID 78 to Notus-R3
ldp add l2-fec customer-id 77 vlan 78 to-peer 223.0.0.3


! -- Send label mapping for customer ID 77 and VLAN ID 79 to Zephir-R2
ldp add l2-fec customer-id 77 vlan 79 to-peer 223.0.0.2


ldp start
```

### 10.10.3  VLAN and LDP Monitoring

Listing 10.7 validates the VLAN configuration of Polydeukes-R12. Ethernet ports et.1.3 and et.1.5 are assigned to blue VLAN with VLAN ID of 79.

### Listing 10.7   Polydeukes-R12 VLAN Information

```
Polydeukes-R12# vlan show
```

| VI | VLAN Name | Used for | Ports D |
|----|-----------|----------|---------|
| 1 | DEFAULT | IP,IPX,ATALK,DEC,SNA,IPv6,L2 | et.1.(1-2,4,6-16), et.2.(1-16), NP.3.(3-4), t3.4.1, NP.4.2 |
| 79 | blue-vlan | IP,IPX,ATALK,DEC,SNA,IPv6,L2 | et.1.(3,5) |

Listing 10.8 validates the VLAN configuration of Aether-R13. Ethernet ports et.1.1 and et.1.5 are assigned to green VLAN with VLAN ID of 78.

### Listing 10.8   Aether-R13 VLAN Information

```
Aether-R13# vlan show
```

| VID | VLAN Name | Used for | Ports |
|-----|-----------|----------|-------|
| 1 | DEFAULT | IP,IPX,ATALK,DEC,SNA,IPv6,L2 | et.1.(2-4,6-16), et.2.(1-16) |
| 78 | green-vlan | IP,IPX,ATALK,DEC,SNA,IPv6,L2 | et.1.(1,5) |

Listing 10.9 validates the VLAN configuration of Iris-R14. Ethernet port et.1.1 is assigned to green VLAN with VLAN ID 78, and port et.1.3 is assigned to blue VLAN with VLAN ID 79. Port et.1.5 is a trunk port.

### Listing 10.9   Iris-R14 VLAN Information

```
Iris-R14# vlan show
```

| VID | VLAN Name | Used for | Ports |
|-----|-----------|----------|-------|
| 1 | DEFAULT | IP,IPX,ATALK,DEC,SNA,IPv6,L2 | et.1.(2,4,6-16), et.2.(1-16), gi.3.(1-2), gi.4.(1-2) |
| 78 | green-vlan | IP,IPX,ATALK,DEC,SNA,IPv6,L2 | et.1.(1,5) |
| 79 | blue-vlan | IP,IPX,ATALK,DEC,SNA,IPv6,L2 | et.1.(3,5) |

Listing 10.10 validates the VLAN configuration of Zephir-R2. Ethernet port et.3.5 is assigned to blue VLAN with VLAN ID of 79.

---

### Listing 10.10    Zephir-R2 VLAN Information

```
Zephir-R2# vlan show
```

| VID | VLAN Name | Used for | Ports |
|-----|-----------|----------|-------|
| 1 | DEFAULT | IP,IPX,ATALK,DEC,SNA,IPv6,L2 | et.3.(1-4,6-14), gi.6.1 |
| 23 | vlan23 | IP,IPX,ATALK,DEC,SNA,IPv6,L2 | gi.4.2 |
| 24 | vlan24 | IP,IPX,ATALK,DEC,SNA,IPv6,L2 | gi.4.1 |
| **79** | **blue-vlan** | **IP,IPX,ATALK,DEC,SNA,IPv6,L2** | **et.3.5** |

---

Listing 10.11 validates the VLAN configuration of Notus-R3. Ethernet port et.7.5 is assigned to green VLAN with VLAN ID of 78.

---

### Listing 10.11    Notus-R3 VLAN Information

```
Notus-R3# vlan show
```

| VID | VLAN Name | Used for | Ports |
|-----|-----------|----------|-------|
| 1 | DEFAULT | IP,IPX,ATALK,DEC,SNA,IPv6,L2 | gi.5.2, et.7.(1-4,6-24) |
| 23 | vlan23 | IP,IPX,ATALK,DEC,SNA,IPv6,L2 | gi.5.4 |
| 34 | vlan34 | IP,IPX,ATALK,DEC,SNA,IPv6,L2 | gi.5.3 |
| **78** | **green-vlan** | **IP,IPX,ATALK,DEC,SNA,IPv6,L2** | **et.7.5** |

---

Listing 10.12 validates the VLAN configuration of Eurus-R4. Ethernet port et.6.5 is a trunk port carrying traffic from both the green and blue VLANs.

---

### Listing 10.12    Eurus-R4 VLAN Information

```
Eurus-R4# vlan show
```

| VID | VLAN Name | Used for | Ports |
|-----|-----------|----------|-------|
| 1 | DEFAULT | IP,IPX,ATALK,DEC,SNA,IPv6,L2 | gi.5.1, et.6.(1-4,6-16) |
| 24 | vlan24 | IP,IPX,ATALK,DEC,SNA,IPv6,L2 | gi.4.2 |
| 34 | vlan34 | IP,IPX,ATALK,DEC,SNA,IPv6,L2 | gi.4.1 |
| **78** | **green-vlan** | **IP,IPX,ATALK,DEC,SNA,IPv6,L2** | **et.6.5** |
| **79** | **blue-vlan** | **IP,IPX,ATALK,DEC,SNA,IPv6,L2** | **et.6.5** |

---

Listing 10.13 illustrates the LDP session state information for Zephir-R2. A remote LDP session is successfully established between local peer Zephir-R2 (223.0.0.2) and remote peer Eurus-R4 (223.0.0.4). The next-hop addresses advertised by Eurus-R4 are 192.168.34.4, 192.168.24.4, and 223.0.0.4, all of which can be used to reach the remote peer (see section 2.7 of [RFC3036] for further details).

---

**Listing 10.13    Zephir-R2 LDP Session State Information**

```
Zephir-R2# ldp show session verbose

Address: 223.0.0.4, State: Operational, Connection: Open, Keepalive
Time: 25
Session operational for 0d, 6h, 50m, 23s
Labels Sent 2, Received total 2, Received filtered 0
Session ID: 223.0.0.2:0—223.0.0.4:0, Remote session
Next keepalive in 8 seconds
Passive, Maximum PDU: 4096, Keepalive Timeout: 30 seconds
Keepalive interval: 10 seconds, Connect retry interval: 14 seconds
Local address: 223.0.0.2, Remote address: 223.0.0.4
Next-hop addresses received:
192.168.34.4
192.168.24.4
223.0.0.4
```

---

Listing 10.14 illustrates the LDP session state information for Notus-R3. A remote LDP session is successfully established between local peer Notus-R3 (223.0.0.3) and remote peer Eurus-R4 (223.0.0.4). The next-hop addresses advertised by Eurus-R4 are 192.168.34.4, 192.168.24.4, and 223.0.0.4, all of which can be used to reach the remote peer.

---

**Listing 10.14    Notus-R3 LDP Session State Information**

```
Notus-R3# ldp show session verbose

Address: 223.0.0.4, State: Operational, Connection: Open, Keepalive
Time: 26
Session operational for 0d, 0h, 50m, 23s
Labels Sent 2, Received total 2, Received filtered 0
Session ID: 223.0.0.3:0—223.0.0.4:0, Remote session
Next keepalive in 3 seconds
Passive, Maximum PDU: 4096, Keepalive Timeout: 30 seconds
```

```
Keepalive interval: 10 seconds, Connect retry interval: 14 seconds
Local address: 223.0.0.3, Remote address: 223.0.0.4
Next-hop addresses received:
192.168.34.4
192.168.24.4
223.0.0.4
```

Listing 10.15 illustrates the LDP session state information for Eurus-R4:

- A remote LDP session is successfully established between local peer Eurus-R4 (223.0.0.4) and remote peer Notus-R3 (223.0.0.3). The next-hop addresses advertised by Notus-R3 are 192.168.34.3, 192.168.23.3, and 223.0.0.3, all of which can be used to reach the remote peer.
- A remote LDP session is successfully established between local peer Eurus-R4 (223.0.0.4) and remote peer Zephir-R2 (223.0.0.2). The next-hop addresses advertised by Zephir-R2 are 192.168.24.2, 192.168.23.2, and 223.0.0.2, all of which can be used to reach the remote peer.

### Listing 10.15   Eurus-R4 LDP Session State Information

```
Eurus-R4# ldp show session verbose

Address: 223.0.0.3, State: Operational, Connection: Open, Keepalive
Time: 23
Session operational for 0d, 0h, 45m, 13s
Labels Sent 2, Received total 2, Received filtered 0
Session ID: 223.0.0.4:0—223.0.0.3:0, Remote session
Next keepalive in 7 seconds
Active, Maximum PDU: 4096, Keepalive Timeout: 30 seconds
Keepalive interval: 10 seconds, Connect retry interval: 28 seconds
Local address: 223.0.0.4, Remote address: 223.0.0.3
Next-hop addresses received:
192.168.34.3
192.168.23.3
223.0.0.3
Address: 223.0.0.2, State: Operational, Connection: Open, Keepalive
Time: 19
Session operational for 0d, 6h, 41m, 29s
Labels Sent 2, Received total 2, Received filtered 0
Session ID: 223.0.0.4:0—223.0.0.2:0, Remote session
Next keepalive in 5 seconds
Active, Maximum PDU: 4096, Keepalive Timeout: 30 seconds
Keepalive interval: 10 seconds, Connect retry interval: 28 seconds
```

```
Local address: 223.0.0.4, Remote address: 223.0.0.2
Next-hop addresses received:
192.168.24.2
192.168.23.2
223.0.0.2
```

Listing 10.16 illustrates the L2 FEC of Zephir-R2, which is customer ID (group ID) 77 and VLAN ID (VC ID) 79. The LDP identifier for remote peer Eurus-R4 is 223.0.0.4:0. It is a 6-byte TLV that identifies the router (first four bytes) and label space (last two bytes). Per-platform label space is announced by setting the label space ID to zero (see section 2.2 of [RFC3036] for more details). A per-platform label space is a single global pool of label values defined for the entire router where a platform-unique label is assigned to a particular FEC and announced to all neighbors. The per-platform label space is typically used in frame-mode MPLS (for instance, MPLS over Ethernet).

The VC label (out-lbl) locally generated by Zephir-R2 and advertised to its neighbor Eurus-R4 is 17, while the VC label (in-lbl) received from Eurus-R4 is 18. Therefore, VC label 18 is pushed onto blue-tagged local Ethernet frames originating from port et.3.5 and heading for the remote blue VLAN serviced by Eurus-R4. The VC LSP established by remote LDP is then tunneled within tunnel LSP "R2-to-R4" (see Listing 9.5) established by RSVP-TE between Zephir-R2 and Eurus-R4 (LDP over RSVP-TE). The corresponding tunnel label 16 is pushed on top of VC label 18, forming a two-level label stack.

### Listing 10.16    Zephir-R2 L2 FEC Information

```
Zephir-R2# ldp show l2-fec verbose

FEC: Forward Equivalence class, in-lbl: Label received, out-lbl: Label
sent

Remote neighbor 223.0.0.4:0

FEC: Customer ID 77, VLAN ID 79
in-lbl: 18, out-lbl: 17
Ports: et.3.5
Transport LSP name/label: R2-to-R4/16
Bytes In: 0, Pkts In: 0, In Pkts Drop: 0
Bytes Out: 0, Pkts Out: 0, Out Pkts Drop: 0
```

Listing 10.17 illustrates the L2 FEC of Notus-R3, which is customer ID (group ID) 77 and VLAN ID (VC ID) 78. The LDP identifier for remote peer Eurus-R4 is 223.0.0.4:0.

The VC label (out-lbl) locally generated by Notus-R3 and advertised to its neighbor Eurus-R4 is 17, and the VC label (in-lbl) received from Eurus-R4 is also 17. Note that the same label can be used on any interface per router provided that it is referring to the same FEC.

VC label 17 is pushed onto green-tagged local Ethernet frames originating from port et.7.5 and heading for the remote green VLAN serviced by Eurus-R4. The VC LSP established by remote LDP is then tunneled within tunnel LSP "R3-to-R4" (see Listing 9.6) established by RSVP-TE between Notus-R3 and Eurus-R4. The corresponding tunnel label 16 is pushed on top of VC label 17, forming a two-level label stack.

---

### Listing 10.17   Notus-R3 L2 FEC Information

```
Notus-R3# ldp show 12-fec verbose

FEC: Forward Equivalence class, in-lbl: Label received, out-lbl: Label
sent

Remote neighbor 223.0.0.4:0

FEC: Customer ID 77, VLAN ID 78
in-lbl: 17, out-lbl: 17
Ports: et.7.5
Transport LSP name/label: R3-to-R4/16
Bytes In: 0, Pkts In: 0, In Pkts Drop: 0
Bytes Out: 0, Pkts Out: 0, Out Pkts Drop: 0
```

---

Listing 10.18 illustrates the two L2 FECs of Eurus-R4. The L2 FEC information for remote peer Notus-R3 is as follows:

- The LDP identifier is 223.0.0.3:0.
- The L2 FEC is customer ID (group ID) 77 and VLAN ID (VC ID) 78.
- The VC label (out-lbl) locally generated by Eurus-R4 and advertised to its neighbor Notus-R3 is 17, and the VC label (in-lbl) received from Notus-R3 is also 17.
- VC label 17 is pushed onto green-tagged local Ethernet frames originating from port et.6.5 and heading for the remote green VLAN serviced by Notus-R3. The VC LSP established by remote LDP is then tunneled within tunnel LSP "R4-to-R3" (see Listing 9.7) established

by RSVP-TE between Eurus-R4 and Notus-R3. The corresponding tunnel label 16 is pushed on top of VC label 17, forming a two-level label stack.

The L2 FEC information for remote peer Zephir-R2 is as follows:

- The LDP identifier is 223.0.0.2:0.
- The L2 FEC is customer ID (group ID) 77 and VLAN ID (VC ID) 79.
- The VC label (out-lbl) locally generated by Eurus-R4 and advertised to its neighbor Zephir-R2 is 18, while the VC label (in-lbl) received from Zephir-R2 is 17.
- VC label 17 is pushed onto blue-tagged local Ethernet frames originating from port et.6.5 and heading for the remote blue VLAN serviced by Zephir-R2. The VC LSP established by remote LDP is then tunneled within tunnel LSP "R4-to-R2" (see Listing 9.7) established by RSVP-TE between Eurus-R4 and Zephir-R2. The corresponding tunnel label 16 is pushed on top of VC label 17, forming a two-level label stack.

---

### Listing 10.18    Eurus-R4 L2 FEC Information

```
Eurus-R4# ldp show l2-fec verbose

FEC: Forward Equivalence class, in-lbl: Label received, out-lbl: Label
sent

Remote neighbor 223.0.0.3:0

FEC: Customer ID 77, VLAN ID 78
in-lbl: 17, out-lbl: 17
Ports: et.6.5
Transport LSP name/label: R4-to-R3/16
Bytes In: 0, Pkts In: 0, In Pkts Drop: 0
Bytes Out: 0, Pkts Out: 0, Out Pkts Drop: 0

Remote neighbor 223.0.0.2:0

FEC: Customer ID 77, VLAN ID 79
in-lbl: 17, out-lbl: 18
Ports: et.6.5
Transport LSP name/label: R4-to-R2/16
Bytes In: 0, Pkts In: 0, In Pkts Drop: 0
Bytes Out: 0, Pkts Out: 0, Out Pkts Drop: 0
```

---

### 10.10.4 VLL Connectivity Verification

As illustrated in Figure 10.4, four IP hosts are connected to the four different VLAN locations to verify whether the VLL connectivity is operational between the respective VLANs. The host-to-VLAN connections are described as follows:

- HOST-7912 is connected to the blue VLAN at port et.1.3 of Polydeukes-R12.
- HOST-7914 is connected to the blue VLAN at port et.1.3 of Iris-R14.
- HOST-7813 is connected to the green VLAN at port et.1.1 of Aether-R13.
- HOST-7814 is connected to the green VLAN at port et.1.1 of Iris-R14.

Listing 10.19 illustrates the successful outcome of the ping test performed from HOST-7912 to HOST-7914, indicating that the VLL connectivity is up and the Martini's point-to-point tunneling is working between the two blue VLAN locations.

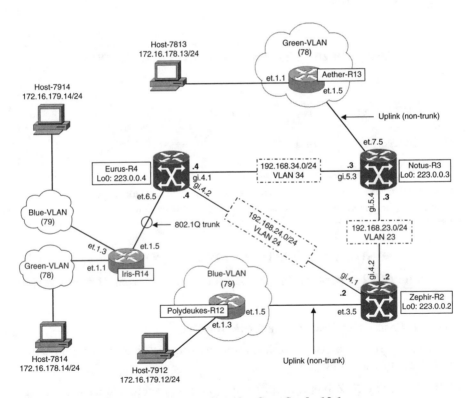

**Figure 10.4   Host-to-VLAN Connections for Case Study 10.1**

---

**Listing 10.19    Ping from HOST-7912 to HOST-7914**

```
HOST-7912# ping 172.16.179.14

PING 172.16.179.14: 36 bytes of data
5 second timeout, 1 repetition
36 bytes from 172.16.179.14: icmp_seq=0 ttl=255 time=0.890 ms

--- 172.16.179.14 ping statistics ---
1 packets transmitted, 1 packets received, 0.00% packet loss
round-trip min/avg/max/dev = 0.890/0.890/0.890/0.000 ms
```

---

Listing 10.20 illustrates the successful outcome of the ping test performed from HOST-7814 to HOST-7813, indicating that the VLL connectivity is up and the Martini's point-to-point tunneling is working between the two green VLAN locations.

---

**Listing 10.20    Ping from HOST-7814 to HOST 7813**

```
HOST-7814# ping 172.16.178.13

PING 172.16.178.13: 36 bytes of data
5 second timeout, 1 repetition
36 bytes from 172.16.178.13: icmp_seq=0 ttl=255 time=4.390 ms

--- 172.16.178.13 ping statistics ---
1 packets transmitted, 1 packets received, 0.00% packet loss
round-trip min/avg/max/dev = 4.390/4.390/4.390/0.000 ms
```

---

## 10.11   CASE STUDY 10.2: VIRTUAL PRIVATE LAN SERVICES

Case study 10.2 describes the provision of VPLS between three different sites.

### 10.11.1  Case Overview and Network Topology

Case study 10.2 is a continuation of case study 10.1, with the following adjustments and add-ons:

- Company XYZ has a main branch in commercial estate 3 and two subbranches: one at commercial estate 4 and the other at commercial estate 2. All three sites must be able to communicate with each other via the MAN as if they are in a single virtual private LAN.

- Company RST has a main office in commercial estate 2 and two branch offices: one at commercial estate 4 and the other at commercial estate 3. All three offices must be able to communicate with each other via the MAN as if they are in a single virtual private LAN.
- Company HIJ has a main office in commercial estate 4 and two branch offices: one at commercial estate 2 and the other at commercial estate 3. All three offices must be able to communicate with each other via the MAN as if they are in a single virtual private LAN.

These companies have existing LAN services in their premises, and they are only interested in scaling these services. Figure 10.5 illustrates the network diagram for this case study. Three VLANs are implemented in this case: blue VLAN for Company RST, green VLAN for Company XYZ, and red VLAN for Company HIJ. These VLANs are spanned across all three districts to provide the VPLS to the respective company sites.

Zephir-R2, Notus-R3, and Eurus-R4 are L3 PE switches, whereas Polydeukes-R12, Aether-R13, and Iris-R14 are L2 CE switches. RSVP-TE is used within the IP/MPLS MAN to establish the tunnel LSP with TE capability

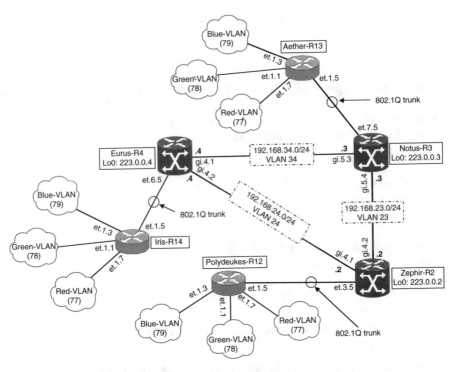

**Figure 10.5  Network Diagram for Case Study 10.2**

339

between the PE switches, while LDP is used to establish the VC LSPs in a point-to-multipoint manner. These LDP LSPs are tunneled within the RSVP LSPs between the PE switches (LDP over RSVP-TE). A total of six VC LSPs are required to emulate the full LAN connectivity between the three sites.

### 10.11.2 Network Configuration

Listing 10.21 illustrates the VLAN configuration for Polydeukes-R12 in commercial estate 2:

- The red VLAN with VLAN ID 77 is created, and Ethernet ports et.1.5 and et.1.7 are added to this VLAN.
- The green VLAN with VLAN ID 78 is created, and Ethernet ports et.1.1 and et.1.5 are added to this VLAN.
- The blue VLAN with VLAN ID 79 is created, and Ethernet ports et.1.3 and et.1.5 are added to this VLAN.

In addition, Ethernet port et.1.5 is configured as an 802.1Q VLAN trunk port that carries traffic belonging to the red, green, and blue VLANs. In this case, traffic from the default VLAN (VLAN ID 1) is excluded.

---

**Listing 10.21   Polydeukes-R12 VLAN Configuration**

```
system set name Polydeukes-R12

vlan make trunk-port et.1.5 exclude-default-vlan

vlan create red-vlan id 77 port-based
vlan add ports et.1.5 to red-vlan
vlan add ports et.1.7 to red-vlan

vlan create green-vlan id 78 port-based
vlan add ports et.1.1 to green-vlan
vlan add ports et.1.5 to green-vlan

vlan create blue-vlan id 79 port-based
vlan add ports et.1.3 to blue-vlan
vlan add ports et.1.5 to blue-vlan
```

---

Listing 10.22 illustrates the VLAN configuration for Aether-R13 in commercial estate 3:

- The red VLAN with VLAN ID 77 is created, and Ethernet ports et.1.5 and et.1.7 are added to this VLAN.

- The green VLAN with VLAN ID 78 is created, and Ethernet ports et.1.1 and et.1.5 are added to this VLAN.
- The blue VLAN with VLAN ID 79 is created, and Ethernet ports et.1.3 and et.1.5 are added to this VLAN.

In addition, Ethernet port et.1.5 is configured as an 802.1Q VLAN trunk port that carries traffic belonging to the red, green, and blue VLANs. In this case, traffic from the default VLAN (VLAN ID 1) is excluded.

---

### Listing 10.22  Aether-R13 VLAN Configuration

```
system set name Aether-R13

vlan make trunk-port et.1.5 exclude-default-vlan

vlan create red-vlan id 77 port-based
vlan add ports et.1.5 to red-vlan
vlan add ports et.1.7 to red-vlan

vlan create green-vlan id 78 port-based
vlan add ports et.1.1 to green-vlan
vlan add ports et.1.5 to green-vlan

vlan create blue-vlan id 79 port-based
vlan add ports et.1.3 to blue-vlan
vlan add ports et.1.5 to blue-vlan
```

---

Listing 10.23 illustrates the VLAN configuration for Iris-R14 in commercial estate 4:

- The red VLAN with VLAN ID 77 is created, and Ethernet ports et.1.5 and et.1.7 are added to this VLAN.
- The green VLAN with VLAN ID 78 is created, and Ethernet ports et.1.1 and et.1.5 are added to this VLAN.
- The blue VLAN with VLAN ID 79 is created, and Ethernet ports et.1.3 and et.1.5 are added to this VLAN.

In addition, Ethernet port et.1.5 is configured as an 802.1Q VLAN trunk port that carries traffic belonging to the red, green, and blue VLANs. In this case, traffic from the default VLAN (VLAN ID 1) is excluded.

---

**Listing 10.23   Iris-R14 VLAN Configuration**

```
system set name Iris-R14

vlan make trunk-port et.1.5 exclude-default-vlan

vlan create red-vlan id 77 port-based
vlan add ports et.1.5 to red-vlan
vlan add ports et.1.7 to red-vlan

vlan create green-vlan id 78 port-based
vlan add ports et.1.1 to green-vlan
vlan add ports et.1.5 to green-vlan

vlan create blue-vlan id 79 port-based
vlan add ports et.1.3 to blue-vlan
vlan add ports et.1.5 to blue-vlan
```

---

Listing 10.24 illustrates the network configuration (with in-line headers and comments) for Zephir-R2, which interconnects Polydeukes-R12. Self-explanatory comments (in italics) are embedded in between the configuration lines to provide clarification to the configuration commands and procedures. The OSPF, MPLS-TE, and RSVP-TE configurations are omitted for brevity, as in case study 10.1. The main focus of this listing is on the VLAN and VPLS configurations:

- The three respective VLANs are created with the "vlan create" and "vlan add ports" commands. Port et.3.5 is made a trunk port that carries traffic from the three VLANs.
- The L2 FEC for each customer is configured in a customer profile with the "mpls set customer-profile" command. The customer profile must include the following:
  - Unique identifier number for each customer: the values 77, 78, and 79 for the red, green, and blue VLAN customers, respectively.
  - One or more physical ports on Zephir-R2 that are assigned to the customers: only trunk port et.3.5 is assigned, which carries traffic from the three different VLANs.
  - The VPN type used with this customer profile, which specifies the source of traffic to be forwarded through the VC LSP: a port-to-VLAN (port-vlan) VPN is specified, indicating that the source of traffic is from trunk port et.3.5. All traffic arriving on this port associated with the specified VLAN is identified as belonging to this customer and is forwarded through the VC LSP.

- Remote LDP peering is achieved through enabling LDP on the loopback interface lo0 of the local LDP peer (Zephir-R2) and specifying the remote LDP peers (Notus-R3 and Eurus-R4) with the "ldp add remote-peer" command.
- Send the label mapping for each customer profile to the appropriate remote peer with the "ldp connect customer-profile" command.
- Set up tunnel LSP with RSVP-TE (see Listing 9.5 for RSVP-TE configuration).

---

### Listing 10.24 Zephir-R2 Configuration

```
system set name Zephir-R2

! -- Configure router loopback
interface add ip lo0 address-netmask 223.0.0.2/32
ip-router global set router-id 223.0.0.2

VLAN configuration
vlan create vlan23 id 23 port-based
vlan create vlan24 id 24 port-based
vlan add ports gi.4.2 to vlan23
vlan add ports gi.4.1 to vlan24
interface create ip port-42 vlan vlan23 address-netmask 192.168.23.2/24
interface create ip port-41 vlan vlan24 address-netmask 192.168.24.2/24

! -- Make et.3.5 a trunk port that excludes traffic from default VLAN
vlan make trunk-port et.3.5 exclude-default-vlan

! -- Configure red VLAN with a VLAN ID of 77
vlan create red-vlan id 77 port-based
vlan add ports et.3.5 to red-vlan

! -- Configure green VLAN with a VLAN ID of 78
vlan create green-vlan id 78 port-based
vlan add ports et.3.5 to green-vlan

! -- Configure blue VLAN with a VLAN ID of 79
vlan create blue-vlan id 79 port-based
vlan add ports et.3.5 to blue-vlan

VPLS configuration

! -- Define L2 customer FEC
```

```
mpls set customer-profile vpls-77 customer_id 77 vlans 77 in-port-list
et.3.5 type port-vlan

mpls set customer-profile vpls-78 customer_id 78 vlans 78 in-port-list
et.3.5 type port-vlan

mpls set customer-profile vpls-79 customer_id 79 vlans 79 in-port-list
et.3.5 type port-vlan


! -- Enable LDP on loopback interface lo0
ldp add interface lo0


! -- Add Eurus-R4 as remote LDP peer
ldp add remote-peer 223.0.0.4


! -- Add Notus-R3 as remote LDP peer
ldp add remote-peer 223.0.0.3


! -- Send label mapping for L2 customer FEC to remote peers
ldp connect customer-profile vpls-77 remote-peer 223.0.0.4
ldp connect customer-profile vpls-77 remote-peer 223.0.0.3
ldp connect customer-profile vpls-78 remote-peer 223.0.0.4
ldp connect customer-profile vpls-78 remote-peer 223.0.0.3
ldp connect customer-profile vpls-79 remote-peer 223.0.0.4
ldp connect customer-profile vpls-79 remote-peer 223.0.0.3


ldp start
```

Listing 10.25 illustrates the network configuration (with in-line headers and comments) for Notus-R3, which interconnects Aether-R13. Self-explanatory comments (in italics) are embedded in between the configuration lines to provide clarification to the configuration commands and procedures. The OSPF, MPLS-TE, and RSVP-TE configurations are omitted for brevity, as in case study 10.1. The main focus of this listing is on the VLAN and VPLS configurations:

- The three respective VLANs are created with the "vlan create" and "vlan add ports" commands. Port et.7.5 is made a trunk port that carries traffic from the three VLANs.
- The L2 FEC for each customer is configured in a customer profile with the "mpls set customer-profile" command. The customer profile must include the following:
  - Unique identifier number for each customer: the values 77, 78, and 79 for the red, green, and blue VLAN customers, respectively.
  - One or more physical ports on Notus-R3 that are assigned to the customers: only trunk port et.7.5 is assigned, which carries traffic from the three different VLANs.

— The VPN type used with this customer profile, which specifies the source of traffic to be forwarded through the VC LSP: a port-to-VLAN (port-vlan) VPN is specified, indicating that the source of traffic is from trunk port et.7.5. All traffic arriving on this port associated with the specified VLAN is identified as belonging to this customer and is forwarded through the VC LSP.

- Remote LDP peering is achieved through enabling LDP on the loopback interface lo0 of the local LDP peer (Notus-R3) and specifying the remote LDP peers (Zephir-R2 and Eurus-R4) with the "ldp add remote-peer" command.
- Send the label mapping for each customer profile to the appropriate remote peer with the "ldp connect customer-profile" command.
- Set up tunnel LSP with RSVP-TE (see Listing 9.6 for RSVP-TE configuration).

---

### Listing 10.25  Notus-R3 Configuration

```
system set name Notus-R3

! -- Configure router loopback
interface add ip lo0 address-netmask 223.0.0.3/32
ip-router global set router-id 223.0.0.3


VLAN configuration
vlan create vlan34 id 34 port-based
vlan create vlan23 id 23 port-based
vlan add ports gi.5.3 to vlan34
vlan add ports gi.5.4 to vlan23
interface create ip port-53 vlan vlan34 address-netmask 192.168.34.3/24
interface create ip port-54 vlan vlan23 address-netmask 192.168.23.3/24


! -- Make et.7.5 a trunk port that excludes traffic from default VLAN
vlan make trunk-port et.7.5 exclude-default-vlan


! -- Configure red VLAN with a VLAN ID of 77
vlan create red-vlan id 77 port-based
vlan add ports et.7.5 to red-vlan


! -- Configure green VLAN with a VLAN ID of 78
vlan create green-vlan id 78 port-based
vlan add ports et.7.5 to green-vlan


! -- Configure blue VLAN with a VLAN ID of 79
vlan create blue-vlan id 79 port-based
```

```
vlan add ports et.7.5 to blue-vlan


VPLS configuration


! -- Define L2 customer FEC
mpls set customer-profile vpls-77 customer_id 77 vlans 77 in-port-list
et.7.5 type port-vlan
mpls set customer-profile vpls-78 customer_id 78 vlans 78 in-port-list
et.7.5 type port-vlan
mpls set customer-profile vpls-79 customer_id 79 vlans 79 in-port-list
et.7.5 type port-vlan


! -- Enable LDP on loopback interface lo0
ldp add interface lo0


! -- Add Eurus-R4 as remote LDP peer
ldp add remote-peer 223.0.0.4


! -- Add Zephir-R2 as remote LDP peer
ldp add remote-peer 223.0.0.2


! -- Send label mapping for L2 customer FEC to remote peers
ldp connect customer-profile vpls-77 remote-peer 223.0.0.2
ldp connect customer-profile vpls-78 remote-peer 223.0.0.2
ldp connect customer-profile vpls-78 remote-peer 223.0.0.4
ldp connect customer-profile vpls-77 remote-peer 223.0.0.4
ldp connect customer-profile vpls-79 remote-peer 223.0.0.2
ldp connect customer-profile vpls-79 remote-peer 223.0.0.4


ldp start
```

Listing 10.26 illustrates the network configuration (with in-line headers and comments) for Eurus-R4, which interconnects Iris-R14. Self-explanatory comments (in italics) are embedded in between the configuration lines to provide clarification to the configuration commands and procedures. The OSPF, MPLS-TE, and RSVP-TE configurations are omitted for brevity, as in case study 10.1. The main focus of this listing is on the VLAN and VPLS configurations:

- The three respective VLANs are created with the "vlan create" and "vlan add ports" commands. Port et.6.5 is made a trunk port that carries traffic from the three VLANs.
- The L2 FEC for each customer is configured in a customer profile with the "mpls set customer-profile" command. The customer profile must include the following:

— Unique identifier number for each customer: the values 77, 78, and 79 for the red, green, and blue VLAN customers respectively.
— One or more physical ports on the Eurus-R4 that are assigned to the customers: only trunk port et.6.5 is assigned, which carries traffic from the three different VLANs.
— The VPN type used with this customer profile, which specifies the source of traffic to be forwarded through the VC LSP: a port-to-VLAN (port-vlan) VPN is specified, indicating that the source of traffic is from trunk port et.6.5. All traffic arriving on this port associated with the specified VLAN is identified as belonging to this customer and is forwarded through the VC LSP.

- Remote LDP peering is achieved through enabling LDP on the loop-back interface lo0 of the local LDP peer (Eurus-R4) and specifying the remote LDP peers (Zephir-R2 and Notus-R3) with the "ldp add remote-peer" command.
- Send the label mapping for each customer profile to the appropriate remote peer with the "ldp connect customer-profile" command.
- Set up tunnel LSP with RSVP-TE (see Listing 9.7 for RSVP-TE config-uration).

---

### Listing 10.26   Eurus-R4 Configuration

```
system set name Eurus-R4

! -- Configure router loopback
interface add ip lo0 address-netmask 223.0.0.4/32
ip-router global set router-id 223.0.0.4

VLAN configuration
vlan create vlan34 id 34 port-based
vlan create vlan24 id 24 port-based
vlan add ports gi.4.1 to vlan34
vlan add ports gi.4.2 to vlan24
interface create ip port-42 vlan vlan24 address-netmask 192.168.24.4/24
interface create ip port-41 vlan vlan34 address-netmask 192.168.34.4/24

! -- Make et.6.5 a trunk port that excludes traffic from default VLAN
vlan make trunk-port et.6.5 exclude-default-vlan

! -- Configure red VLAN with a VLAN ID of 77
vlan create red-vlan id 77 port-based
vlan add ports et.6.5 to red-vlan

! -- Configure green VLAN with a VLAN ID of 78
```

```
vlan create green-vlan id 78 port-based
vlan add ports et.6.5 to green-vlan

! -- Configure blue VLAN with a VLAN ID of 79
vlan create blue-vlan id 79 port-based
vlan add ports et.6.5 to blue-vlan


VPLS configuration

! -- Define L2 customer FEC
mpls set customer-profile vpls-77 customer_id 77 vlans 77 in-port-list
et.6.5 type port-vlan
mpls set customer-profile vpls-78 customer_id 78 vlans 78 in-port-list
et.6.5 type port-vlan
mpls set customer-profile vpls-79 customer_id 79 vlans 79 in-port-list
et.6.5 type port-vlan

! -- Enable LDP on loopback interface lo0
ldp add interface lo0

! -- Add Notus-R3 as remote LDP peer
ldp add remote-peer 223.0.0.3

! -- Add Zephir-R2 as remote LDP peer
ldp add remote-peer 223.0.0.2

! -- Send label mapping for L2 customer FEC to remote peers
ldp connect customer-profile vpls-77 remote-peer 223.0.0.2
ldp connect customer-profile vpls-77 remote-peer 223.0.0.3
ldp connect customer-profile vpls-78 remote-peer 223.0.0.2
ldp connect customer-profile vpls-78 remote-peer 223.0.0.3
ldp connect customer-profile vpls-79 remote-peer 223.0.0.3
ldp connect customer-profile vpls-79 remote-peer 223.0.0.2

ldp start
```

### 10.11.3  VLAN and LDP Monitoring

Listing 10.27 validates the VLAN configuration of Polydeukes-R12:

- Ethernet ports et.1.5 and et.1.7 are assigned to red VLAN with VLAN ID of 77.
- Ethernet ports et.1.1 and et.1.5 are assigned to green VLAN with VLAN ID of 78.
- Ethernet ports et.1.3 and et.1.5 are assigned to blue VLAN with VLAN ID of 79.

- Because port et.1.5 is a trunk port that carries traffic from the three different VLANs, it is assigned in all three VLANs.

---

**Listing 10.27   Polydeukes-R12 VLAN Information**

```
Polydeukes-R12# vlan show
```

| VID | VLAN Name | Used for | Ports |
|-----|-----------|----------|-------|
| 1 | DEFAULT | IP,IPX,ATALK,DEC,SNA,IPv6,L2 | et.1.(2,4,6,8-16), et.2.(1-16), NP.3.(3-4), t3.4.1, NP.4.2 |
| 77 | red-vlan | IP,IPX,ATALK,DEC,SNA,IPv6,L2 | et.1.(5,7) |
| 78 | green-vlan | IP,IPX,ATALK,DEC,SNA,IPv6,L2 | et.1.(1,5) |
| 79 | blue-vlan | IP,IPX,ATALK,DEC,SNA,IPv6,L2 | et.1.(3,5) |

---

Listing 10.28 validates the VLAN configuration of Aether-R13, which is identical to that of Polydeukes-R12 in Listing 10.27.

---

**Listing 10.28   Aether-R13 VLAN Information**

```
Aether-R13# vlan show
```

| VID | VLAN Name | Used for | Ports |
|-----|-----------|----------|-------|
| 1 | DEFAULT | IP,IPX,ATALK,DEC,SNA,IPv6,L2 | et.1.(2,4,6,8-16), et.2.(1-16) |
| 77 | red-vlan | IP,IPX,ATALK,DEC,SNA,IPv6,L2 | et.1.(5,7) |
| 78 | green-vlan | IP,IPX,ATALK,DEC,SNA,IPv6,L2 | et.1.(1,5) |
| 79 | blue-vlan | IP,IPX,ATALK,DEC,SNA,IPv6,L2 | et.1.(3,5) |

---

Listing 10.29 validates the VLAN configuration of Iris-R14, which is identical to that of Polydeukes-R12 in Listing 10.27.

---

**Listing 10.29   Iris-R14 VLAN Information**

```
Iris-R14# vlan show
```

| VID | VLAN Name | Used for | Ports |
|-----|-----------|----------|-------|
| 1 | DEFAULT | IP,IPX,ATALK,DEC,SNA,IPv6,L2 | et.1.(2,4,6,8-16), et.2.(1-16), gi.3.(1-2), gi.4.(1-2) |
| 77 | red-vlan | IP,IPX,ATALK,DEC,SNA,IPv6,L2 | et.1.(5,7) |
| 78 | green-vlan | IP,IPX,ATALK,DEC,SNA,IPv6,L2 | et.1.(1,5) |
| 79 | blue-vlan | IP,IPX,ATALK,DEC,SNA,IPv6,L2 | et.1.(3,5) |

Listing 10.30 validates the VLAN configuration of Zephir-R2 in which Ethernet port et.3.5 is made a trunk port carrying traffic from the three different VLANs.

**Listing 10.30   Zephir-R2 VLAN Information**

```
Zephir-R2# vlan show
```

| VID | VLAN Name | Used for | Ports |
|-----|-----------|----------|-------|
| 1 | DEFAULT | IP,IPX,ATALK,DEC,SNA,IPv6,L2 | et.3.(1-4,6-14), gi.6.1 |
| 23 | vlan23 | IP,IPX,ATALK,DEC,SNA,IPv6,L2 | gi.4.2 |
| 24 | vlan24 | IP,IPX,ATALK,DEC,SNA,IPv6,L2 | gi.4.1 |
| 77 | red-vlan | IP,IPX,ATALK,DEC,SNA,IPv6,L2 | et.3.5 |
| 78 | green-vlan | IP,IPX,ATALK,DEC,SNA,IPv6,L2 | et.3.5 |
| 79 | blue-vlan | IP,IPX,ATALK,DEC,SNA,IPv6,L2 | et.3.5 |

Listing 10.31 validates the VLAN configuration of Notus-R3 in which Ethernet port et.7.5 is made a trunk port carrying traffic from the three different VLANs.

**Listing 10.31   Notus-R3 VLAN Information**

```
Notus-R3# vlan show
```

| VID | VLAN Name | Used for | Ports |
|-----|-----------|----------|-------|
| 1 | DEFAULT | IP,IPX,ATALK,DEC,SNA,IPv6,L2 | gi.5.2, et.7.(1-4,6-24) |
| 23 | vlan23 | IP,IPX,ATALK,DEC,SNA,IPv6,L2 | gi.5.4 |
| 34 | vlan34 | IP,IPX,ATALK,DEC,SNA,IPv6,L2 | gi.5.3 |
| 77 | red-vlan | IP,IPX,ATALK,DEC,SNA,IPv6,L2 | et.7.5 |
| 78 | green-vlan | IP,IPX,ATALK,DEC,SNA,IPv6,L2 | et.7.5 |
| 79 | blue-vlan | IP,IPX,ATALK,DEC,SNA,IPv6,L2 | et.7.5 |

Listing 10.32 validates the VLAN configuration of Eurus-R4 in which Ethernet port et.6.5 is made a trunk port carrying traffic from the three different VLANs.

#### Listing 10.32  Eurus-R4 VLAN Information

```
Eurus-R4# vlan show


VID  VLAN Name   Used for                       Ports
1    DEFAULT     IP,IPX,ATALK,DEC,SNA,IPv6,L2   gi.5.1, et.6.(1-4,6-16)
24   vlan24      IP,IPX,ATALK,DEC,SNA,IPv6,L2   gi.4.2
34   vlan34      IP,IPX,ATALK,DEC,SNA,IPv6,L2   gi.4.1
77   red-vlan    IP,IPX,ATALK,DEC,SNA,IPv6,L2   et.6.5
78   green-vlan  IP,IPX,ATALK,DEC,SNA,IPv6,L2   et.6.5
79   blue-vlan   IP,IPX,ATALK,DEC,SNA,IPv6,L2   et.6.5
```

Listing 10.33 illustrates the LDP session state information for Zephir-R2:

- A remote LDP session is successfully established between local peer Zephir-R2 (223.0.0.2) and remote peer Eurus-R4 (223.0.0.4). The next-hop addresses advertised by Eurus-R4 are 192.168.34.4, 192.168.24.4, and 223.0.0.4, all of which can be used to reach the remote peer.
- A remote LDP session is successfully established between local peer Zephir-R2 (223.0.0.2) and remote peer Notus-R3 (223.0.0.3). The next-hop addresses advertised by Notus-R3 are 192.168.34.3, 192.168.23.3, and 223.0.0.3, all of which can be used to reach the remote peer.

#### Listing 10.33  Zephir-R2 LDP Session State Information

```
Zephir-R2# ldp show session verbose


Address: 223.0.0.4, State: Operational, Connection: Open, Keepalive
Time: 21
Session operational for 0d, 0h, 35m, 46s
Labels Sent 4, Received total 4, Received filtered 0
Session ID: 223.0.0.2:0—223.0.0.4:0, Remote session
Next keepalive in 9 seconds
Passive, Maximum PDU: 4096, Keepalive Timeout: 30 seconds
Keepalive interval: 10 seconds, Connect retry interval: 14 seconds
Local address: 223.0.0.2, Remote address: 223.0.0.4
Next-hop addresses received:
192.168.34.4
```

```
192.168.24.4
223.0.0.4
Address: 223.0.0.3, State: Operational, Connection: Open, Keepalive
Time: 29
Session operational for 0d, 1h, 15m, 35s
Labels Sent 4, Received total 4, Received filtered 0
Session ID: 223.0.0.2:0-223.0.0.3:0, Remote session
Next keepalive in 0 seconds
Passive, Maximum PDU: 4096, Keepalive Timeout: 30 seconds
Keepalive interval: 10 seconds, Connect retry interval: 14 seconds
Local address: 223.0.0.2, Remote address: 223.0.0.3
Next-hop addresses received:
192.168.34.3
192.168.23.3
223.0.0.3
```

Listing 10.34 illustrates the LDP session state information for Notus-R3:

- A remote LDP session is successfully established between local peer Notus-R3 (223.0.0.3) and remote peer Zephir-R2 (223.0.0.2). The next-hop addresses advertised by Zephir-R2 are 192.168.24.2, 192.168.23.2, and 223.0.0.2, all of which can be used to reach the remote peer.
- A remote LDP session is successfully established between local peer Notus-R3 (223.0.0.3) and remote peer Eurus-R4 (223.0.0.4). The next-hop addresses advertised by Eurus-R4 are 192.168.34.4, 192.168.24.4, and 223.0.0.4, all of which can be used to reach the remote peer.

### Listing 10.34    Notus-R3 LDP Session State Information

```
Notus-R3# ldp show session verbose

Address: 223.0.0.2, State: Operational, Connection: Open, Keepalive
Time: 27
Session operational for 0d, 1h, 21m, 2s
Labels Sent 4, Received total 4, Received filtered 0
Session ID: 223.0.0.3:0-223.0.0.2:0, Remote session
Next keepalive in 8 seconds
Active, Maximum PDU: 4096, Keepalive Timeout: 30 seconds
Keepalive interval: 10 seconds, Connect retry interval: 14 seconds
Local address: 223.0.0.3, Remote address: 223.0.0.2
Next-hop addresses received:
192.168.24.2
192.168.23.2
```

```
223.0.0.2
Address: 223.0.0.4, State: Operational, Connection: Open, Keepalive
Time: 26
Session operational for 0d, 2h, 43m, 32s
Labels Sent 4, Received total 4, Received filtered 0
Session ID: 223.0.0.3:0—223.0.0.4:0, Remote session
Next keepalive in 8 seconds
Passive, Maximum PDU: 4096, Keepalive Timeout: 30 seconds
Keepalive interval: 10 seconds, Connect retry interval: 14 seconds
```
**Local address: 223.0.0.3, Remote address: 223.0.0.4**
**Next-hop addresses received:**
**192.168.34.4**
**192.168.24.4**
**223.0.0.4**

---

Listing 10.35 illustrates the LDP session state information for Eurus-R4:

- A remote LDP session is successfully established between local peer Eurus-R4 (223.0.0.4) and remote peer Zephir-R2 (223.0.0.2). The next-hop addresses advertised by Zephir-R2 are 192.168.24.2, 192.168.23.2, and 223.0.0.2, all of which can be used to reach the remote peer.
- A remote LDP session is successfully established between local peer Eurus-R4 (223.0.0.4) and remote peer Notus-R3 (223.0.0.3). The next-hop addresses advertised by Notus-R3 are 192.168.34.3, 192.168.23.3, and 223.0.0.3, all of which can be used to reach the remote peer.

---

### Listing 10.35   Eurus-R4 LDP Session State Information

```
Eurus-R4# ldp show session verbose

Address: 223.0.0.2, State: Operational, Connection: Open, Keepalive
Time: 29
Session operational for 0d, 0h, 46m, 1s
Labels Sent 4, Received total 4, Received filtered 0
Session ID: 223.0.0.4:0—223.0.0.2:0, Remote session
Next keepalive in 5 seconds
Active, Maximum PDU: 4096, Keepalive Timeout: 30 seconds
Keepalive interval: 10 seconds, Connect retry interval: 14 seconds
```
**Local address: 223.0.0.4, Remote address: 223.0.0.2**
**Next-hop addresses received:**
**192.168.24.2**
**192.168.23.2**
**223.0.0.2**

```
Address: 223.0.0.3, State: Operational, Connection: Open, Keepalive
Time: 28
Session operational for 0d, 2h, 48m, 21s
Labels Sent 4, Received total 4, Received filtered 0
Session ID: 223.0.0.4:0—223.0.0.3:0, Remote session
Next keepalive in 8 seconds
Active, Maximum PDU: 4096, Keepalive Timeout: 30 seconds
Keepalive interval: 10 seconds, Connect retry interval: 14 seconds
Local address: 223.0.0.4, Remote address: 223.0.0.3
Next-hop addresses received:
192.168.34.3
192.168.23.3
223.0.0.3
```

Listing 10.36 illustrates the L2 FECs of Zephir-R2 that are advertised to its remote peers. The L2 FEC information for remote peer Eurus-R4 is as follows:

- The LDP identifier is 223.0.0.4:0.
- Customer ID 77 and VLAN ID 77 are the L2 FECs used to classify traffic from the red VLAN. For the red VLAN:
  - The VC label (out-lbl) locally generated by Zephir-R2 and advertised to its neighbor Eurus-R4 is 19, while the VC label (in-lbl) received from Eurus-R4 is 17.
  - VC label 17 is pushed onto red-tagged local Ethernet frames originating from port et.3.5 and heading for the remote red VLAN serviced by Eurus-R4. The VC LSP established by remote LDP is then tunneled within tunnel LSP "R2-to-R4" (see Listing 9.5) established by RSVP-TE between Zephir-R2 and Eurus-R4. The corresponding tunnel label 16 is pushed on top of VC label 17, forming a two-level label stack.
- Customer ID 78 and VLAN ID 78 are the L2 FECs used to classify traffic from the green VLAN. For the green VLAN:
  - The VC label (out-lbl) locally generated by Zephir-R2 and advertised to its neighbor Eurus-R4 is 18, while the VC label (in-lbl) received from Eurus-R4 is 19.
  - VC label 19 is pushed onto green-tagged local Ethernet frames originating from port et.3.5 and heading for the remote green VLAN serviced by Eurus-R4. The VC LSP established by remote LDP is then tunneled within tunnel LSP "R2-to-R4" (see Listing 9.5) established by RSVP-TE between Zephir-R2 and Eurus-R4. The corresponding tunnel label 16 is pushed on top of VC label 19, forming a two-level label stack.

- Customer ID 79 and VLAN ID 79 are the L2 FECs used to classify traffic from the blue VLAN. For the blue VLAN:
  — The VC label (out-lbl) locally generated by Zephir-R2 and advertised to its neighbor Eurus-R4 is 17, while the VC label (in-lbl) received from Eurus-R4 is 22.
  — VC label 22 is pushed onto blue-tagged local Ethernet frames originating from port et.3.5 and heading for the remote blue VLAN serviced by Eurus-R4. The VC LSP established by remote LDP is then tunneled within tunnel LSP "R2-to-R4" (see Listing 9.5) established by RSVP-TE between Zephir-R2 and Eurus-R4. The corresponding tunnel label 16 is pushed on top of VC label 22, forming a two-level label stack.

The L2 FEC information for remote peer Notus-R3 is as follows:

- The LDP identifier is 223.0.0.3:0.
- Customer ID 77 and VLAN ID 77 are the L2 FECs used to classify traffic from the red VLAN. For the red VLAN:
  — The VC label (out-lbl) locally generated by Zephir-R2 and advertised to its neighbor Notus-R3 is 22, while the VC label (in-lbl) received from Notus-R3 is 17.
  — VC label 17 is pushed onto red-tagged local Ethernet frames originating from port et.3.5 and heading for the remote red VLAN serviced by Notus-R3. The VC LSP established by remote LDP is then tunneled within tunnel LSP "R2-to-R3" (see Listing 9.5) established by RSVP-TE between Zephir-R2 and Notus-R3. The corresponding tunnel label 16 is pushed on top of VC label 17, forming a two-level label stack.
- Customer ID 78 and VLAN ID 78 are the L2 FECs used to classify traffic from the green VLAN. For the green VLAN:
  — The VC label (out-lbl) locally generated by Zephir-R2 and advertised to its neighbor Notus-R3 is 21, while the VC label (in-lbl) received from Notus-R3 is 18.
  — VC label 18 is pushed onto green-tagged local Ethernet frames originating from port et.3.5 and heading for the remote green VLAN serviced by Notus-R3. The VC LSP established by remote LDP is then tunneled within tunnel LSP "R2-to-R3" (see Listing 9.5) established by RSVP-TE between Zephir-R2 and Notus-R3. The corresponding tunnel label 16 is pushed on top of VC label 18, forming a two-level label stack.
- Customer ID 79 and VLAN ID 79 are the L2 FECs used to classify traffic from the blue VLAN. For the blue VLAN:
  — The VC label (out-lbl) locally generated by Zephir-R2 and advertised to its neighbor Notus-R3 is 20, while the VC label (in-lbl) received from Notus-R3 is 21.

— VC label 21 is pushed onto blue-tagged local Ethernet frames originating from port et.3.5 and heading for the remote blue VLAN serviced by Notus-R3. The VC LSP established by remote LDP is then tunneled within tunnel LSP "R2-to-R3" (see Listing 9.5) established by RSVP-TE between Zephir-R2 and Notus-R3. The corresponding tunnel label 16 is pushed on top of VC label 21, forming a two-level label stack.

---

### Listing 10.36   Zephir-R2 L2 FEC Information

```
Zephir-R2# ldp show 12-fec verbose

FEC: Forward Equivalence class, in-lbl: Label received, out-lbl: Label
sent

Remote neighbor 223.0.0.4:0

FEC: Customer ID 77, VLAN ID 77
in-lbl: 17, out-lbl: 19
Ports: et.3.5
Transport LSP name/label: R2-to-R4/16
Bytes In: 0, Pkts In: 0, In Pkts Drop: 0
Bytes Out: 0, Pkts Out: 0, Out Pkts Drop: 0

FEC: Customer ID 78, VLAN ID 78
in-lbl: 19, out-lbl: 18
Ports: et.3.5
Transport LSP name/label: R2-to-R4/16
Bytes In: 0, Pkts In: 0, In Pkts Drop: 0
Bytes Out: 0, Pkts Out: 0, Out Pkts Drop: 0

FEC: Customer ID 79, VLAN ID 79
in-lbl: 22, out-lbl: 17
Ports: et.3.5
Transport LSP name/label: R2-to-R4/16
Bytes In: 0, Pkts In: 0, In Pkts Drop: 0
Bytes Out: 0, Pkts Out: 0, Out Pkts Drop: 0

Remote neighbor 223.0.0.3:0

FEC: Customer ID 77, VLAN ID 77
in-lbl: 17, out-lbl: 22
Ports: et.3.5
Transport LSP name/label: R2-to-R3/16
```

```
Bytes In: 0, Pkts In: 0, In Pkts Drop: 0
Bytes Out: 0, Pkts Out: 0, Out Pkts Drop: 0


FEC: Customer ID 78, VLAN ID 78
in-lbl: 18, out-lbl: 21
Ports: et.3.5
Transport LSP name/label: R2-to-R3/16
Bytes In: 0, Pkts In: 0, In Pkts Drop: 0
Bytes Out: 0, Pkts Out: 0, Out Pkts Drop: 0


FEC: Customer ID 79, VLAN ID 79
in-lbl: 21, out-lbl: 20
Ports: et.3.5
Transport LSP name/label: R2-to-R3/16
Bytes In: 0, Pkts In: 0, In Pkts Drop: 0
Bytes Out: 0, Pkts Out: 0, Out Pkts Drop: 0
```

Listing 10.37 illustrates the L2 FECs of Notus-R3 that are advertised to its remote peers. The L2 FEC information for remote peer Zephir-R2 is as follows:

- The LDP identifier is 223.0.0.2:0.
- Customer ID 77 and VLAN ID 77 are the L2 FECs used to classify traffic from the red VLAN. For the red VLAN:
  — The VC label (out-lbl) locally generated by Notus-R3 and advertised to its neighbor Zephir-R2 is 17, while the VC label (in-lbl) received from Zephir-R2 is 22.
  — VC label 22 is pushed onto red-tagged local Ethernet frames originating from port et.7.5 and heading for the remote red VLAN serviced by Zephir-R2. The VC LSP established by remote LDP is then tunneled within tunnel LSP "R3-to-R2" (see Listing 9.6) established by RSVP-TE between Notus-R3 and Zephir-R2. The corresponding tunnel label 16 is pushed on top of VC label 22, forming a two-level label stack.
- Customer ID 78 and VLAN ID 78 are the L2 FECs used to classify traffic from the green VLAN. For the green VLAN:
  — The VC label (out-lbl) locally generated by Notus-R3 and advertised to its neighbor Zephir-R2 is 18, while the VC label (in-lbl) received from Zephir-R2 is 21.
  — VC label 21 is pushed onto green-tagged local Ethernet frames originating from port et.7.5 and heading for the remote green VLAN serviced by Zephir-R2. The VC LSP established by remote LDP is then tunneled within tunnel LSP "R3-to-R2" (see Listing 9.6) established by RSVP-TE between Notus-R3 and Zephir-R2.

The corresponding tunnel label 16 is pushed on top of VC label 21, forming a two-level label stack.

- Customer ID 79 and VLAN ID 79 are the L2 FECs used to classify traffic from the blue VLAN. For the blue VLAN:
  — The VC label (out-lbl) locally generated by Notus-R3 and advertised to its neighbor Zephir-R2 is 21, while the VC label (in-lbl) received from Zephir-R2 is 20.
  — VC label 20 is pushed onto blue-tagged local Ethernet frames originating from port et.7.5 and heading for the remote blue VLAN serviced by Zephir-R2. The VC LSP established by remote LDP is then tunneled within tunnel LSP "R3-to-R2" (see Listing 9.6) established by RSVP-TE between Notus-R3 and Zephir-R2. The corresponding tunnel label 16 is pushed on top of VC label 20, forming a two-level label stack.

The L2 FEC information for remote peer Eurus-R4 is as follows:

- The LDP identifier is 223.0.0.4:0.
- Customer ID 77 and VLAN ID 77 are the L2 FECs used to classify traffic from the red VLAN. For the red VLAN:
  — The VC label (out-lbl) locally generated by Notus-R3 and advertised to its neighbor Eurus-R4 is 20, while the VC label (in-lbl) received from Eurus-R4 is 18.
  — VC label 18 is pushed onto red-tagged local Ethernet frames originating from port et.7.5 and heading for the remote red VLAN serviced by Eurus-R4. The VC LSP established by remote LDP is then tunneled within tunnel LSP "R3-to-R4" (see Listing 9.6) established by RSVP-TE between Notus-R3 and Eurus-R4. The corresponding tunnel label 16 is pushed on top of VC label 18, forming a two-level label stack.
- Customer ID 78 and VLAN ID 78 are the L2 FECs used to classify traffic from the green VLAN. For the green VLAN:
  — The VC label (out-lbl) locally generated by Notus-R3 and advertised to its neighbor Eurus-R4 is 19, while the VC label (in-lbl) received from Eurus-R4 is 20.
  — VC label 20 is pushed onto green-tagged local Ethernet frames originating from port et.7.5 and heading for the remote green VLAN serviced by Eurus-R4. The VC LSP established by remote LDP is then tunneled within tunnel LSP "R3-to-R4" (see Listing 9.6) established by RSVP-TE between Notus-R3 and Eurus-R4. The corresponding tunnel label 16 is pushed on top of VC label 20, forming a two-level label stack.
- Customer ID 79 and VLAN ID 79 are the L2 FECs used to classify traffic from the blue VLAN. For the blue VLAN:

— The VC label (out-lbl) locally generated by Notus-R3 and advertised to its neighbor Eurus-R4 is 22, while the VC label (in-lbl) received from Eurus-R4 is 21.

— VC label 21 is pushed onto blue-tagged local Ethernet frames originating from port et.7.5 and heading for the remote blue VLAN serviced by Eurus-R4. The VC LSP established by remote LDP is then tunneled within tunnel LSP "R3-to-R4" (see Listing 9.6) established by RSVP-TE between Notus-R3 and Eurus-R4. The corresponding tunnel label 16 is pushed on top of VC label 21, forming a two-level label stack.

---

### Listing 10.37   Notus-R3 L2 FEC Information

```
Notus-R3# ldp show 12-fec verbose

FEC: Forward Equivalence class, in-lbl: Label received, out-lbl: Label
sent

Remote neighbor 223.0.0.2:0

FEC: Customer ID 77, VLAN ID 77
in-lbl: 22, out-lbl: 17
Ports: et.7.5
Transport LSP name/label: R3-to-R2/16
Bytes In: 0, Pkts In: 0, In Pkts Drop: 0
Bytes Out: 0, Pkts Out: 0, Out Pkts Drop: 0

FEC: Customer ID 78, VLAN ID 78
in-lbl: 21, out-lbl: 18
Ports: et.7.5
Transport LSP name/label: R3-to-R2/16
Bytes In: 0, Pkts In: 0, In Pkts Drop: 0
Bytes Out: 0, Pkts Out: 0, Out Pkts Drop: 0

FEC: Customer ID 79, VLAN ID 79
in-lbl: 20, out-lbl: 21
Ports: et.7.5
Transport LSP name/label: R3-to-R2/16
Bytes In: 0, Pkts In: 0, In Pkts Drop: 0
Bytes Out: 0, Pkts Out: 0, Out Pkts Drop: 0

Remote neighbor 223.0.0.4:0

FEC: Customer ID 77, VLAN ID 77
```

```
in-lbl: 18, out-lbl: 20
Ports: et.7.5
Transport LSP name/label: R3-to-R4/16
Bytes In: 0, Pkts In: 0, In Pkts Drop: 0
Bytes Out: 0, Pkts Out: 0, Out Pkts Drop: 0

FEC: Customer ID 78, VLAN ID 78
in-lbl: 20, out-lbl: 19
Ports: et.7.5
Transport LSP name/label: R3-to-R4/16
Bytes In: 0, Pkts In: 0, In Pkts Drop: 0
Bytes Out: 0, Pkts Out: 0, Out Pkts Drop: 0

FEC: Customer ID 79, VLAN ID 79
in-lbl: 21, out-lbl: 22
Ports: et.7.5
Transport LSP name/label: R3-to-R4/16
Bytes In: 0, Pkts In: 0, In Pkts Drop: 0
Bytes Out: 0, Pkts Out: 0, Out Pkts Drop: 0
```

Listing 10.38 illustrates the L2 FECs of Eurus-R4 that are advertised to its remote peers. The L2 FEC information for remote peer Zephir-R2 is as follows:

- The LDP identifier is 223.0.0.2:0.
- Customer ID 77 and VLAN ID 77 are the L2 FECs used to classify traffic from the red VLAN. For the red VLAN:
  — The VC label (out-lbl) locally generated by Eurus-R4 and advertised to its neighbor Zephir-R2 is 17, while the VC label (in-lbl) received from Zephir-R2 is 19.
  — VC label 19 is pushed onto red-tagged local Ethernet frames originating from port et.6.5 and heading for the remote red VLAN serviced by Zephir-R2. The VC LSP established by remote LDP is then tunneled within tunnel LSP "R4-to-R2" (see Listing 9.7) established by RSVP-TE between Eurus-R4 and Zephir-R2. The corresponding tunnel label 16 is pushed on top of VC label 19, forming a two-level label stack.
- Customer ID 78 and VLAN ID 78 are the L2 FECs used to classify traffic from the green VLAN. For the green VLAN:
  — The VC label (out-lbl) locally generated by Eurus-R4 and advertised to its neighbor Zephir-R2 is 19, while the VC label (in-lbl) received from Zephir-R2 is 18.
  — VC label 18 is pushed onto green-tagged local Ethernet frames originating from port et.6.5 and heading for the remote green

VLAN serviced by Zephir-R2. The VC LSP established by remote LDP is then tunneled within tunnel LSP "R4-to-R2" (see Listing 9.7) established by RSVP-TE between Eurus-R4 and Zephir-R2. The corresponding tunnel label 16 is pushed on top of VC label 18, forming a two-level label stack.

- Customer ID 79 and VLAN ID 79 are the L2 FECs used to classify traffic from the blue VLAN. For the blue VLAN:
  — The VC label (out-lbl) locally generated by Eurus-R4 and advertised to its neighbor Zephir-R2 is 22, while the VC label (in-lbl) received from Zephir-R2 is 17.
  — VC label 17 is pushed onto blue-tagged local Ethernet frames originating from port et.6.5 and heading for the remote blue VLAN serviced by Zephir-R2. The VC LSP established by remote LDP is then tunneled within tunnel LSP "R4-to-R2" (see Listing 9.7) established by RSVP-TE between Eurus-R4 and Zephir-R2. The corresponding tunnel label 16 is pushed on top of VC label 17, forming a two-level label stack.

The L2 FEC information for remote peer Notus-R3 is as follows:

- The LDP identifier is 223.0.0.3:0.
- Customer ID 77 and VLAN ID 77 are the L2 FECs used to classify traffic from the red VLAN. For the red VLAN:
  — The VC label (out-lbl) locally generated by Eurus-R4 and advertised to its neighbor Notus-R3 is 18, while the VC label (in-lbl) received from Notus-R3 is 20.
  — VC label 20 is pushed onto red-tagged local Ethernet frames originating from port et.6.5 and heading for the remote red VLAN serviced by Notus-R3. The VC LSP established by remote LDP is then tunneled within tunnel LSP "R4-to-R3" (see Listing 9.7) established by RSVP-TE between Eurus-R4 and Notus-R3. The corresponding tunnel label 16 is pushed on top of VC label 20, forming a two-level label stack.
- Customer ID 78 and VLAN ID 78 are the L2 FECs used to classify traffic from the green VLAN. For the green VLAN:
  — The VC label (out-lbl) locally generated by Eurus-R4 and advertised to its neighbor Notus-R3 is 20, while the VC label (in-lbl) received from Notus-R3 is 19.
  — VC label 19 is pushed onto green-tagged local Ethernet frames originating from port et.6.5 and heading for the remote green VLAN serviced by Notus-R3. The VC LSP established by remote LDP is then tunneled within tunnel LSP "R4-to-R3" (see Listing 9.7) established by RSVP-TE between Eurus-R4 and Notus-R3. The corresponding tunnel label 16 is pushed on top of VC label 19, forming a two-level label stack.

- Customer ID 79 and VLAN ID 79 are the L2 FECs used to classify traffic from the blue VLAN. For the blue VLAN:
  — The VC label (out-lbl) locally generated by Eurus-R4 and advertised to its neighbor Notus-R3 is 21, while the VC label (in-lbl) received from Notus-R3 is 22.
  — VC label 22 is pushed onto blue-tagged local Ethernet frames originating from port et.6.5 and heading for the remote blue VLAN serviced by Notus-R3. The VC LSP established by remote LDP is then tunneled within tunnel LSP "R4-to-R3" (see Listing 9.7) established by RSVP-TE between Eurus-R4 and Notus-R3. The corresponding tunnel label 16 is pushed on top of VC label 22, forming a two-level label stack.

---

### Listing 10.38    Eurus-R4 L2 FEC Information

```
Eurus-R4# ldp show 12-fec verbose

FEC: Forward Equivalence class, in-lbl: Label received, out-lbl: Label
sent

Remote neighbor 223.0.0.2:0


FEC: Customer ID 77, VLAN ID 77
in-lbl: 19, out-lbl: 17
Ports: et.6.5
Transport LSP name/label: R4-to-R2/16
Bytes In: 0, Pkts In: 0, In Pkts Drop: 0
Bytes Out: 0, Pkts Out: 0, Out Pkts Drop: 0


FEC: Customer ID 78, VLAN ID 78
in-lbl: 18, out-lbl: 19
Ports: et.6.5
Transport LSP name/label: R4-to-R2/16
Bytes In: 0, Pkts In: 0, In Pkts Drop: 0
Bytes Out: 0, Pkts Out: 0, Out Pkts Drop: 0


FEC: Customer ID 79, VLAN ID 79
in-lbl: 17, out-lbl: 22
Ports: et.6.5
Transport LSP name/label: R4-to-R2/16
Bytes In: 0, Pkts In: 0, In Pkts Drop: 0
Bytes Out: 0, Pkts Out: 0, Out Pkts Drop: 0


Remote neighbor 223.0.0.3:0
```

```
FEC: Customer ID 77, VLAN ID 77
in-lbl: 20, out-lbl: 18
Ports: et.6.5
Transport LSP name/label: R4-to-R3/16
Bytes In: 0, Pkts In: 0, In Pkts Drop: 0
Bytes Out: 0, Pkts Out: 0, Out Pkts Drop: 0


FEC: Customer ID 78, VLAN ID 78
in-lbl: 19, out-lbl: 20
Ports: et.6.5
Transport LSP name/label: R4-to-R3/16
Bytes In: 0, Pkts In: 0, In Pkts Drop: 0
Bytes Out: 0, Pkts Out: 0, Out Pkts Drop: 0


FEC: Customer ID 79, VLAN ID 79
in-lbl: 22, out-lbl: 21
Ports: et.6.5
Transport LSP name/label: R4-to-R3/16
Bytes In: 0, Pkts In: 0, In Pkts Drop: 0
Bytes Out: 0, Pkts Out: 0, Out Pkts Drop: 0
```

## 10.12  SUMMARY

VPLS can be used to deliver Ethernet multipoint services that span across geographically dispersed locations and provide connectivity between multiple sites as if these sites were attached to the same Ethernet LAN. VPLS uses the IP/MPLS network infrastructure, which provides better scalability over an infrastructure composed of only legacy Ethernet switches. Within the metro SP's MAN, the use of IP/MPLS routing protocols and procedures instead of the Spanning Tree Protocol, and MPLS labels instead of VLAN IDs, results in significant improvements in the scalability of VPLS as a service. The two case studies in this chapter examine the setup of a point-to-point L2 VPN service (VLL) and a multipoint L2 VPN service (VPLS).

VPLS offers a solution that addresses high-speed, secure, any-to-any forwarding at L2. It fills in the L2 multipoint VPN gap between customer requirements and existing L2 VPN point-to-point technologies such as Martini's point-to-point tunneling. The requirement to forward frames at L2 is crucial, as many new applications and services dictate that the service be transparent to upper-layer protocols (for instance, IP) or may lack L3 addressing altogether (for instance, NetBEUI). A variant of VPLS, H-VPLS provides a more scalable solution to delivering Ethernet multipoint services over MPLS using Ethernet-based IEEE 802.1AD provider bridges.

In short, VPLS expands the service portfolio of existing MPLS networks to include multipoint L2 VPN services with any-to-any connectivity, as well as extends the physical reach of Ethernet to that of WAN access. Metro SPs should fully utilize the strings of benefits offered by VPLS and add it to their arsenal of new-age services powered by MPLS.

# Part 5
# Quality-of-Service Aspect of Metropolitan Area Networks

# Chapter 11
# QoS and MPLS

## 11.1 THE NEED FOR DIFFERENT SERVICE CLASSES

If the metro service providers (MSPs) only offer a single homogeneous service, they can only charge a fixed rate for bandwidth and cannot take full advantage of the new revenue opportunities created by differentiated services. Bandwidth has become such a commodity in the metro space that it is no longer enough for MSPs to offer plain bandwidth. In this new-age era, metro subscribers would like to have the benefit of selecting a whole range of services that can fulfill each and every one of their individual requirements and preferences.

User differentiation is achievable through layer-2 (L2) or layer-3 (L3) Multi-Protocol Label Switching (MPLS) virtual private networks (VPNs), but in order for MSPs to differentiate their service offerings, service classes need to be incorporated into the IP/MPLS metropolitan area network (MAN). A service class is essentially a specific set of quality-of-service (QoS) parameters grouped together to achieve a particular type of traffic classification and handling. To gain a greater competitive edge and to realize higher revenue on top of basic VPN and Internet services, MSPs must offer different classes of service (CoS) to their customers. By grouping user traffic into different CoS, the MSPs can also have better control over service levels. A differentiated pricing strategy can be formulated from different service classes as well. For instance, an MSP can offer gold, silver, and bronze service classes to support different application needs (see Figure 11.1).

Gold service, which provides guaranteed latency and delivery, can be used to support mission-critical transaction processing and interactive real-time applications, ranging from enterprise resource planning (ERP) programs to IP telephony. Silver service, which provides guaranteed delivery, can be reserved for video streaming, two-tier customer applications, and other intermediate priority traffic. Bronze service would be the equivalent of best-effort (BE) service for which the delivery is nonguaranteed. Bronze service can be used for bulk data transfers (such as data backup, file transfer, and video download) and to transport noncritical data, such as e-mail. For customers who typically have their own specific business and application needs, such differentiated services will be very appealing.

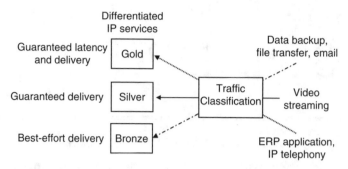

**Figure 11.1   Different Classes of Service**

## 11.2   TERMINOLOGY

This section gives a general conceptual overview of the terms used in this chapter. Some of these terms are more specifically defined in the later sections.

- *Behavior aggregate (BA)*: A collection of packets with the same DSCP crossing a link in a particular direction (and thus receiving the same PHB). The terms *aggregate* and *behavior aggregate* are used interchangeably in this chapter.
- *Differentiated services boundary (DS boundary)*: The edge of a DS domain, where classifiers and traffic conditioners are likely to be deployed. A DS boundary can be further subdivided into ingress and egress nodes, where the ingress/egress nodes are the downstream/upstream nodes of a boundary link in a given traffic direction.
- *Differentiated services code point (DSCP)*: The value in the DiffServ field of the IP header, indicating which PHB to be applied to the packet.
- *Differentiated services domain (DS domain)*: A contiguous portion of the network over which consistent sets of differentiated services policies are administered in a coordinated manner.
- *Ordered aggregate (OA)*: The set of BAs that have an ordering constraint ("must go into the same queue"). In short, an OA is a single DiffServ class.
- *Per-hop behavior (PHB)*: The QoS treatment (scheduling/dropping) applied to a BA at a DiffServ-compliant node.
- *PHB mechanism*: A specific algorithm or operation (for instance, queuing discipline) that is implemented in a node to realize a set of one or more PHBs.
- *PHB scheduling class (PSC)*: The set of one or more PHB(s) that are applied to the BA(s) belonging to a given OA.

- *Traffic flow*: A single instance of an application-to-application flow of packets that is identified by source address, source port, destination address, destination port, and protocol ID.

## 11.3 QoS MODELS FOR MPLS

MPLS does not define a new QoS architecture. Most of the work on MPLS QoS has focused on supporting the two existing IP QoS architectures: integrated services (IntServ) and differentiated services (DiffServ).

### 11.3.1 IntServ Model

The IntServ model [RFC1633] requires resources such as bandwidth to be reserved beforehand for a given traffic flow to ensure that the QoS requested by that traffic flow is fulfilled. This model includes additional components beyond those used in the BE model, such as packet classifiers, packet schedulers, and admission control. A packet classifier is used to identify flows that are to receive a specific level of service. A packet scheduler handles the scheduling of service to different packet flows, ensuring that QoS commitments are satisfied. Admission control is used to determine whether a router has the necessary resources to accept a new flow. In a nutshell, the IntServ model defines per-flow QoS and is used for providing QoS/service-level agreement (SLA) by admission control.

In addition, two services have been defined under the IntServ model: guaranteed service [RFC2212] and controlled-load service [RFC2211]. Guaranteed service ensures that packets arrive within a stipulated delivery time by controlling the maximum queuing delay. It is intended for use by real-time applications such as interactive voice and video, financial transactions, and other delay-sensitive applications. Controlled-load service, on the other hand, was designed to provide the equivalent of BE service (in which no target values of delay or packet loss are specified) on an unloaded or lightly loaded network. It is intended for applications such as non-real-time audio and video that require a high percentage of their packets to be delivered with minimum transit delay.

The IntServ model requires explicit signaling of QoS requirements from end systems (hosts) to routers [RFC2753]. Standard Resource Reservation Protocol (RSVP) [RFC2205] [RFC2208] [RFC2210] performs this signaling function and is used by hosts to maintain an end-to-end QoS for an individual traffic flow (a sequence of packets that have the same source and destination). As RSVP is also used to reserve network resources; the RSVP routers would need to maintain a state for each reservation. Therefore, the number of states in the involved RSVP routers increases in proportion to the number of concurrent reservations. This poses a serious scalability issue to both IntServ [RFC2998] and RSVP [RFC2961], particularly in large public IP networks that may potentially have millions of active traffic flows in transit concurrently.

Standard RSVP was extended to RSVP with traffic engineering (TE) extensions (see chapter 6) to serve as a label distribution protocol for MPLS. RSVP-TE is used to implement resource reservation for flow aggregates (collections of traffic flows) in MPLS-TE as well. MPLS-TE does not deploy RSVP-TE for per-flow QoS due to the limited scalability mentioned earlier. Besides, the two CoS provided by IntServ pose a great hindrance to the maximum number of service classes that can be supported in an MPLS domain.

### 11.3.2   DiffServ Model

Contrary to the IntServ's per-flow-based reservation model, the DiffServ model [RFC2475] defines a QoS architecture based on flow aggregates. The DiffServ model allows network traffic (data packets) to be classified at the network edge (DS boundary) into a small amount of aggregated flows or classes known as behavior aggregates (BAs). The BA has a color or marking associated with it so that each network node within the network core (DS domain) along the traffic path can provide appropriate QoS treatment, such as scheduling and dropping to packets belonging to different BAs. Without classification, all packets will be given the same QoS treatment and the hope of implementing different CoS with different service levels diminishes.

The markings made at the DS boundary indicate the type of QoS treatment that a packet should receive at each node. The different per-hop QoS treatments applied to packets associated with each BA inside the DS domain are discussed in the following sections.

### 11.3.3   Definition of the DiffServ Field

The DiffServ (DS) field [RFC2474] is composed of six bits of the part of the IP header formerly known as the type of service (ToS) octet [RFC791] and is used for packet classification and marking at the DS boundary. The six bits represent a mark or DiffServ code point (DSCP) that is used to indicate the QoS treatment or per-hop behavior (PHB) a packet should experience at each node. A 2-bit currently unused (CU) field is reserved for explicit congestion notification (ECN) [RFC3168]. DS-compliant nodes ignore the CU field when determining the DSCP value that is associated with the PHB for a received packet. Figure 11.2 illustrates the DS field structure.

Put another way, traffic entering a DS domain is classified into several BAs at the DS boundary and each BA is assigned a corresponding DSCP value.

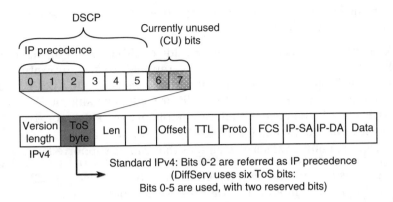

DSCP

Currently unused
(CU) bits

IP precedence

| 0 | 1 | 2 | 3 | 4 | 5 | 6 | 7 |

| Version length | ToS byte | Len | ID | Offset | TTL | Proto | FCS | IP-SA | IP-DA | Data |

IPv4

Standard IPv4: Bits 0-2 are referred as IP precedence
(DiffServ uses six ToS bits:
Bits 0-5 are used, with two reserved bits)

**Figure 11.2　The DS Field Structure**

### 11.3.4　Per-Hop Behavior

A PHB refers to the packet scheduling (queuing), dropping, policing, or shaping behavior of a node on any given packet belonging to a BA configured in accordance to a service-level agreement (SLA). Using PHBs, several classes of services can be defined using different classification, policing, shaping, scheduling, and dropping rules. Within the DS domain, traffic is handled according to its DSCP value, which indicates the desired PHB. The following subsections describe the four available standard PHBs:

- Default PHB
- Class-selector PHB
- Assured forwarding PHB (substitutes the controlled-load service provided by the IntServ model)
- Expedited forwarding PHB (substitutes the guaranteed service provided by the IntServ model)

**11.3.4.1　Default PHB.** The default PHB [RFC2474] essentially specifies that a packet marked with a recommended DSCP value of 000000 should receive the traditional BE service from a DS-compliant node. In addition, if a packet arrives at a DS-compliant node and the DSCP value does not map to any other PHB, the packet will be mapped to the default PHB.

**11.3.4.2　Class-Selector PHB.** To maintain backward compatibility with any existing IP precedence-based classification and forwarding scheme deployed in the network, DiffServ has defined a DSCP value in the form XXX000, where XXX corresponds to the 3-bit IP precedence value [RFC1122]. These DSCP values are referred to as class-selector code points, and the PHB associated with a class-selector code point is a class-selector PHB [RFC2474]. These class-selector PHBs retain most of the forwarding behavior as nodes that implement IP precedence-based classification and forwarding.

For instance, packets with a DSCP value of 10100 (the equivalent of the IP precedence value of 101) have preferential forwarding treatment for scheduling, dropping, and so on, compared to packets with a DSCP value of 011000 (the equivalent of the lower IP precedence value of 011). These class-selector PHBs ensure that DS-compliant nodes can coexist with IP precedence-based nodes.

**11.3.4.3 Assured Forwarding PHB.** Assured forwarding (AF) PHB [RFC2597] is an alternative to the controlled-load service [RFC2211] available in the IntServ model. The AF PHB defines four AF classes: AF1, AF2, AF3, and AF4. Each class is allocated a specific amount of buffer space and port bandwidth in accordance with the SLA. It is worthwhile to note that the classes are really queues. Within each AF class, there are three levels of drop priority (DP): 1 (low), 2 (medium), and 3 (high).

AF PHB can be expressed as AFny, where n represents the AF class number (1, 2, 3, or 4) and y represents the DP value (1, 2, or 3) within the AFn class. Suppose three AF classes—AF11, AF12, and AF13—are used. When congestion is experienced in the network core, packets in the AF13 class will be dropped before packets in the AF12 class, which in turn will be dropped before packets in the AF11 class. This dropping scheme can be used to protect user traffic conforming to an agreed-upon guarantee rate while increasing the odds of packets exceeding the contracted rate of being dropped if congestion is experienced in the network core. The four AF classes are assigned the following DSCP values (where dd = drop priority level):

- AF Class 1: 001dd0
- AF Class 2: 010dd0
- AF Class 3: 011dd0
- AF Class 4: 100dd0

Because of the three levels of drop priority, each AF class has up to 3 DSCP values, so a total of 12 DSCP values are defined for the four AF classes. The DSCP value given to a BA will determine which AF class this BA is assigned, and each AF class is independently forwarded with its guaranteed/allocated bandwidth.

Because AF PHB guarantees/allocates available bandwidth to each of its four classes and allows access to extra bandwidth if it is available, the class-based weighted fair queuing scheme (see section 11.3.5.3) can be used to implement this functionality. As congestion avoidance is also required within each class, the weighted (class-based) random early detection scheme (see section 11.3.5.4) can be used for this purpose.

**11.3.4.4 Expedited Forwarding PHB.** Mission-critical applications such as Voice-over-IP (VoIP), videoconferencing, and e-commerce would usually require some form of guaranteed bandwidth service. Even though standard RSVP of the IntServ model can be used to provide a guaranteed bandwidth service for these applications, its scalability constraint is still a great concern. The other alternative is to use the expedited forwarding (EF) PHB [RFC3246] defined by the DiffServ model. Through the implementation of the priority queuing scheme along with rate limiting (an enhanced form of traffic policing that can be applied to both ingress and egress traffic) on the class (or BA), the EF PHB is able to provide the low packet loss, low latency, low/constant jitter, and guaranteed bandwidth service typically required by most mission-critical applications (particularly VoIP).

EF PHB is considered a premium service in a DiffServ setup. For optimal efficiency, EF PHB should be reserved for only the most critical applications. Treating all or most traffic as high priority will defeat the usefulness of EF PHB, especially in instances of traffic congestion. The recommended DSCP value for EF PHB is 101110.

### 11.3.5  QoS Components of the DiffServ Model

The DiffServ model essentially deals with traffic management issues on a per-hop basis and has the following QoS components (or PHB mechanisms):

- Classification and marking
- Policing and shaping
- Queue scheduling
- Queue management (or drop policy)

**11.3.5.1  Classification and Marking.** Classification is the primary component of DiffServ provisioning. It segregates traffic into multiple priority levels, or CoS. Hereafter, other QoS components can be used to specify the appropriate traffic handling policies for each traffic class. Traffic can be classified based on source destination, source port, destination address, destination port, port ranges, protocol number, DSCP values, IP precedence values, IEEE 802.1P priority values, and so on. Packets can also be classified by external sources, such as by a customer or by a downstream network provider.

Marking is used to identify and differentiate packet flows. Some of the common markings include:

- IP precedence value
- IP DSCP value
- IEEE 802.1P priority value
- MPLS experimental (Exp) value (see section 11.4)

The service provider usually provides the marking information for the customer edge (CE) and provider edge (PE) devices, which includes which devices and ports to mark. These markings can be used to identify traffic within the customer network. Markings such as IP precedence and DSCP can be used in the following QoS components:

- *Policing*: To determine how packets should be treated when they conform to or exceed the transmission rate.
- *Queue scheduling*: To determine how packets should be buffered and scheduled.
- *Queue management*: To determine how packets should be treated with a packet dropping mechanism such as weighted random early detection (WRED) during congestion situations.

**11.3.5.2  Policing and Shaping.** When traffic exceeds the capacity of a router port, the excess packets must be handled properly to minimize packet loss and maintain the desired service level. The two ways to alleviate this are policing and shaping. Policing is used on ingress traffic, or traffic coming into the router. Policing relies on rate parameters such as mean or peak rate and burst sizes to execute a corresponding policing action. In general, traffic that conforms to these rate parameters is forwarded, while traffic that exceeds these parameters is dropped or still sent out, but with a reduced priority. In other words, policing typically limits incoming traffic flow to a set of preconfigured rate parameters and drops excess traffic in the presence of congestion so that it can stay within the imposed rate limit.

Policing is applied at the input to prevent any one user from exceeding its allocation and thus defeating the provisioning for the service class as a whole. If a user exceeds its traffic contract, the excess packets can be discarded or re-marked as "overcontract" and forwarded out again. Subsequently, routers throughout the network core can preferably discard overcontract packets when congestion arises to ensure that such packets do not affect the service guarantees intended for other packets that actually conform to the traffic contract.

Shaping is similar to policing except that it is used on egress traffic. Furthermore, shaping does not drop excess traffic like policing does. It simply regulates outgoing traffic flow to a preconfigured bit rate and burst size. In the presence of congestion, shaping delays excess outgoing traffic using a buffer, or queuing mechanism, so that it can stay within the imposed rate limit.

**11.3.5.3  Queue Scheduling.** Because egress traffic is assigned into multiple queues based on packet classification, each queue must be serviced so that its packets can be sent to the next-hop neighbor in the correct sequence. Queue scheduling determines the order of packet transmission by controlling which packets are placed in which queue and how queues are serviced with respect to each other. Different types of queuing algorithms are available for queue scheduling, each allowing the creation of a different number of queues, the differentiation of traffic, and the order arrangement of outgoing packets. Two commonly used queuing techniques are priority queuing and weighted fair queuing.

With priority queuing (PQ), a queue is serviced only when all queues of a higher priority are empty. This strict priority scheduling guarantees that high-priority queues will always be serviced first. PQ is useful for supporting the delivery of a high-throughput, low- or predictable-delay, and low-packet-loss service such as interactive voice or video. Packets assigned to a high-priority queue are not affected by congestion in lower-priority queues. Higher-priority queues should be rate limited to prevent queue starvation where a lower-priority queue is never serviced because a higher-priority queue always contains packets.

Weighted fair queuing (WFQ) offers a more lenient scheduling approach to PQ through fair allocation of bandwidth among queues. It is based on a relative bandwidth applied to each of the queues. Class-based weighted fair queuing (CBWFQ) extends the standard WFQ functionality to guarantee bandwidth or throughput to classes. In other words, a queue is reserved for each class, and traffic belonging to a class is directed to that class queue. CBWFQ guarantees bandwidth according to weights assigned to traffic classes. Active classes can also access unused bandwidth based on their weights.

**11.3.5.4 Queue Management.** In queue scheduling, traffic is assigned into a queue and excess packets are buffered until they can be sent again. If all queue buffers become full, subsequent packets are usually dropped. This behavior is known as tail drop because the packets at the end of the queue are discarded first.

Tail drop, however, has some adverse effects on a network. Because tail drop can only discard excess packets and cannot free out buffer space in queues, the queues are continually full and packets are always being discarded. In addition, tail drop can cause packets of different Transmission Control Protocol (TCP) sessions using the same congested queue to be dropped at the same time, thus leading to global TCP synchronization, where all affected TCP sessions synchronize their congestion control behavior. This can cause a corresponding increase or decrease in traffic, resulting in the inefficient use of bandwidth. Tail drops also cause TCP sessions to go into slow-start [RFC2001], where TCP speakers gradually, yet incessantly, increase their transmission rate, depleting all available bandwidth and slowing down only when they detect packet loss.

One method of avoiding all these undesirable behaviors is to use a technique known as random early detection (RED). RED is a congestion avoidance mechanism that randomly drops packets even before a queue is full. RED drops packets with increasing probability, and as a result, TCP sessions slow down to the approximate rate of the output port bandwidth and the necessary average queue size is reduced.

RED uses a packet drop profile to determine when to drop a packet. Three modes are defined:

- *No drop*: When the average queue occupancy is between 0 and the minimum threshold.
- *Random drop*: When the average queue occupancy is between the minimum and maximum thresholds.
- *Full drop (tail drop)*: When the average queue occupancy is at maximum threshold or above.

Random drops prevent congestion as well as tail drops. Figure 11.3 shows a sample RED profile. When the queue occupancy is within its minimum threshold of 25 percent, there is 0 percent probability that a packet will be dropped. As the queue reaches 50 percent occupancy, there is about a 25 percent chance for a packet drop. At the maximum threshold, there is a 60 percent chance for a packet drop, which is the maximum drop probability in this example. When the queue occupancy surpasses its maximum threshold of 75 percent, there is a 100 percent chance that a packet will be dropped.

Weighted random early detection (WRED) is similar to RED, except that packets can be dropped selectively based on IP precedence or DSCP value, which becomes the weight. A different RED profile can be used for each

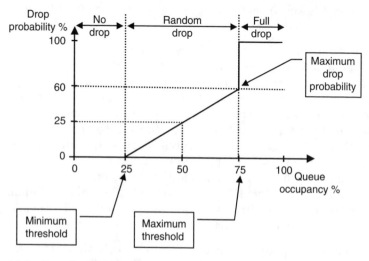

**Figure 11.3   Sample RED Profile**

weight. The weighted aspect of WRED ensures that high-precedence traffic has lower packet loss rates than other traffic during congestion.

### 11.3.6 Comparison between BE, IntServ, and DiffServ Services

In general, there are three types of service models:

- *BE*: The delivery for this service category has no QoS guarantee whatsoever and no state is maintained. In other words, no QoS is applied to packets (default behavior).
- *IntServ*: In this model, applications signal to the network that they require special QoS. The QoS is guaranteed end to end through resource reservation and admission control mechanisms on a per-flow basis. A per-flow state is maintained as well. IntServ's guaranteed service is the direct opposite of BE service.
- *DiffServ*: In this model, the network differentiates traffic that requires special QoS into classes. Resources are allocated on a per-class basis and the amount of state information is proportional to the number of classes rather than to the number of application flows. DiffServ offers a provisioned, hop-by-hop approach to QoS and maintains an aggregated state (per-flow aggregate) rather than a per-flow state. The DiffServ model can be considered a compromise between the BE and IntServ models. Its objective is to seek simplicity and scale over IntServ's per-flow QoS handling with a per-class-per-hop QoS treatment to packets, thus ensuring that a specific service level can be achieved for a class.

As can be seen from the above comparison between the three different services, the DiffServ model provides a more generalized QoS architecture than the IntServ model. By provisioning QoS in a hop-by-hop manner (PHB) in the network core to traffic that is classified and marked at the network edge (into BA), the DiffServ model mitigates the scalability issues encountered with the IntServ model's per-flow QoS implementation. As such, MPLS QoS is implemented using the DiffServ architecture for scalability reasons.

### 11.4 DIFFSERV AND MPLS

There is a seamless fit in the implementation of DiffServ and MPLS because they do the following:

- Aggregate traffic at the network edge
- Process the aggregate only in the network core

The similarities between both models are summarized in Table 11.1.

Nevertheless, if the two models were to be incorporated as one, the DSCP field would not be directly visible to the core MPLS label switch

**Table 11.1    The Similarities between DiffServ and MPLS**

| At/Model | DiffServ | MPLS |
|---|---|---|
| Network edge | Aggregation at DS boundary<br>Multiple flows associated with a class or BA (marked with DSCP) | Aggregation at edge or ingress point<br>Multiple flows associated with an FEC (marked with label) |
| Network core | Aggregated processing within DS domain<br>Scheduling/dropping (PHB) based on DSCP value | Aggregated processing within MPLS domain<br>Forwarding based on label value |

routers (LSRs) because these routers forward frames based solely on the MPLS shim header (see section 2.2.3). Therefore, information on DiffServ must be made available to the core LSRs using specific fields found in the MPLS shim header, such as the Exp field or the label value. MPLS supports this capability using two different types of label switched paths (LSPs):

- Exp-inferred PHB scheduling class LSPs (E-LSPs)
- Label-only-inferred PHB scheduling class LSPs (L-LSPs)

E-LSPs and L-LSPs are not intrinsically different from standard LSPs. They do not specify new MPLS label distribution protocols or change the label swapping operation defined for MPLS packets. They are LSPs on which a certain PHB mechanism implementing DiffServ QoS is being used.

### 11.4.1    E-LSP

For an E-LSP [DRAFT-MPLS-DIFF-EXT], each core LSR along the path uses the label to determine where a packet should be forwarded and the Exp field to infer the PHB (includes both the PHB scheduling class and drop precedence) that should be applied to the packet during the MPLS forwarding process. Because the QoS treatment or PHB is inferred from the Exp field, several classes of traffic can be multiplexed onto a single LSP (using the same label).

A single E-LSP can be used to support one or more ordered aggregates (OAs). Because the Exp field is a 3-bit field, such LSPs can support up to a maximum of eight BAs for a particular forwarding equivalence class (FEC), no matter how many OAs the BAs contain. The maximum number of classes would be less after setting aside some values for control plane traffic or if some of the classes have a drop precedence associated with them. Put simply, an E-LSP infers the PHB of a packet directly from the value of the Exp field found in the MPLS shim header.

Figure 11.4 illustrates how an E-LSP supports three different PHBs (EF, AF1y, and BE) when it transports an aggregate flow from the ingress LSR to the egress LSR. The ingress LSR is responsible for inspecting all IP packets

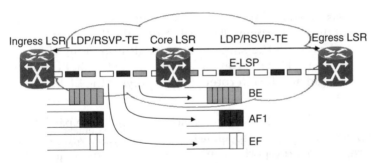

**Figure 11.4   An E-LSP Supporting Three PHBs**

arriving on customer-facing ports, assigning a label that places the packet onto the right E-LSP at the output port, and correctly setting the Exp bits in the MPLS shim header. Note that the EF, AF1, and BE packets travel on a single LSP with a single label but are enqueued in different queues based on different Exp values. The queue scheduling and drop precedence are selected according to the Exp values.

E-LSPs can be established by the various available MPLS signaling and label distribution protocols such as Label Distribution Protocol (LDP) or RSVP-TE. No new signaling protocol is required. With these protocols, the setting of the Exp bits for a given LSP either is explicitly signaled during LSP setup without modification for default PHB-to-Exp mapping or relies on a preconfigured mapping for specific PHB-to-Exp mapping. Because the 6-bit DS field in the IP header specifies a maximum of 64 PHBs while the 3-bit Exp field specifies only a maximum of 8 PHBs, it is necessary for the ingress LSR to maintain a PHB-to-Exp table that can map up to 64 different PHBs to only 8 Exp values. The PHB-to-Exp mapping is typically implemented at the ingress LSR using existing default markings (such as IP precedence or 802.1P priority) or by manual configuration (for DSCP values). Table 11.2 illustrates a sample PHB-to-Exp mapping at an ingress LSR.

**Table 11.2   Sample PHB-to-Exp Mapping**

| DiffServ PHB | Exp Value |
| --- | --- |
| Network control (111000) | 111 |
| EF (101110) | 101 |
| AF1y | 001 |
| AF2y | 010 |
| AF3y | 011 |
| BE (000000) | 000 |

Regardless of what PHB mechanisms are used to define the PHB-to-Exp mappings at the ingress point of the MPLS domain, the interpretation of the Exp value should be the same at every LSR hop along the E-LSP. For instance, if Exp = 000 is interpreted as identifying the BE PHB on one LSR, Exp = 000 should also be interpreted as identifying the BE PHB on all LSRs along the same E-LSP.

Each core LSR must be configured so that it can place packets with specific Exp values onto the respective queues associated with this value. Furthermore, to fulfill the performance requirements for each service class, the ratio of output port bandwidth allocated to each queue, the maximum queue size, and the specific RED drop profile should be configured accordingly on each LSR.

### 11.4.2   L-LSP

An L-LSP [DRAFT-MPLS-DIFF-EXT] is an LSP on which each core LSR along the path infers the PHB scheduling class (PSC) or queue scheduling from the MPLS label, as well as the corresponding optional drop precedence from the MPLS Exp bits (or the cell-loss priority bit for cell-mode MPLS), both of which can be applied to packets traversing the MPLS domain. In this approach, an ingress LSR can be configured to establish multiple LSPs between itself and a given egress LSR. Each LSP carries traffic belonging to a single <FEC, OA> pair. For example, three separate LSPs will be established to a single destination (FEC) if there are packets belonging to three different PSCs bound for that destination.

The PSC associated with an L-LSP needs to be signaled explicitly during label establishment so that after label establishment, the LSR can infer exclusively from the label value which PSC to be applied to an MPLS packet. The RSVP-TE DIFFSERV object and the LDP DiffServ type/length/value (TLV) are defined for this purpose (see [DRAFT-MPLS-DIFF-EXT]). This also implies that each LSP can be traffic engineered in the RSVP-TE case to support the specific requirements of the aggregated flow it carries. Put simply, an L-LSP can fully infer the PHB (only the PSC, not the drop precedence) of an MPLS packet from the value of the label field regardless of the value carried in the Exp field.

Figure 11.5 illustrates the establishment of three separate L-LSPs supporting three different PHBs (EF, AF1y, and BE) that need to be transported across the MPLS domain to the same FEC. EF, AF1, and BE packets travel on separate LSPs and are enqueued in different queues (queue scheduling) based on different label values. The drop precedence can be selected with the optional Exp field.

The ingress LSR inspects all native IP packets arriving on customer-facing ports and, based on the destination IP address (FEC) as well as the

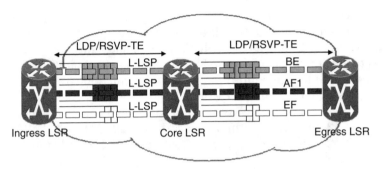

**Figure 11.5 Three L-LSPs Supporting Three PHBs**

DSCP value carried in the native IP packet's header, assigns a label that places the packet onto the correct L-LSP. In other words, the ingress LSR establishes a binding between a label (an L-LSP), an FEC, and a PHB. The ingress LSR needs to apply packet filtering to classify packets, assign a label, and place the packets onto the respective L-LSPs. Successive core LSRs along the LSP should be configured to place packets arriving on this LSP into the appropriate queue. In addition, the ratio of output port bandwidth allocated to each queue should be configured along with the queue size and the RED profile for the queue.

The deployment of L-LSPs with RSVP-TE can provide specific performance guarantees because each L-LSP carries only a single class of traffic and the explicit route can be defined such that it only includes those LSR hops that have been explicitly provisioned with the required resources and configured to handle the anticipated traffic load for this particular class of traffic.

### 11.4.3 E-LSP versus L-LSP

The key advantage of E-LSP is its manageability and simplicity. The support for multiple PHBs over a single E-LSP simplifies network management by minimizing the total number of LSPs that need to be established to support specific application requirements. However, the E-LSP faces the "do not put all your eggs in one basket" dilemma. When an existing E-LSP is preempted by another, higher-priority LSP, all service classes traversing the preempted E-LSP are affected. Moreover, E-LSPs are not an option for an Asynchronous Transfer Mode (ATM) LSR because cell-mode MPLS uses the original ATM cell encapsulation and no Exp field exists in the cell header for the LSR to infer the PHB. Nevertheless, the main limitation of E-LSP is the maximum number of PHBs it can support per FEC, which tallies at eight.

The L-LSP eases the limit on the number of PHBs by supporting up to 64 PHBs. This is possible because the L-LSP infers the PHB from the label value rather than from the Exp bits. Because an L-LSP does not rely on a PHB-to-Exp mapping to determine the PHB, it can be used impeccably with an ATM LSR. Unlike E-LSPs, which use traffic-engineered paths to fulfill the constraints of an aggregate FEC containing multiple PHBs, an L-LSP traverses a traffic-engineered path that is optimized to meet the constraint of a single PHB. In other words, one LSP is constructed per CoS or PHB in the L-LSP approach. With thorough planning and proper configuration, an L-LSP can provide the necessary throughput, delay, jitter, and packet loss guarantees to support mission-critical and guaranteed bandwidth services.

The only setback for the L-LSP approach is during the establishment of a binding between a label (an L-LSP), an FEC, and a PHB, which requires the definition of new extensions for signaling and label distribution protocols such as RSVP-TE and LDP (see [DRAFT-MPLS-DIFF-EXT]). In addition, the deployment of L-LSPs when supporting various application requirements can complicate the network management process, as the total number of LSPs tends to increase in this case. As for which option is better, it really depends on the specification of the QoS design, such as the number of classes, drop precedence values, and whether ATM LSRs are deployed.

### 11.4.4 DiffServ Tunneling Models over MPLS

The DiffServ specification [DRAFT-MPLS-DIFF-EXT] for MPLS defines three different tunneling models with specific rules of interaction between the original DiffServ marking of an IP packet and the marking used within the MPLS domain. These rules are independent of the MPLS application used. The three tunnel models are uniform, pipe, and short pipe. Using these tunnel models, an MPLS network can transparently tunnel the DiffServ marking of packets from one edge of an MPLS domain to the other edge of the same MPLS domain. These tunnel models only affect the behavior of edge and penultimate LSRs where labels are pushed or popped. They do not affect the label swapping process at intermediate (or core) LSRs. An MSP can choose different tunneling models for each metro subscriber, depending on which is the most appropriate in fulfilling that subscriber's needs.

**11.4.4.1 Uniform Model.** In the uniform model, there is only one DiffServ marking that is pertinent to a packet when it traverses the MPLS domain. If the DiffServ marking of the packet is revised within the MPLS domain (due to traffic conditioning tasks such as policing and shaping), the egress LSR will act in accordance with the modified marking. In other words, the marking in the packet may be manipulated to reflect the MSP's DiffServ marking in the core. Any adjustment to the packet marking within the MPLS domain is permanent and gets propagated when the packet exits from the MPLS domain.

**11.4.4.2 Pipe and Short-Pipe Models.** In the pipe model, two markings are pertinent to a packet when it traverses the MPLS domain: (1) the DiffServ marking that is used by the intermediate LSRs along the LSP, including the egress LSR, and (2) the original DiffServ marking carried by the packet before entering the MPLS domain, which will be retained and used once the packet exits from the MPLS domain. Any adjustment to the packet marking within the MPLS domain is not permanent and does not get propagated when the packet exits the MPLS domain.

Note that in the pipe model, the egress LSR still uses the marking that was used by the intermediate LSRs. On the other hand, it has to remove all labels imposed on the original packet. To preserve this marking carried in the labels, the egress LSR maintains an internal copy of the marking before popping the labels. This internal copy is used to classify the packet on the outbound port (facing the CE connected to this port) once the labels are removed.

The short-pipe model is a slight variation of the pipe model. The only distinction is that the egress LSR uses the original packet marking instead of the marking used by the intermediate LSRs. The pipe and short-pipe models provide QoS transparency in which the metro subscriber's DiffServ marking in the packet is preserved.

## 11.5 DIFFSERV-AWARE MPLS-TE

Even though MPLS-TE and MPLS DiffServ can run in parallel and deliver their respective benefits—MPLS-TE distributes aggregate load, while MPLS DiffServ provides differentiation—their functionalities are totally independent of each other. Thus, MPLS-TE cannot provide its benefit on a per-class basis such as per-class constraint-based routing (CBR) with different bandwidth constraints and per-class call admission control (CAC) over different bandwidth pools (specifically, bandwidth that is allocated to each class queue). Likewise, due to its simplicity, MPLS DiffServ is strong in delivering scalable multi-class services in IP networks, but it is still weak on guarantees because it does not provide any topology-aware per-class admission control mechanism.

Nevertheless, performing TE at a per-class level instead of an aggregated level is required in networks where fine optimization of resources is required to further enhance service performance and efficiency. DiffServ-aware MPLS-TE (DS-TE) [DRAFT-MPLS-DIFF-TE-REQMT] is an enhancement to MPLS-TE that introduces the concept of classes (or class types, to be exact). Each participating link advertises the amount of available bandwidth of each class type on that link. When the CBR process is initiated for a new TE tunnel, a bandwidth constraint of a particular class type can be defined as one of the criteria to be used for the path selection. The admission control process using RSVP-TE at each hop is performed against the available bandwidth of the particular class type. The RSVP-TE CLASSTYPE object is defined for this purpose (see [DRAFT-MPLS-DIFF-TE-EXT]).

DS-TE not only allows the configuration of a global pool for bandwidth accounting, but also provides a restrictive subpool (which is a portion of the global pool) configuration that can be used for high-priority network traffic such as voice or other real-time applications. This capability to fulfill a more restrictive bandwidth constraint would result in achieving higher QoS (in terms of delay, jitter, and packet loss) and better bandwidth guarantee for traffic using the subpool.

DS-TE also involves extending open shortest path first (OSPF) [DRAFT-DIFF-TE-OSPF] and integrated intermediate system to intermediate system (ISIS) [DRAFT-DIFF-TE-ISIS] so that the available subpool bandwidth at each preemption level is advertised together with the available global pool bandwidth at each preemption level. DS-TE further modifies CBR [DRAFT-MPLS-DIFF-TE-REQMT] to take this additional advertised information into consideration during path computation. CBR is performed with bandwidth constraints on a per-class-type basis. As such, each LSR is now able to keep track of the available bandwidth when admitting new LSPs for high-priority traffic. In this way, MSPs can have the flexibility to oversubscribe lower-priority classes or even undersubscribe higher-priority traffic to meet tight QoS/SLA requirements.

A typical application for DS-TE would be voice trunking/toll bypass or IP virtual leased line services, where a strict (hard) point-to-point guarantee is required in terms of bandwidth, delay, jitter, and packet loss. The delay/jitter bounds and the packet loss probability are satisfied along the LSP using queuing and dropping policies, while the bandwidth requirements/constraints per class type are fulfilled by DS-TE. The admission control process in DS-TE also ensures that the queuing and dropping mechanisms at each LSR hop are not overcommitted and the QoS guarantees (delay, jitter, packet loss) and bandwidth guarantees are met.

At this juncture, it is worthwhile to mention that DS-TE and L-LSP are both mechanisms that can be used to provide different LSPs for different CoS. However, they have different characteristics. For instance, DS-TE provides CBR capabilities that are DiffServ-aware, while L-LSPs could be used on an MPLS domain that does not implement TE. Though they serve different purposes, they are not mutually exclusive. A DS-TE setup could potentially use L-LSP.

## 11.6 PER-VPN QoS SERVICE MODELS

Two MPLS QoS service models can be used to implement QoS guarantees to VPN customers. They are the point-to-cloud model (also known as the hose model) and the point-to-point model (also known as the pipe model).

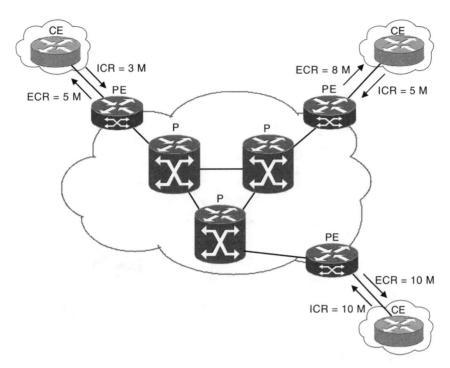

**Figure 11.6   Point-to-Cloud Model**

Note that these models are of a different categorization from those mentioned earlier in section 11.4.4. The point-to-cloud and point-to-point QoS models are only relevant in the context of MPLS VPNs and refer to how the QoS guarantees between sites are implemented.

In the point-to-cloud model, each site receives a single QoS guarantee for traffic sent to and received from all other VPN sites. Two parameters are defined for this purpose: ingress committed rate (ICR) and egress committed rate (ECR). In a VPN with three sites (see Figure 11.6), each site will have a single QoS guarantee for all incoming traffic, irrespective of the traffic source, and all outgoing traffic, irrespective of the traffic destination.

In the point-to-point model, each VPN site has specific QoS guarantees to other remote VPN sites. In a VPN with three sites (see Figure 11.7), each site will have specific QoS guarantees to the other two sites, all associated with the same VPN membership. This approach can be used to offer hard QoS guarantees similar to those found in Frame Relay or ATM networks.

These two models are not mutually exclusive. They can be offered in a hybrid manner where some VPN sites require a hard (point-to-point) guarantee to some other VPN sites, while other VPN sites only require a soft (point-to-cloud) guarantee.

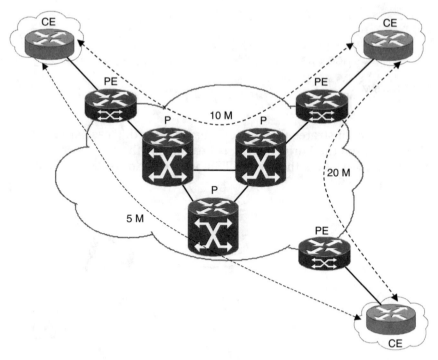

**Figure 11.7    Point-to-Point Model**

## 11.7    CASE STUDY 11.1: QoS AND L2 MPLS VPN

Case study 11.1 describes the deployment of MPLS DiffServ in an L2 MPLS VPN environment.

### 11.7.1    Case Overview and Network Topology

The setting for case study 11.1 is extracted straight from case study 10.2 with the following modifications and add-ons:

- The three virtual local area networks (VLANs) do not correspond to company sites anymore. Instead, they now represent the various services to be offered as a whole to a particular customer who has branch offices spanning across the three different districts.
- The VoIP service has been ported onto the "green" VLAN.
- The Internet service has been ported onto the "blue" VLAN.
- The transparent LAN service (or virtual private LAN services (VPLS)) has been ported onto the "red" VLAN.
- VoIP traffic is given the top priority (due to its delay-sensitive characteristics), followed by Internet access and then transparent LAN service (TLS). This means that VoIP packets will always be scheduled

for transmission earliest, followed by Internet packets and then TLS packets.

- The VoIP service (VLAN ID 78) is given a priority value of 4, the Internet service (VLAN ID 79) is assigned a priority value of 2, and the TLS (VLAN ID 77) is assigned a priority value of 0.
- The Ethernet link between Eurus-R4 and Iris-R14 has been upgraded to Gigabit Ethernet.

Figure 11.8 illustrates the network diagram for this case study. Traffic originating from the Eurus district is heading for the other two districts. Therefore, Eurus-R4 is the ingress LSR, whereas Notus-R3 and Zephir-R2 are the egress LSRs.

## 11.7.2   Network Configuration

Listing 11.1 illustrates the configuration of the L2 MPLS DiffServ ingress policy for Eurus-R4, which is built on top of Listing 10.26. Self-explanatory comments (in italics) are embedded in between the configuration lines to provide clarification to the configuration commands and procedures.

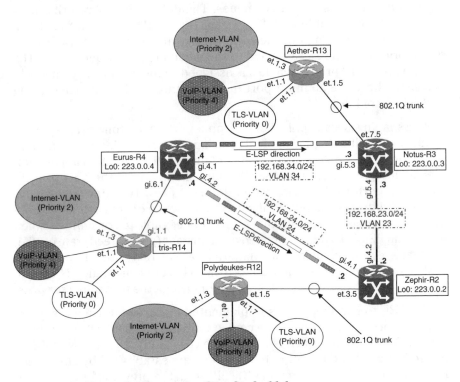

**Figure 11.8   Network Diagram for Case Study 11.1**

**Table 11.3    Default 802.1P Priority Mappings**

| 802.1P Priority Values | Internal Priority Queue |
|---|---|
| 0, 1 | Low |
| 2, 3 | Medium |
| 4, 5 | High |
| 6, 7 | Control |

Because the case study involves L2 MPLS VPNs, packet classification is achieved via VLAN IDs. Specifically, VoIP traffic has a VLAN ID of 78, Internet traffic has a VLAN ID of 79, and TLS traffic has a VLAN ID of 77. Over at the customer-facing or ingress port (gi.6.1) of Eurus-R4, Ethernet frames that contain VLAN ID 78 are marked with an 802.1P priority value of 4. Those with VLAN ID 79 are marked with an 802.1P priority value of 2, and those with VLAN ID 77 are marked with an 802.1P priority value of 0.

These markings in turn determine the ingress queue scheduling and rate shaping. Riverstone maintains an internal hardware-buffering scheme (in each of the ports of its metro routers), which imposes strict priority queuing with four different priority queues. The default 802.1P priority mappings are shown in Table 11.3. This means that VoIP traffic with 802.1P priority value 4 is assigned to the high-priority queue, Internet traffic with 802.1P priority value 2 is assigned to the medium-priority queue, and TLS traffic with 802.1P priority value 0 is assigned to the low-priority queue. In this case, these are all input port queues acting on the customer-facing port (gi.6.1) of Eurus-R4.

Rate shaping is also applied to this input port. Riverstone's rate shaping is enhanced traffic shaping applied at an input port or on ingress traffic and is supported only on line cards that contain its fifth-generation application specific integration circuits (ASICs). Rate shaping uses the port priority queues to perform buffering when different priority traffic exceeds its respective burst (maximum) rates in the presence of congestion. To control the amount of bandwidth that each service is able to consume between the CE and PE, VoIP traffic is shaped at a constant (guaranteed) rate of 10M, Internet access is shaped at a committed rate of 10M with a burst rate of 20M, and TLS traffic can burst up to 100M. This implementation is equivalent to a soft (point-to-cloud) guarantee for each of the respective services.

The PHB-to-Exp mapping is implemented at Eurus-R4 using the given 802.1P priority values, which are copied onto the Exp field without modification. The E-LSP approach is used in the case study, and the two E-LSPs, one to Notus-R3 and the other to Zephir-R2, are established by LDP. These E-LSPs are really virtual circuit (VC) LSPs from the previous VPLS setup in Listing 10.26. Strict priority queuing is still used to schedule packets onto the E-LSP—high-priority packets first, followed by medium-priority packets and finally low-priority packets.

---

### Listing 11.1  Eurus-R4 L2 MPLS DiffServ Ingress Policy Configuration

```
system set name Eurus-R4

! -- Set the 802.1P priority value to 4 for traffic from the VoIP VLAN
(VLAN ID 78)
! -- By default, a value of 4 will assign VoIP packets to the high-
priority internal queue
qos set l2 name VoIP dest-mac any vlan 78 in-port-list gi.6.1 priority 4

! -- Set the 802.1P priority value to 2 for traffic from the Internet
VLAN (VLAN ID 79)
! -- By default, a value of 2 will assign Internet packets to the medium-
priority internal queue
qos set l2 name Internet dest-mac any vlan 79 in-port-list gi.6.1
priority 2

! -- Set the 802.1P priority value to 0 for traffic from the TLS VLAN
(VLAN ID 77)
! -- By default, a value of 0 will assign TLS packets to the low-priority
internal queue
qos set l2 name TLS dest-mac any vlan 77 in-port-list gi.6.1 priority 0

! -- Maintain a constant (committed) rate of 10M for VoIP service
service VoIP create rate-shape input rate 10000000 burst 0
service VoIP apply rate-shape port gi.6.1 queue high

! -- Provide a minimum rate of 10M and a maximum rate of 20M for Internet
access
service Internet create rate-shape input rate 10000000 burst 20000000
service Internet apply rate-shape port gi.6.1 queue medium

! -- Allow TLS to burst up to 100M
service TLS create rate-shape input rate 0 burst 100000000
service TLS apply rate-shape port gi.6.1 queue low

! -- Set up E-LSP to Zephir-R2
! -- Apply MPLS DiffServ ingress policy to traffic traversing this LSP
ldp set l2-tls customer-id 77 copy-1p-to-exp vlan 77 to-peer 223.0.0.2
ldp set l2-tls customer-id 78 copy-1p-to-exp vlan 78 to-peer 223.0.0.2
ldp set l2-tls customer-id 79 copy-1p-to-exp vlan 79 to-peer 223.0.0.2
```

```
! -- Set up E-LSP to Notus-R3
! -- Apply MPLS DiffServ ingress policy to traffic traversing this LSP
ldp set l2-tls customer-id 77 copy-1p-to-exp vlan 77 to-peer 223.0.0.3
ldp set l2-tls customer-id 78 copy-1p-to-exp vlan 78 to-peer 223.0.0.3
ldp set l2-tls customer-id 79 copy-1p-to-exp vlan 79 to-peer 223.0.0.3
```

Listing 11.2 illustrates the configuration of the L2 MPLS DiffServ egress policy for Notus-R3, which is built on top of Listing 10.25. As VLAN traffic traverses the E-LSP, the Exp field in the MPLS shim header carries the 802.1P priority value belonging to the different CoS (VoIP, Internet, or TLS). When the VLAN frame leaves the E-LSP at Notus-R3, the Exp value is copied back to the 802.1P field in its 802.1Q header.

### Listing 11.2   Notus-R3 L2 MPLS DiffServ Egress Policy Configuration

```
system set name Notus-R3

mpls set egress-l2-diffserv-policy copy-exp-to-1p
```

Listing 11.3 illustrates the configuration of the L2 MPLS DiffServ egress policy for Zephir-R2, which is built on top of Listing 10.24. When a VLAN frame leaves the E-LSP at Zephir-R2, the Exp value is copied back to the 802.1P field in its 802.1Q header.

### Listing 11.3   Zephir-R2 L2 MPLS DiffServ Egress Policy Configuration

```
system set name Zephir-R2

mpls set egress-l2-diffserv-policy copy-exp-to-1p
```

### 11.7.3   Monitoring

Listing 11.4 verifies the QoS information configured for Eurus-R4:

- Incoming traffic from VLAN 78 (VoIP VLAN) is given an 802.1P priority value of 4.
- Incoming traffic from VLAN 79 (Internet VLAN) is given an 802.1P priority value of 2.

- Incoming traffic from VLAN 77 (TLS VLAN) is given an 802.1P priority value of 0.

---

### Listing 11.4    Eurus-R4 QoS Information

```
Eurus-R4# qos show 12 all-destination
```

| | |
|---|---|
| **Name:** | **VoIP** |
| Direction: | destination |
| **VLAN:** | **78** |
| Source MAC: | 000000:000000 |
| Source MAC Mask: | FFFFFF:FFFFFF |
| Dest MAC: | any |
| Dest MAC Mask: | FFFFFF:FFFFFF |
| **In-List ports:** | **gi.6.1** |
| **Priority:** | **4** |
| 802.1Q: | honor incoming 802.1Q value |

| | |
|---|---|
| **Name:** | **Internet** |
| Direction: | destination |
| **VLAN:** | **79** |
| Source MAC: | 000000:000000 |
| Source MAC Mask: | FFFFFF:FFFFFF |
| Dest MAC: | any |
| Dest MAC Mask: | FFFFFF:FFFFFF |
| **In-List ports:** | **gi.6.1** |
| **Priority:** | **2** |
| 802.1Q: | honor incoming 802.1Q value |

| | |
|---|---|
| **Name:** | **TLS** |
| Direction: | destination |
| **VLAN:** | **77** |
| Source MAC: | 000000:000000 |
| Source MAC Mask: | FFFFFF:FFFFFF |
| Dest MAC: | any |
| Dest MAC Mask: | FFFFFF:FFFFFF |
| **In-List ports:** | **gi.6.1** |
| **Priority:** | **0** |
| 802.1Q: | honor incoming 802.1Q value |

---

Listing 11.5 illustrates the default 802.1P priority mappings at Eurus-R4. Based on the previous configuration in Listing 11.1:

- VoIP traffic (VLAN ID 78) with an 802.1P priority value of 4 will be mapped to the high-priority queue.
- Internet traffic (VLAN ID 79) with an 802.1P priority value of 2 will be mapped to the medium-priority queue.
- TLS traffic (VLAN ID 77) with an 802.1P priority value of 0 will be mapped to the low-priority queue.

---

### Listing 11.5   Eurus-R4 Default 802.1P Priority Mappings

```
Eurus-R4# qos show priority-map name default
```

| 802.1p Priority | Internal Priority |
|:---:|:---|
| **0** | **Low** |
| 1 | Low |
| **2** | **Medium** |
| 3 | Medium |
| **4** | **High** |
| 5 | High |
| 6 | Control |
| 7 | Control |

---

Listing 11.6 verifies the L2 MPLS DiffServ policy configured for Eurus-R4. The ingress DiffServ policy for the three different service traffic streams traversing the two E-LSPs (one to Zephir-R2 and the other to Notus-R3) is to copy the given 802.1P priority value onto the Exp field without modification.

---

### Listing 11.6   Eurus-R4 L2 MPLS DiffServ Policy

```
Eurus-R4# ldp show 12-fec verbose

FEC: Forward Equivalence class, in-lbl: Label received, out-lbl: Label
sent

Remote neighbor 223.0.0.2:0

FEC: Customer ID 77, VLAN ID 77
in-lbl: 22, out-lbl: 22
Ports: gi.6.1
Transport LSP name/label: R4-to-R2/16
MPLS Diff-Serv: Copy 1p to exp
```

Bytes In: 0, Pkts In: 0, In Pkts Drop: 0
Bytes Out: 0, Pkts Out: 0, Out Pkts Drop: 0

**FEC: Customer ID 78, VLAN ID 78**
in-lbl: 21, out-lbl: 21
Ports: gi.6.1
Transport LSP name/label: R4-to-R2/16
**MPLS Diff-Serv: Copy 1p to exp**
Bytes In: 0, Pkts In: 0, In Pkts Drop: 0
Bytes Out: 0, Pkts Out: 0, Out Pkts Drop: 0

**FEC: Customer ID 79, VLAN ID 79**
in-lbl: 20, out-lbl: 20
Ports: gi.6.1
Transport LSP name/label: R4-to-R2/16
**MPLS Diff-Serv: Copy 1p to exp**
Bytes In: 0, Pkts In: 0, In Pkts Drop: 0
Bytes Out: 0, Pkts Out: 0, Out Pkts Drop: 0

**Remote neighbor 223.0.0.3:0**

**FEC: Customer ID 77, VLAN ID 77**
in-lbl: 21, out-lbl: 19
Ports: gi.6.1
Transport LSP name/label: R4-to-R3/16
**MPLS Diff-Serv: Copy 1p to exp**
Bytes In: 0, Pkts In: 0, In Pkts Drop: 0
Bytes Out: 0, Pkts Out: 0, Out Pkts Drop: 0

**FEC: Customer ID 78, VLAN ID 78**
in-lbl: 22, out-lbl: 18
Ports: gi.6.1
Transport LSP name/label: R4-to-R3/16
**MPLS Diff-Serv: Copy 1p to exp**
Bytes In: 0, Pkts In: 0, In Pkts Drop: 0
Bytes Out: 0, Pkts Out: 0, Out Pkts Drop: 0

**FEC: Customer ID 79, VLAN ID 79**
in-lbl: 20, out-lbl: 17
Ports: gi.6.1
Transport LSP name/label: R4-to-R3/16
**MPLS Diff-Serv: Copy 1p to exp**
Bytes In: 0, Pkts In: 0, In Pkts Drop: 0
Bytes Out: 0, Pkts Out: 0, Out Pkts Drop: 0

Listing 11.7 verifies the L2 MPLS DiffServ policy configured for Zephir-R2 and Notus-R3. Both LSRs have been configured with an egress DiffServ policy that copies the Exp value back to the 802.1P field in the 802.1Q header of the respective VLAN frames from the three different service classes (VoIP, Internet, and TLS).

---

**Listing 11.7    L2 MPLS DiffServ Policy at Zephir-R2 and Notus-R3**

```
Zephir-R2# mpls show egress-diffserv-policy 12

L2 Egress Diff-serv Policy:
Policy: copy exp to 1p

Notus-R3# mpls show egress-diffserv-policy 12

L2 Egress Diff-serv Policy:
Policy: copy exp to 1p
```

---

## 11.8    CASE STUDY 11.2: QoS AND L3 MPLS VPN

Case study 11.2 describes the deployment of MPLS DiffServ in an L3 MPLS VPN environment.

### 11.8.1    Case Overview and Network Topology

The setting for case study 11.2 is extracted straight from case study 9.2 with the following modifications and add-ons:

- Only the two green VPN sites are involved in this case study.
- Aeolus-R33 is attached to subnets 10.33.1.0/24 and 10.33.2.0/24. These are the source subnets and are discovered via RIPv2.
- Iris-R14 is attached to subnet 10.14.14.0/24. This is the destination subnet that is also discovered via RIPv2.
- Traffic flows from either one of the source subnets to the same destination subnet. However, a higher priority is given to the traffic from subnet 10.33.1.0/24.
- Traffic originating from source subnet 10.33.1.0/24, heading for destination subnet 10.14.14.0/24, is assigned an IP precedence value of 4 and given a high priority.
- Traffic originating from source subnet 10.33.2.0/24, heading for destination subnet 10.14.14.0/24, is assigned an IP precedence value of 0 and given a low priority.

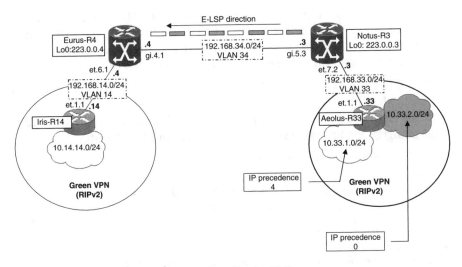

**Figure 11.9   Network Diagram for Case Study 11.2**

Figure 11.9 illustrates the network diagram for this case study. Because traffic flows from Notus-R3 to Eurus-R4, Notus-R3 is the ingress LSR, whereas Eurus-R4 is the egress LSR.

### 11.8.2   Network Configuration

Listing 11.8 illustrates the configuration of the DiffServ marking for Aeolus-R33, which is built on top of Listing 9.28. The packet classification is based on traffic originating from source subnets 10.33.1.0/24 and 10.33.2.0/24, heading for the same destination subnet 10.14.14.0/24. Packets from subnet 10.33.1.0/24 are marked with an IP precedence value of 4 and given a high priority, while packets from subnet 10.33.2.0/24 are marked with an IP precedence value of 0 and given a low priority. The queue scheduling is based on the given priority values of the packets; those with higher priority go first. The internal priority queue will still be used. In this case, the high-priority packet is assigned to the high-priority queue and is sent out first, while the low-priority packet is assigned to the low-priority queue and is sent out last.

---

**Listing 11.8   Aeolus-R33 DiffServ Marking Configuration**

```
system set name Aeolus-R33


qos set ip Aeolus high 10.33.1.0/24 10.14.14.0/24 any any any any any
any 4 any
qos set ip Aeolus low 10.33.2.0/24 10.14.14.0/24 any any any any any
0 any
```

---

Listing 11.9 illustrates the configuration of the L3 MPLS DiffServ ingress policy for Notus-R3, which is built on top of Listing 9.31. The PHB-to-Exp mapping is implemented at Notus-R3 using the given IP precedence values, which are copied onto the Exp field without modification.

The E-LSP approach is used in the case study, and the E-LSP to Eurus-R4 is established by RSVP-TE based on the previous MPLS-TE setup in Listing 9.31. Strict priority queuing is still used to schedule packets onto the E-LSP; high-priority packets go first and low-priority packets go last.

---

**Listing 11.9    Notus-R3 L3 MPLS DiffServ Ingress Policy Configuration**

```
system set name Notus-R3

routing-instance green vrf set copy-tosprec-to-exp
```

---

Listing 11.10 illustrates the configuration of the L3 MPLS DiffServ egress policy for Eurus-R4, which is built on top of Listing 9.32. As IP traffic traverses the E-LSP, the Exp field in the MPLS shim header carries the IP precedence value belonging to the different source subnets (10.33.1.0/24 and 10.33.2.0/24). When the packet leaves the E-LSP at Eurus-R4, the Exp value is copied back to the IP precedence field in its IP header.

---

**Listing 11.10    Eurus-R4 L3 MPLS DiffServ Egress Policy Configuration**

```
system set name Eurus-R4

mpls set egress-13-diffserv-policy copy-exp-to-tosprec
```

---

### 11.8.3    Monitoring

Listing 11.11 verifies the QoS information configured for Aeolus-R33:

- Traffic from source subnet 10.33.1.0/24, heading for destination subnet 10.14.14.0/24, is given an IP precedence value of 4 and a high priority. With high priority, IP packets from this particular service class will be mapped onto the high-priority queue.
- Traffic from source subnet 10.33.2.0/24, heading for destination subnet 10.14.14.0/24, is given an IP precedence value of 0 and a low

priority. With low priority, IP packets from this particular service class will be mapped onto the low-priority queue.

---

### Listing 11.11    Aeolus-R33 QoS Information

```
Aeolus-R33# qos show ip
```

| | |
|---|---|
| Name: | Aeolus |
| **Priority:** | **high** |
| Interface: | any |
| TOS: | any |
| TOSMask: | 0x030 |
| Protocol: | any |
| **SourceIP:** | **10.33.1.0** |
| SourceMask: | 255.255.255.0 |
| **DestinationIP:** | **10.14.14.0** |
| DestinationMask: | 255.255.255.0 |
| SourcePort: | any |
| DestPort: | any |
| **TosPrecedence:** | **4** |
| TosByte: | any |

| | |
|---|---|
| Name: | Aeolus |
| **Priority:** | **low** |
| Interface: | any |
| TOS: | any |
| TOSMask: | 0x030 |
| Protocol: | any |
| **SourceIP:** | **10.33.2.0** |
| SourceMask: | 255.255.255.0 |
| **DestinationIP:** | **10.14.14.0** |
| DestinationMask: | 255.255.255.0 |
| SourcePort: | any |
| DestPort: | any |
| **TosPrecedence:** | **0** |
| TosByte: | any |

---

Listing 11.12 verifies the L3 MPLS DiffServ policy configured for Notus-R3. The ingress DiffServ policy for the two different service traffic streams traversing the E-LSP (established by RSVP-TE) to Eurus-R4 is to copy the given IP precedence value onto the Exp field without modification.

---

### Listing 11.12  Notus-R3 L3 MPLS DiffServ Policy

```
Notus-R3# routing-instance show instance green
No Prefix policy configured for any VRF


green
Type                 :  vrf
Route-Distinguisher:  65000:101
QoS Policy           :  copy-tosprec-to-exp
Interfaces           :  Aeolus lo0
vrf-import           :  green-import, sequence 1
                           permit, sequence 10
                             Match clauses
                               community list green-import
                               Action Sequence  Count  Community List
                               ====== ========  =====  ==============
                               permit 10    1   target:65000:101
                             Set clauses

vrf-export           :  green-export, sequence 1
                           permit, sequence 10
                             Match clauses
                             Set clauses
                               set community target:65000:101


Default route        :  not present
Default route active:  0
```

---

Listing 11.13 verifies the L3 MPLS DiffServ policy configured for Eurus-R4. This LSR has been configured with an egress DiffServ policy that copies the Exp value back to the IP precedence field in the IP header of the respective IP packets from the two different source subnets.

---

### Listing 11.13  Eurus-R4 L3 MPLS DiffServ Policy

```
Eurus-R4# mpls show egress-diffserv-policy 13


L3 Egress Diff-Serv Policy:
Policy: copy exp to tosprec
```

---

## 11.9   SUMMARY

Even though MPLS VPNs can be used to provide user differentiation, they are still not refined enough to offer a clear distinction between the various types of user services and applications. The DiffServ model is used to implement such service/application classification. It allows network traffic to be classified at the network edge into a small number of aggregated flows or classes. Each class has a color or marking associated with it so that each network node within the network core along the traffic path can deliver appropriate QoS treatment, such as scheduling and dropping packets according to their assigned priorities and classes, thus achieving different QoS levels for different CoS.

Because DiffServ aggregates traffic at the network edge and processes aggregates only in the network core, its operations parallel that of MPLS. MPLS supports DiffServ using two LSP options. One option is to use the E-LSP approach, whereby a single LSP carries multiple classes of traffic, each of which is explicitly marked for a particular queue scheduling and drop treatment. The other is to use the L-LSP approach, whereby multiple LSPs are set up, each of which provides a particular QoS treatment for the specific class of traffic it carries.

MPLS DiffServ is strong in delivering scalable multi-class services in IP/MPLS networks, but it is still weak on guarantees because it does not provide any admission control mechanism. MPLS-TE, on the other hand, distributes aggregate load and provides admission control, but only at an aggregated level rather than at a per-class basis. DiffServ-aware MPLS-TE resolves this dilemma by combining the best of both worlds (MPLS DiffServ and MPLS-TE)—providing CoS, per-class-type resource reservation, per-class-type admission control, all the TE functionalities (such as protection and restoration mechanisms) inherited from MPLS-TE, and the PHB mechanisms (such as policing, shaping, queue scheduling, and packet dropping) derived from MPLS DiffServ.

When it comes to QoS and MPLS VPN, two service models are defined: the point-to-cloud model and the point-to-point model. The point-to-cloud model is used when VPN sites require a guarantee between the PE and CE. The point-to-point model is deployed when an end-to-end guarantee is required between all VPN sites. The two case studies in this chapter adopt the point-to-cloud model and use the E-LSP approach. One uses an L2 MPLS VPN scenario, while the other takes on an L3 MPLS VPN environment.

To sum up, MPLS QoS offers substantial benefits to MSPs and metro subscribers alike. It further enhances MPLS networks in terms of performance and efficiency, along with the support for real-time applications and services such as IP telephony and interactive gaming.

QoS-based services are usually difficult to implement consistently across multiple service provider environments. The lack of standardized QoS and the need for service providers to negotiate QoS handling at each service provider edge pose some major stumbling blocks. Overcoming these hindrances is both time-consuming and challenging. However, the MSPs have the advantage at this juncture.

MSPs typically own the MANs, and thus they can easily deploy MPLS QoS mechanisms without any impediment across those MANs to support QoS-sensitive applications. The key benefit of such a single-provider approach is that the MSP has total control over the QoS to be delivered across the MAN. MANs are also a natural venue for content localization as well as direct customer connectivity. The main advantage of localization is that it offers complete control over QoS as well.

By focusing on metro-oriented solutions, deploying QoS mechanisms on their own MANs, and localizing content, MSPs can now start to reap the benefits of MPLS QoS together with the full suite of other value-added services made possible by MPLS.

# Bibliography

[DRAFT-BGP-EXT-COMM] Sangli et al., "BGP Extended Communities Attribute," draft-ietf-idr-bgp-ext-communities-05.txt, May 2002.

[DRAFT-DIFF-TE-ISIS] Le Faucheur et al., "Extension to ISIS for Support of DiffServ-Aware MPLS Traffic Engineering," draft-lefaucheur-diff-te-isis-01.txt, November 2000.

[DRAFT-DIFF-TE-OSPF] Le Faucheur et al., "Extension to OSPF for Support of DiffServ-Aware MPLS Traffic Engineering," draft-lefaucheur-diff-te-ospf-01.txt, November 2000.

[DRAFT-ISCSI] Satran et al., "iSCSI," draft-ietf-ips-iscsi-20.txt, January 2003.

[DRAFT-ISIS-TE] Smit and Li, "IS-IS Extensions for Traffic Engineering," draft-ietf-isis-traffic-04.txt, December 2002.

[DRAFT-L2VPN-FRAMEWORK] Andersson et al., "L2VPN Framework," draft-ietf-l2vpn-l2-framework-03.txt, October 2003.

[DRAFT-L2VPN-SIGNALING] Rosen and Radoaca, "Provisioning Models and Endpoint Identifiers in L2VPN Signaling," draft-ietf-l2vpn-signaling-00.txt, September 2003.

[DRAFT-MARTINI-ENCAP] Martini et al., "Encapsulation Methods for Transport of Layer 2 Frames over IP and MPLS Networks," draft-martini-l2circuit-encap-mpls-05.txt, April 2003.

[DRAFT-MARTINI-TRANSP] Martini et al., "Transport of Layer 2 Frames over MPLS," draft-martini-l2circuit-trans-mpls-11.txt, April 2003.

[DRAFT-MPLS-DIFF-EXT] Le Faucheur et al., "MPLS Support of DiffServ," draft-ietf-mpls-diff-ext-09.txt, April 2001.

[DRAFT-MPLS-DIFF-TE-EXT] Le Faucheur et al., "Extensions to RSVP and CR-LDP for Support of DiffServ-Aware MPLS Traffic Engineering," draft-ietf-mpls-diff-te-ext-01.txt, February 2001.

[DRAFT-MPLS-DIFF-TE-REQMT] Le Faucheur et al., "Requirements for Support of DiffServ-Aware MPLS Traffic Engineering," draft-ietf-mpls-diff-te-reqts-00.txt, July 2000.

[DRAFT-MPLS-RSVP-LSP-FASTREROUTE] Pan et al., "Fast Reroute Extensions to RSVP-TE for LSP Tunnels," draft-ietf-mpls-rsvp-lsp-fastreroute-02.txt, August 2003.

[DRAFT-OSPF-2547-DNBIT] Rosen et al., "Using an LSA Options Bit to Prevent Looping in BGP/MPLS IP VPNs," draft-ietf-ospf-2547-dnbit-00.txt, June 2003.

[DRAFT-OSPF-BGP-MPLS] Rosen et al., "OSPF as the PE/CE Protocol in BGP/MPLS VPNs," draft-rosen-vpns-ospf-bgp-mpls-06.txt, February 2003.

[DRAFT-OSPF-TE] Katz et al., "Traffic Engineering Extensions to OSPF Version 2," draft-katz-yeung-ospf-traffic-10.txt, June 2003.

[DRAFT-PWE3-CONTROL-PROTO] Martini et al., "Pseudowire Setup and Maintenance Using LDP," draft-ietf-pwe3-control-protocol-05.txt, December 2003.

[DRAFT-PWE3-ETHER-ENCAP] Martini et al., "Encapsulation Methods for Transport of Ethernet Frames over IP/MPLS Networks," draft-ietf-pwe3-ethernet-encap-05.txt, December 2003.

[DRAFT-VPLS-BGP] Kompella et al., "Virtual Private LAN Service," draft-ietf-l2vpn-vpls-bgp-00.txt, May 2003.

[DRAFT-VPLS-LDP] Lasserre et al., "Virtual Private LAN Services over MPLS," draft-ietf-ppvpn-vpls-ldp-01.txt, November 2003.

[DRAFT-VPLS-REQMT] Augustyn et al., "Requirements for Virtual Private LAN Services (VPLS)," draft-ietf-l2vpn-vpls-requirements-00.txt, October 2002.

[REF01] Nam-Kee Tan, *Building VPNs with IPSec and MPLS*, McGraw-Hill, 2003.

[REF02] Agilent Technologies, Inc., "Unlocking Business Opportunities in the Metro Market," White Paper, 2002.

[REF03] Agilent Technologies, Inc., "Ethernet in Telecommunications Transmission Networks," White Paper, 2002.

[REF04] Montañez, "Deploying QoS in the Enterprise," *Packet Magazine*, Vol. 14, No. 4, 2002.

[REF05] Cisco Systems, Inc., "Advanced Topics in MPLS-TE Deployment," White Paper, 2002.

[REF06] Juniper Networks, Inc., "RSVP Signaling Extensions for MPLS Traffic Engineering," White Paper, 2000.

[REF07] Riverstone Networks, Inc., "RS Switch Router User Guide, Release 9.3," 2002.

[REF08] Juniper Networks, Inc., "IP Dependability: Network Link and Node Protection," White Paper, 2002.

[REF09] Cisco Systems, Inc., "Cisco IOS Switching Services Configuration Guide, Release 12.2," 2001.

[REF10] Juniper Networks, Inc., "VPLS: Scalable Transparent LAN Services," White Paper, 2003.

[REF11] Cisco Systems, Inc., "Cisco IOS MPLS Virtual Private LAN Service," Application Note, 2004.

[REF12] Cisco Systems, Inc., "Cisco IOS MPLS Virtual Private LAN Service," Q & A, 2004.

[REF13] Cisco Systems, Inc., "Quality of Service for MPLS Networks," Q & A, 2001.

[REF14] Cisco Systems, Inc., "Cisco IOS MPLS Quality of Service," White Paper, 2001.

[REF15] Juniper Networks, Inc., "Supporting Differentiated Service Classes: Multiprotocol Label Switching (MPLS)," White Paper, 2002.

[RFC791] Information Sciences Institute, University of Southern California, "Internet Protocol," RFC 791, September 1981.

[RFC1122] Braden, "Requirements for Internet Hosts: Communication Layers," RFC 1122, October 1989.

[RFC1321] Rivest, "The MD5 Message-Digest Algorithm," RFC 1321, April 1992.

[RFC1633] Braden, "Integrated Services in the Internet Architecture: An Overview," RFC 1633, June 1994.

[RFC1771] Rekhter and Li, "A Border Gateway Protocol 4 (BGP-4)," RFC 1771, March 1995.

[RFC1997] Chandra et al., "BGP Communities Attribute," RFC 1997, August 1996.

[RFC2001] Stevens, "TCP Slow Start, Congestion Avoidance, Fast Retransmit, and Fast Recovery Algorithms," RFC 2001, January 1997.

[RFC2113] Katz, "IP Router Alert Option," RFC 2113, February 1997.

[RFC2205] Braden et al., "Resource Reservation Protocol: Version 1 Functional Specification," RFC 2205, September 1997.

[RFC2208] Mankin et al., "Resource Reservation Protocol (RSVP) Version 1 Applicability Statement: Some Guidelines on Deployment," RFC 2208, September 1997.

[RFC2210] Wroclawski, "The Use of RSVP with Integrated Services," RFC 2210, September 1997.

[RFC2211] Wroclawski, "Specification of Controlled-Load Network Element Service," RFC 2211, September 1997.

[RFC2212] Shenker et al., "Specification of Guaranteed Quality of Service," RFC 2212, September 1997.

[RFC2328] Moy, "OSPF Version 2," RFC 2328, April 1998.

[RFC2370] Coltun, "The OSPF Opaque LSA Option," RFC 2370, July 1998.

[RFC2386] Crawley et al., "A Framework for QoS-Based Routing in the Internet," RFC 2386, August 1998.

[RFC2474] Nichols et al., "Definition of the Differentiated Services Field (DS Field) in the IPv4 and IPv6 Headers," RFC 2474, December 1998.

[RFC2475] Blake et al., "An Architecture for Differentiated Services," RFC 2475, December 1998.

[RFC2547] Rosen and Rekhter, "BGP/MPLS VPNs," RFC 2547, March 1999.

[RFC2547bis] Rosen et al., "BGP/MPLS IP VPNs," draft-ietf-ppvpn-rfc2547bis-04.txt, May 2003.

[RFC2597] Heinanen et al., "Assured Forwarding PHB Group," RFC 2597, June 1999.

[RFC2702] Awduche et al., "Requirements for Traffic Engineering Over MPLS," RFC 2702, September 1999.

[RFC2747] Baker, "RSVP Cryptographic Authentication," RFC 2747, January 2000.

[RFC2751] Herzog, "Signaled Preemption Priority Policy Element," RFC 2751, January 2000.

[RFC2753] Yavatkar et al., "A Framework for Policy-Based Admission Control," RFC 2753, January 2000.

[RFC2796] Bates et al., "BGP Route Reflection: An Alternative to Full Mesh IBGP," RFC 2796, April 2000.

[RFC2858] Bates et al., "Multiprotocol Extensions for BGP-4," RFC 2858, June 2000.

[RFC2961] Berger et al., "RSVP Refresh Overhead Reduction Extensions," RFC 2961, April 2001.

[RFC2998] Bernet et al., "A Framework for Integrated Services Operation over DiffServ Networks," RFC 2998, November 2000.

[RFC3031] Rosen et al., "Multiprotocol Label Switching Architecture," RFC 3031, January 2001.

[RFC3032] Rosen et al., "MPLS Label Stack Encoding," RFC 3032, January 2001.

[RFC3036] Andersson et al., "LDP Specification," RFC 3036, January 2001.

[RFC3037] Thomas and Gray, "LDP Applicability," RFC 3037, January 2001.

[RFC3097] Braden and Zhang, "RSVP Cryptographic Authentication: Updated Message Type Value," RFC 3097, April 2001.

[RFC3107] Rekhter and Rosen, "Carrying Label Information in BGP-4," RFC 3107, May 2001.

[RFC3168] Ramakrishnan et al., "The Addition of Explicit Congestion Notification (ECN) to IP," RFC 3168, September 2001.

[RFC3209] Awduche et al., "RSVP-TE: Extensions to RSVP for LSP Tunnels," RFC 3209, December 2001.

[RFC3210] Awduche et al., "Applicability Statement for Extensions to RSVP for LSP-Tunnels," RFC 3210, December 2001.

[RFC3212] Jamoussi et al., "Constraint-Based LSP Setup Using LDP," RFC 3212, January 2002.

[RFC3213] Ash et al., "Applicability Statement for CR-LDP," RFC 3213, January 2002.

[RFC3246] Davie et al., "An Expedited Forwarding PHB (Per-Hop Behavior)," RFC 3246, March 2002.

[RFC3272] Awduche et al., "Overview and Principles of Internet Traffic Engineering," RFC 3272, May 2002.

[RFC3346] Boyle et al., "Applicability Statement for Traffic Engineering with MPLS," RFC 3346, August 2002.

# Index

## A

ABRs 281, 283
abstract nodes 74, 76, 82, 93
acceptable loss rate 11
access technologies 4
ACK_Desired bit 111
Activate Operation 48
adaptability attributes 47
additive cost metric 52
administrative group sub-TLV 61
administrative groups 54, 61, 73
administrative policies 104
admission control 101, 104, 369
ADSL 4
ADSPEC object 86
AF PHB 372
aggregate flow 378
aggregate reservation 76
aggregate state *See per-flow aggregate*
anticipatory TE policies 44
any-to-any connectivity 12, 36, 239, 247
any-to-any L2 VPN services 316
application flows 75
application-oriented services 5
area border routers *See ABRs*
area-local scope 58
AS 58, 76, 82
AS external routes 58, 282, 283
AS path 243
ASBRs 283
AS-wide scope 58
Assured Forwarding PHB *See AF PHB*
Asymmetric Digital Subscriber Line *See ADSL*
Asynchronous Transfer Mode *See ATM*
ATM 12, 18, 36, 39, 44, 84, 247, 250, 322, 381
ATM LSRs 381, 382
attachment circuits 318
attribute filters 85
autodiscovery 319
autonomous system *See AS*
autonomous system border routers *See ASBRs*
autoroute announce feature 103
Avoid Node ID 200

## B

BAs 368, 370, 378
backbone area 281
backbone IGPs 244, 246
backup LSPs 26, 187, 189
backup of last resort 198
bandwidth overprovisioning 34
bandwidth oversubscription 5
bandwidth protection 191
bandwidth reservation 78
behavior aggregates *See BAs*
best-effort delivery 248
best-effort LSPs 23
Best-Effort Service Model 31, 367, 369, 371, 377
BGP 241, 319
BGP extended communities 242, 243, 282
BGP next-hop 244
BGP next-hop label 244
BGP route-reflectors 243
BGP standard communities 243
block-level access 9
block-level data 9
Border Gateway Protocol *See BGP*
bulk data transfers 367
bundle message extension 110
burst sizes 374
bypass tunnels 26, 187, 192, 194

## C

cable modems 4
call admission control 38, 383
capacity management 31
capacity planning 31
CBR 39, 40, 47, 51, 52, 55, 64, 65, 81, 203, 383
CBWFQ 375
CE routers 240, 243, 247
cell rate conformance 44
Cell-Mode MPLS 18, 380, 381
chronic local congestion 33, 34
Class-Based WRQ *See CBWFQ*
Classes of Service *See CoS*
Classification 373
Class-Num 74, 80
Class-Selector PHB 371